Vehicle-to-Grid Technology in Power Distribution Systems

Other related titles:

You may also like

- PBPO198 | Van den Bossche and Moghaddam | Battery Management Systems and Inductive Balancing | 2021
- PBPO238 | Wang, Liu, Wang, Stroe, Fernandez, and Guerrero | AI for Status Monitoring of Utility Scale Batteries | 2022
- PBPO111 | Zobaa, Ribeiro, Aleem, and Afifi | Energy Storage at Different Voltage Levels: Technology, integration, and market aspects | 2018
- PBTR002 | Chau | Energy Systems for Electric and Hybrid Vehicles | 2016
- PBPO079 | Hossain | Vehicle-to-Grid: Linking electric vehicles to the smart grid | 2015

We also publish a wide range of books on the following topics:
Computing and Networks
Control, Robotics and Sensors
Electrical Regulations
Electromagnetics and Radar
Energy Engineering
Healthcare Technologies
History and Management of Technology
IET Codes and Guidance
Materials, Circuits and Devices
Model Forms
Nanomaterials and Nanotechnologies
Optics, Photonics and Lasers
Production, Design and Manufacturing
Security
Telecommunications
Transportation

All books are available in print via https://shop.theiet.org or as eBooks via our Digital Library https://digital-library.theiet.org.

IET ENERGY ENGINEERING 269

Vehicle-to-Grid Technology in Power Distribution Systems

Edited by
Ali Moradi Amani, Junwei Lu, Mahdi Jalili and Feifei Bai

The Institution of Engineering and Technology

About the IET

This book is published by the Institution of Engineering and Technology (The IET).

We inspire, inform and influence the global engineering community to engineer a better world. As a diverse home across engineering and technology, we share knowledge that helps make better sense of the world, to accelerate innovation and solve the global challenges that matter.

The IET is a not-for-profit organisation. The surplus we make from our books is used to support activities and products for the engineering community and promote the positive role of science, engineering and technology in the world. This includes education resources and outreach, scholarships and awards, events and courses, publications, professional development and mentoring, and advocacy to governments.

To discover more about the IET please visit https://www.theiet.org/.

About IET books

The IET publishes books across many engineering and technology disciplines. Our authors and editors offer fresh perspectives from universities and industry. Within our subject areas, we have several book series steered by editorial boards made up of leading subject experts.

We peer review each book at the proposal stage to ensure the quality and relevance of our publications.

Get involved

If you are interested in becoming an author, editor, series advisor, or peer reviewer please visit https://www.theiet.org/publishing/publishing-with-iet-books/ or contact author_support@theiet.org.

Discovering our electronic content

All of our books are available online via the IET's Digital Library. Our Digital Library is the home of technical documents, eBooks, conference publications, real-life case studies and journal articles. To find out more, please visit https://digital-library.theiet.org.

In collaboration with the United Nations and the International Publishers Association, the IET is a Signatory member of the SDG Publishers Compact. The Compact aims to accelerate progress to achieve the Sustainable Development Goals (SDGs) by 2030. Signatories aspire to develop sustainable practices and act as champions of the SDGs during the Decade of Action (2020–30), publishing books and journals that will help inform, develop, and inspire action in that direction.

In line with our sustainable goals, our UK printing partner has FSC accreditation, which is reducing our environmental impact to the planet. We use a print-on-demand model to further reduce our carbon footprint.

British Library Cataloguing in Publication Data

A catalogue record for this product is available from the British Library

ISBN 978-1-83724-026-5 (hardback)
ISBN 978-1-83724-027-2 (PDF)

Typeset in India by MPS Limited

Cover image credit: narvikk/E+ via Getty Images

Contents

List of abbreviations

Abbreviations	Full form
AC	Alternating Current
ADL	Alexander Dennis Ltd
BAU	Business-as-Usual
BEB	Battery Electric Bus
BESS	Battery Energy Storage System
BEVs	Battery electric vehicles
BSS	Battery Swapping Station
CAISO	California Independent System Operator
CCS	Combined Charging System
CEC	California Energy Commission
CHAdeMO	CHArge de MOve
CO_2	Carbon dioxide
DC	Direct Current
DER	Distributed Energy Resources
DG	Distributed Generation
DLMP	Distribution Locational Marginal Price
DMO	Distribution Market Operator
DNO	Distribution Network Operator
DOE	Dynamic Operating Envelope
DPM	Dynamic Power Management
DR	Demand Response
DSO	Distribution System Operator
DSR	Demand Side Response
DWPT	Dynamic Wireless Power Transfer
EB	Electric Bus
ECOS	Environmental Citizens' Organisation for Standardisation
EDP	Energias de Portugal
EMI	ElectroMagnetic Interference
EMS	Energy Management System
EU	European Union
EV	Electric Vehicle
EVSE	Electric Vehicle Supply Equipment
FBMS	Fleet and battery management system
FCS	Fast Charging Station
FERC	Federal Energy Regulatory Commission
FIT	Feed-in-Tariff
FREEDM	Future Renewable Electric Energy Delivery and Management
G2V	Grid-to-Vehicle
GB/T	Guobiao Standards
GHG	Greenhouse gas
HEMS	Home Energy Management System

(Continues)

(*Continued*)

Abbreviations	Full form
ICE	Internal Combustion Engine
IEC	International Electrotechnical Commission
IEEE	Institute of Electrical and Electronics Engineers
IPT	Inductive Power Transfer
IRP	Integrated Resource Provider
ISO	International Organization for Standardization
KKT	Karush-Kuhn-Tucker
LEM	Local Energy Market
LLC	Inductor-Inductor-Capacitor
LV	Low Voltage
MCC	Multi-step Constant Current
MV	Medium Voltage
NEC	National Electrical Code
NEDO	New Energy and Industrial Technology Development Organization
NEM	National Electricity Market
NFPA	National Fire Protection Association
NREL	National Renewable Energy Laboratory
OBC	On-Board Charger
OCPI	Open Charge Point Interface
OCPP	Open Charge Point Protocol
P2P	Peer-to-Peer
PG&E	Pacific Gas and Electric
PLL	Phase-Locked Loop
PSFB	Phase-Shift Full-Bridge
PV	Photovoltaic
RES	Renewable Energy Sources
REVC	Reconfigurable EV Charger
RMI	Rocky Mountain Institute
SAE	Society of Automotive Engineers
SCADA	Supervisory control and data acquisition
SEP	Smart Energy Profile
SM	Smart meter
SoC	State of Charge
TCO	Total Cost of Ownership
TEPCO	Tokyo Electric Power Company
THD	Total Harmonic Distortion
THDI	Total Harmonic Current Distortion
THDV	Total Harmonic Voltage Distortion
TOU	Time-of-Use
TSO	Transmission System Operator
UK	United Kingdom of Great Britain and Northern Ireland
UL	Underwriters Laboratories
UPS	Uninterruptible Power Systems
USA	United States of America
V2B	Vehicle-to-Building
V2G	Vehicle-to-Grid
V2H	Vehicle-to-Home
V2L	Vehicle-to-Load

(*Continues*)

(*Continued*)

Abbreviations	Full form
V2V	Vehicle-to-Vehicle
V2X	Vehicle-to-Everything
VAR	Volt-Ampere Reactive
VGI	Vehicle-to-Grid Integration
VPP	Virtual Power Plant
WEM	Wholesale Electricity Market (Western Australia)
XFC	Extreme Fast Charger
ZEV	Zero Emission Vehicle
ZPA	Zero Phase Angle
ZVS	Zero Voltage Switching

About the editors

Ali Moradi Amani is a senior research fellow and facility manager of Australia's first EV Living Lab at the Royal Melbourne Institute of Technology, Australia. His research focuses on resilient and reliable power grid control in the presence of renewables and electrification. A senior IEEE member with 10+ years of industry experience, he collaborates with international collaborators from institutions in Hong Kong, Vietnam, Japan, and Spain on clean energy and complex systems solutions.

Junwei Lu is a foundation professor in Electrical and Electronic Engineering at the School of Engineering and Built Environment of Griffiths University in Australia. He is also an IEEE Life senior member and ICS board member. His prior positions include State Grid in China and Toyama and Kanazawa Universities in Japan. His research interests include EMC computer modeling and simulation, transformers, WPT for EV and AGV, and V2G. He has published three books and over 400 papers and holds over ten international patents.

Mahdi Jalili is a professor of AI and electrical engineering at the Royal Melbourne Institute of Technology, Australia, and a senior member of IEEE. He earned a PhD degree at the Swiss Federal Institute of Technology Lausanne (EPFL), Switzerland. His research interests include network science, clean energy and electrification, and data analytics. He is the director of the Australian Research Council Industrial Transformation Training Centre in Electrifying Australia for Net-zero Future and holds an Australian Research Council Mid-Career Industry Fellowship. Previously, he was an Australian Research Council DECRA fellow and an RMIT Vice-Chancellor Research fellow.

Feifei Bai is a senior lecturer at the University of Queensland, Australia. She was awarded the Advance Queensland Fellow in 2018. Her research interests are renewable energy integration into the power grid, PMU applications, power system oscillation analysis and control. She collaborates with Australian industries through a range of projects, such as Energy Queensland, NOJA Power, EPEC Group, Iberdrola Australia, Powerlink, and AEMO. One project received an Australian Engineering Excellence Award in 2020.

Chapter 1

Introduction to V2G

Ali Moradi Amani[1], Junwei Lu[2], Mahdi Jalili[1]
and Feifei Bai[3]

The transport sector stands as a significant contributor to global greenhouse gas (GHG) emissions, accounting for around one-fifth of the total emissions. According to the Intergovernmental Panel on Climate Change (IPCC), in 2019, direct GHG emissions from the transport sector were a staggering 8.7 gigatonnes of carbon dioxide equivalent ($GtCO_2$-eq), a significant increase from 5.0 $GtCO_2$-eq in 1990. This translates to 23% of global energy-related CO_2 emissions at that time. Similarly, national breakdowns paint a concerning picture. In Australia, for example, transportation was the third largest source of GHG emissions in 2023, contributing 21% of national emissions [1]. The situation is even worse in the United States, where the Environmental Protection Agency (EPA) reported that transportation accounted for the largest portion (28%) of total US GHG emissions in 2022 [2]. These statistics highlight the urgent need for innovative solutions to decarbonize the transportation sector and mitigate its impact on climate change.

There are two primary options for decarbonizing the transport sector: electrification and hydrogen fuel cell vehicles. While hydrogen offers potential benefits, such as long-range capabilities and quick refueling times, it faces significant challenges, including high production costs, limited infrastructure, and energy-intensive production processes. On the other hand, electrification has matured significantly in recent years, with advancements in battery technology and charging infrastructure making it a more viable and cost-effective solution. Electric passenger vehicles are already gaining traction in many markets, and the technology is rapidly expanding to heavy-duty vehicles like trucks and buses. Rail transport, a relatively low-carbon mode of transportation, offers a sustainable alternative for long-distance freight and passenger transport. By electrifying rail networks and powering them with renewable energy sources, we can further reduce emissions and improve the overall sustainability of the transportation system.

[1]School of Engineering, RMIT University, Australia
[2]School of Engineering and Built Environment, Griffith University, Australia
[3]School of Electrical Engineering and Computer Science, University of Queensland, Australia

Electric vehicle (EV) sales continue to surge, with projections indicating a potential of around 17 million units sold globally in 2024 [3]. This represents a significant market share, exceeding one-fifth of total car sales. While factors like fluctuating battery rare material prices, inflationary pressures, and the phasing out of purchase incentives in certain regions may influence growth rates, the overall trend remains positive. In the first quarter of 2024, electric car sales experienced a substantial year-on-year increase of approximately 25%, mirroring the growth observed in the same period of 2022. Key factors driving this growth include intensified competition among manufacturers, decreasing battery and vehicle prices, and sustained government support. Notably, China, Europe, and the United States are leading the charge, with projected market shares for electric cars reaching up to 45%, 25%, and 11%, respectively.

The rapid growth of EVs has necessitated the development of robust charging infrastructure, known as electric vehicle supply equipment (EVSE). EVSEs come in various types, ranging from simple home chargers to powerful public fast-charging stations. In recent years, significant advancements have been made in EVSE technology, and several vendors have manufactured very high-capacity EV chargers. Faster charging speeds, increased reliability, and smarter charging solutions have emerged; however, we are not there yet. As with charge point availability, 42% of hesitant EV buyers state that they will not purchase an EV until the battery capacity and driving range improve [4]. Similar objections have been reported in other countries.

On the other hand, electricity distributors are experiencing significant technical challenges as most of the EV charging demand should be delivered in power distribution grids, which were not originally established to manage this demand. Innovative solutions, such as demand response and smart charging, have been proposed, but there is still a long way to fully implement them. Integration of smart grid technologies enables intelligent charging, optimizing energy consumption and reducing peak demand on the electrical grid. Optimal sizing and location of charging infrastructure is another complex challenge that requires careful consideration of factors such as EV adoption rates, traffic patterns, and grid capacity. A well-planned charging network can ensure efficient utilization of energy resources and minimize grid congestion. This interdisciplinary problem involves expertise from various fields, including electrical engineering, urban planning, and economics.

1.1   What is V2G?

Even within a complex and demanding energy landscape, EVs offer considerable potential for enhancing the flexibility of future power grids. They can be charged using renewable energy sources during off-peak periods, promoting sustainable energy use, particularly during daylight hours when solar generation is abundant. The stored energy within EV batteries can then be utilized to supply consumers or the grid as needed. This energy transfer can take various forms, including vehicle-to-vehicle (V2V), vehicle-to-infrastructure (V2I), vehicle-to-home (V2H), vehicle-to-grid (V2G), vehicle-to-network

(V2N), vehicle-to-building (V2B), or vehicle-to-load (V2L). These diverse interactions are often collectively termed V2X, signifying the interconnectedness of vehicles with other entities and the smart city infrastructure. This book focuses specifically on the interaction of EVs with the power grid, namely V2G, V2H, and V2L. In this context, EVs are viewed as mobile energy storage units – essentially batteries on wheels – capable of providing localized support to the grid by balancing supply and demand. Like stationary battery storage, these technologies can contribute to grid stability through load leveling, voltage and frequency regulation, and peak shaving.

V2G technology offers a promising solution to address grid challenges, particularly in the era of electrification [5]. As clean energy technologies like electric heating and cooking gain momentum, electricity demand is surging faster than the existing grid infrastructure can adapt. This rapid increase strains existing distribution networks, potentially leading to overloading and the need for costly upgrades. Distribution network operators (DNOs) face the dilemma of either investing heavily in grid expansion or exploring innovative solutions. Battery energy storage systems (BESS) and V2G, as flexible resources, can help bridge this gap. By storing excess energy from renewable sources, such as solar PV, during off-peak hours and releasing it during peak demand periods, these technologies can alleviate grid stress and reduce the need for immediate infrastructure upgrades. This strategic approach enables DNOs to optimize grid operations, defer significant capital expenditures, and ultimately lower electricity costs for consumers.

Another significant potential of V2G is to enhance grid resilience in remote and rural areas prone to frequent power outages, often caused by extreme weather conditions. These areas face unique challenges due to their vast geographic expanse and dispersed infrastructure, making centralized grid upgrades prohibitively expensive. Traditional power systems in these regions, often supporting only a few households and essential services, require substantial investments for electrification. V2G, along with other V2X technologies, can provide a decentralized solution to address these challenges. By enabling EVs to store and discharge energy, these technologies can support local energy demands, reducing the reliance on centralized grid infrastructure. This decentralized approach eliminates the need for DNOs to invest heavily in grid upgrades and manage complex operational and maintenance tasks. Instead, DNOs can focus on creating a supportive framework that incentivizes EV owners to optimize energy consumption through V2L, V2V, and V2G. This framework should include clear communication channels, robust control systems, and fair compensation mechanisms for energy fed back into the grid.

1.2   The costs and benefits of V2G

While offering significant potential, V2G technology presents challenges in balancing the flow of energy, information, and revenue between EV owners and power grid operators to achieve a stable supply–demand equilibrium. V2G promises to optimize energy exchange between EVs and the grid, providing beneficial services to both EV owners and grid operators through bidirectional energy flow. The V2G market encompasses a broad range of stakeholders (EV owners, manufacturers, aggregators, electricity generators, network operators, and the public), each with distinct cost and benefit considerations. A key challenge lies in addressing the inherent trade-offs between the costs and benefits experienced by EV owners and DNOs to ensure a sustainable implementation of V2G. The existing body of research on stakeholder costs and benefits reflects this complexity, with variations in methodology and approach often leading to differing conclusions.

1.2.1   V2G for grid operators

V2G technology offers significant advantages for DNOs. By strategically managing EV charging and discharging, V2G can contribute to minimizing operational expenses within active distribution networks, especially those incorporating distributed energy resources (DERs). This cost reduction stems from addressing challenges like solar intermittency and peak demand. Furthermore, optimized V2G operation can mitigate battery degradation by minimizing charge–discharge cycles. For DNOs, the potential cost savings associated with V2G can offset the expenses related to infrastructure upgrades. However, a comprehensive cost–benefit analysis over the long term is crucial to determine the return on investment.

EVs participating in V2G schemes offer several advantages, including energy savings, improved supply management, and enhanced utilization of renewable energy storage. Both DNOs and EV owners benefit from V2G's ability to alleviate grid congestion. Furthermore, financial incentives can encourage EV owners to make their vehicles available for V2G services. Key benefits for grid operators include,

- **Peak demand reduction:** V2G can help alleviate peak demand by allowing EVs to discharge energy back into the grid during periods of high consumption, reducing the strain on the power system and potentially deferring costly grid upgrades.
- **Grid stability:** V2G can support grid stability by providing ancillary services such as frequency regulation and voltage control, ensuring a reliable and efficient power supply.
- **Renewable energy integration:** By storing excess renewable energy during periods of low demand and releasing it when needed, V2G can enhance the integration of variable renewable energy sources, such as solar and wind power, into the grid.

New charging infrastructure, along with the associated data communications and management systems, will likely represent a significant cost for DNOs, who may also face additional regulatory, administrative, and algorithm development expenses.

1.2.2   V2G benefits for EV owners

Integrating large-scale V2G with comprehensive controlled-charging regimes has the potential to deliver significant energy and cost savings while simultaneously extending battery lifespan [6]. The benefits for EV owners are contingent on the tariff structure and the extent of battery degradation. EV owners' earnings should offset any battery degradation costs. Fair implementation of smart discharging strategies is crucial to prevent substantial battery degradation. Research indicates that an efficient, smart, bi-directional charging algorithm, from a battery degradation perspective, is highly dependent on the EV's daily usage patterns and the ambient temperature, necessitating local data-driven studies. One approach to mitigating distribution network congestion during peak periods is to mandate time of use (TOU) tariffs for households charging EVs. However, with increasing EV penetration, a secondary peak may occur. Consequently, incentivizing EV owners to adopt real-time tariffs would be more effective. A UK trial demonstrated that such smart charging reduced peak electricity consumption by up to 47% [7]. Research in Melbourne, Australia, examined the economic benefits of an EV combined with a home energy management system under various scenarios [8]. The study showed that V2H implementation reduced monthly electricity bills by 11.6% compared to unmanaged unidirectional charging. However, many studies fail to consider battery degradation [9].

From the EV owner's perspective, V2G can generate additional income by returning excess energy to the grid during peak demand periods, effectively turning their vehicles into mobile energy storage assets. When combined with renewable resources, EV owners can charge their cars at the workplace or in public parking during the day, e.g., using surplus solar generation, and sell electricity back to the grid when they return to home or depot. By optimizing charging schedules and utilizing TOU pricing, EV owners can lower their electricity bills, potentially saving significant costs over time. EV owners may encounter several costs, including the installation of bi-directional charging equipment, a loss of charging flexibility, and the potential for increased battery degradation.

While V2G offers numerous benefits, several challenges must be addressed to ensure its successful implementation:

- **Battery degradation**: Frequent bidirectional charging can accelerate battery degradation, potentially reducing the lifespan of EV batteries.
- **Economic viability**: The economic viability of V2G depends on factors such as electricity tariffs, battery costs, and government incentives.
- **Technical complexity**: Implementing V2G requires advanced control systems and communication protocols to ensure seamless integration with the grid and EV charging infrastructure.
- **Regulatory framework**: Clear and supportive regulatory frameworks are essential to facilitate the deployment of V2G technologies and incentivize participation from both EV owners and grid operators.

Various V2X technologies offer distinct potential benefits. V2L can provide emergency backup power to essential loads during power cuts or serve as an alternative electricity source to diesel generators in off-grid locations. V2H is slightly more complex than V2L, aiming to utilize an EV as a whole-home power backup, as well as reducing household grid energy consumption during peak periods. V2H is most effective when combined with rooftop solar, where TOU or peak/off-peak tariffs are in place. However, economic analysis and simulations are required to evaluate the financial benefits of V2H, taking into account the costs of the bi-directional charger, electricity tariff, rooftop solar array size, and battery degradation. This option may appeal to a limited number of consumers seeking grid independence. V2B is similar to V2H, but aggregates multiple EV batteries to optimize energy use in larger buildings (e.g., apartment blocks or office buildings). Beyond demand reduction and peak-time electricity cost savings, V2B can benefit buildings through power factor correction, reactive power control, and voltage regulation.

To fully realize the potential of V2G, it is crucial to conduct thorough economic analyses, technical feasibility studies, and field trials to assess the long-term benefits and costs associated with this technology. By addressing these challenges and fostering collaboration between various stakeholders, V2G can play a significant role in shaping a sustainable and resilient energy future.

1.2.3 Battery degradation

Battery degradation is a key obstacle to the wider adoption of V2G programs, and further research is crucial to fully understand its impact. Generally, V2X services involving small fluctuations in the battery's state of charge (SoC) incur lower degradation costs compared to those requiring substantial energy throughput. Consequently, power-focused (capacity) value streams, such as frequency regulation, which demand less energy throughput, are likely to be the most viable for V2X, as they minimize battery degradation. Resource adequacy and network deferral – where V2X reduces future peak system loads – represent other high-revenue, power-based value streams. Network deferral, also known as non-wire alternatives, enabled by V2X, allows the grid to accommodate projected load growth, potentially avoiding or delaying (for up to 2–3 years) costly upgrades to the distribution network or individual substations.

Several key factors influence battery degradation. Depth of discharge (DoD) plays a crucial role, with deeper discharges leading to greater stress and accelerated capacity fade. Each charge/discharge cycle contributes to wear, and the extent of this wear is amplified by the DoD. Similarly, high charging and discharging rates can increase internal battery temperatures and exacerbate degradation mechanisms. Temperature, both ambient and within the battery itself, significantly impacts battery health. Elevated temperatures accelerate chemical reactions that contribute to degradation, while extremely low temperatures can also hinder performance and potentially cause damage. Studies have shown a strong correlation between temperature and battery lifespan during V2G operations, highlighting the need for thermal management strategies. Furthermore, storing batteries at a high SoC for extended periods, even when not actively cycling, can accelerate calendar degradation. This is particularly relevant for V2G, where vehicles may spend considerable time connected to the grid at a high SoC, ready to provide ancillary services.

To mitigate battery degradation in V2G scenarios, several strategies can be employed. Smart charging algorithms can optimize charging and discharging profiles to minimize stress on the battery. These algorithms can consider factors like DoD, charging/discharging rates, and temperature to dynamically adjust the charging process. Battery management systems (BMS) play a vital role in monitoring battery health and controlling charging/discharging to prolong lifespan. The BMS can track individual cell voltages, temperatures, and SoC and use this information to balance cell charge, prevent overcharging or deep discharging, and implement thermal management strategies. Optimized V2G strategies, such as limiting the DoD during V2G operations and avoiding extreme temperatures, can also contribute to minimizing battery wear. Despite these progresses, there is still a big gap of knowledge on battery degradation. In fact, future research should focus on developing more sophisticated battery degradation models that accurately predict long-term performance under various V2G operating conditions. Further investigation into advanced cooling and thermal management techniques for batteries in V2G applications is also needed.

1.2.4   Supporting integration of renewables into the grid

For V2G/V2H/V2L to be truly effective, EVs should ideally be charged using clean, renewable energy resources. This encourages EV owners to utilize rooftop solar panels for home charging or to charge their vehicles daily using electricity generated primarily by solar or wind farms. By predicting energy supply and utilizing the flexibility afforded by bi-directional EV charging, network operators can address challenges related to low energy demand, improve network and renewable energy utilization, and gain greater flexibility in using stored EV battery energy. This requires DNOs to establish a comprehensive energy storage management system, incorporating flexible loads such as BESS and EVs. It is also essential to develop strategies that integrate V2G with existing DERs within low-voltage networks to capitalize on the associated opportunities and mitigate potential risks.

1.3   Technical challenges and opportunities

The potential of V2G is influenced by several factors, including EV availability, which itself depends on owner acceptance of the technology, driving habits, willingness to participate in V2G schemes, system readiness (e.g., charger availability), technical constraints (e.g., battery degradation), market maturity, technical standards, and regulations. If not managed effectively, V2G can negatively impact power quality. Injecting extra power into the grid via a converter can lead to voltage compliance issues, harmonics, and transformer overloading. Traditionally, utilities have used devices like voltage regulators, capacitor banks, and transformer tap changers to maintain power quality. Smart bi-directional EV charging offers a flexible alternative for mitigating these issues in DER- and EV-rich power grids. A US study indicates that personal vehicles are only used for travel approximately 4%–5% of the time, spending the remaining 95%–96% parked at home or elsewhere [10]. Charging infrastructure is a critical component of EV adoption. The widespread adoption of EVs hinges on the development of a robust and accessible charging infrastructure. Charging stations come in various types, categorized by power output and charging speeds, and two well-known standards have been developed for them: IEC 61851-1 and SAE J1772. Four modes of charging have been introduced in IEC 61851-1 (Table 1.1).

Table 1.1   EV charging modes based on the IEC 61851-1 standard

Mode	Specification	Description
Mode 1	1-Phase, 16 A	There is a direct connection between the power outlet and EV. This mode is rarely used and not allowed in many countries due to the lack of communication between EVs and the power grid and safety concerns
Mode 2	1-Phase, up to 32 A	A portable control box, typically placed between the power outlet and the EV's charging port, ensures electrical safety and communicates with the vehicle's on-board charger. This method is suitable for slow, overnight charging at home and is particularly common in plug-in hybrid electric vehicles (PHEVs)
Mode 3	1-Phase or 3-phase, up to 32 A	EVSE is fixed (not portable), e.g., wall mounted, and includes charging control and protection features. Specific connectors using the pulse-width modulation safety system with slow or fast charging must be used
Mode 4	DC charging	Fast and ultra-fast charging

Another EVSE categorization is based on the power capacity and was originally defined in SAE J1772. Based on that, we have Level 1 chargers (up to 7 kW) mainly for residential applications, Level 2 chargers (up to 20 kW) for both residential and public places such as shopping malls, workplaces, and along highways, and finally Level 3 (DC fast chargers) with up to 100 kW capacity that are primarily located on major highways and in urban and interstate areas. With the advancement of EVSE technologies, DC chargers with a capacity of 500 kW are now available that can make a flat EV, with an 80 kWh battery, fully charged in less than 10 min. Thus, these levels of charging have been modified for different vendors (Figure 1.1). (Figure 1.1 provides a general overview of connecting EV to the grid for Grid-to-Vehicle (G2V) and V2G cases.

1.3.1   Active and reactive power regulation

Active power exchange for load leveling is a highly effective V2G service. Since EV charging often coincides with peak demand periods, strategic V2G scheduling can play a crucial role in mitigating peak load issues. Optimization techniques, often employed by aggregators, are used to create effective V2G schedules. These schedules take into account EV arrival and departure times, the initial SoC, and the potential energy contribution from V2G-enabled vehicles. Accurate forecasting of V2G potential requires predicting EV arrival times and discharge capacity, which can be achieved through in-depth data analysis and advanced machine learning methods.

1.3.2   Frequency control ancillary service

Energy storage systems offer the potential to provide rapid and accurate compensation for frequency deviations. Traditionally, conventional generators have provided this

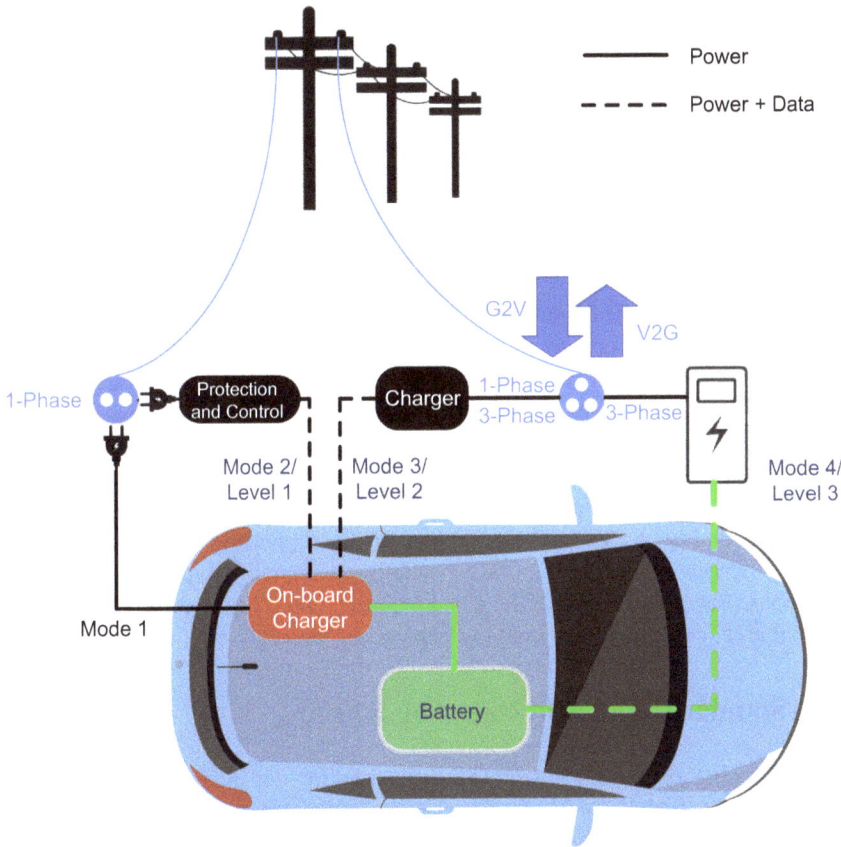

Figure 1.1   Different modes and levels of EV charging

service without direct cost, but with the increasing decommissioning of large-scale fossil fuel generation, DERs will need to assume this role. Frequency regulation is becoming increasingly complex due to declining system inertia and greater volatility caused by the growing proportion of renewable energy resources. V2G-enabled EVs can participate in the frequency control ancillary services market through aggregated EV behavior. Providing this service requires appropriate reserve capacity and response times (particularly for fast response services), increased visibility and controllability of the distribution grid for DNOs, and a policy framework that safeguards owner benefits while ensuring reliable grid services.

Droop control, a traditional local frequency control method used in generation units, adjusts injected active power linearly in proportion to frequency variations. A similar control strategy can be applied to energy storage systems. Droop control, combined with a dead band to minimize battery degradation, can serve as a primary frequency control mechanism. Given the fast response inverters required, appropriate standards settings should be considered. The dead band and droop slopes must be determined based on system characteristics, owner willingness to participate in frequency regulation and battery specifications. Market participation requires aggregators to maintain sufficient V2G reserve. Aggregators can submit bids based on forecasts of EV charging patterns, predicted V2G capacity, and contractual agreements with EV owners. These contracts will specify a minimum number of hours (daily or monthly) for EVs to provide secondary reserves and may define control ranges for each EV, such as the option to engage V2G mode or operate solely as controllable loads.

1.3.3   Operating reserves

Operating reserves refer to generation capacity that is available and can be brought online quickly to respond to fluctuations in demand or unexpected outages. Its about having a buffer of readily accessible power. Operating reserves are capacity-based services activated as required. Spinning reserve resources must reach full generation capacity within a specified timeframe upon receiving a dispatch signal. V2G-enabled EVs can contribute to operating reserves, providing positive reserves by exporting excess power to the grid or negative reserves by increasing system load. The mobility of EVs can be advantageous for supporting remote communities during emergencies or generation shortfalls, potentially avoiding costly grid extensions. For instance, a V2G-enabled electric school bus, typically parked in a regional area, could function as a local operating reserve.

1.4   Standards and grid codes for EVs

Efficient operation of EV-integrated infrastructure depends on comprehensive standardization. Five key areas for EV charging standardization are EVSE, data and communication systems, grid integration, safety, and energy/service market participation. V2G integration requires standardization of hardware, safety, and communication protocols. EV and grid system compatibility and interoperability are addressed by four sets of standards: plug standards, communication protocols, charging standards, and safety standards. Table 1.2 provides an overview of the principal EV-related standards.

1.4.1   EV charging interfaces

The widespread adoption of EVs hinges on a crucial factor: consistent charging standards. Currently, the world lacks a single, universal standard, creating confusion and inefficiency. Different regions have their dominant players:

- **US:** SAE, IEEE, and the recently introduced Tesla's North American Charging Standard (NACS);
- **Europe:** IEC (with the Combined Charging System Type 2 (CCS2) connector: a single connector for both AC and DC charging);
- **Japan:** CHAdeMO (a faster DC charging standard, though less common globally, capable of higher charging speeds than earlier standards);
- **China:** Guobiao (GB/T).

Table 1.2   EV charging modes based on the IEC 61851-1 standard

Standard	Title
Plug	
IEC 62196-1	Plugs, socket outlets, vehicle connectors, and vehicle inlets – General requirements for conductive charging of EVs
IEC 62196-2	Plugs, socket outlets, vehicle connectors, and vehicle inlets – Dimensional compatibility and interchangeability requirements for AC pin and contact-tube accessories
IEC 62196-3	Plugs, socket outlets, vehicle connectors, and vehicle inlets – Dimensional compatibility and interchangeability requirements for DC and AC/DC pin and contact-tube vehicle couplers
Communication	
ISO 15118-20	Vehicle-to-grid communication interface
IEC 61850	Communication protocols for electrical substation systems and networks
Charging	
IEC 61851-1	Electrical vehicle conductive charging system – General requirements for different charging modes
IEC 61851-21	Electrical vehicle conductive charging system – Requirements for conductive connection to AC/DC supply for both on-board and off-board charging systems
IEC 61851-22	Electrical vehicle conductive charging system – AC charging station
IEC 61851-23	Electrical vehicle conductive charging system – DC charging station
UL 9741	Standard for bi-directional EV charging system equipment
Safety	
IEC 61140	Protection against electric shock – Common aspects of installation and equipment
IEC 62040	Uninterruptible power systems
IEC 60529	Classification and ratings for degrees of protection provided by enclosures

These variations manifest in the design of charging ports and connectors. For instance, the US uses SAE J1772 and, recently, Tesla's NACS connector, while Europe favors the CCS2 connector, which offers both AC and DC charging capabilities in a single unit. CCS1 is another type of combo connector primarily used in North America. Australia recognizes the need for standardization to drive EV adoption and infrastructure development. The Federal Chamber of Automotive Industries has pledged to harmonize national standards, focusing on the IEC 61851-1, IEC 62196-2, and IEC 62196-3 standards from 2020. This ensures compatibility between car manufacturers, charging stations, and consumers. Beyond functionality, safety is paramount. Standards Australia is addressing battery safety with the AS/NZ 5139 standard, focusing on fixed batteries in buildings. However, it does not explicitly encompass EV batteries themselves. Mandatory application requires referencing it in the Australian Wiring Standards (AS3000) or state legislation.

1.4.2   Data and communication infrastructures

The successful implementation of V2G necessitates a robust communication infrastructure that seamlessly connects EVs, aggregators, and DNOs. This infrastructure should facilitate EV owner participation in energy markets through V2G by enabling a streamlined process like this: (i) *EV Owner Initiative*: EV owners express their interest in V2G participation by sending requests and relevant settings to an aggregator, (ii) *Aggregator Optimization*: Aggregators, in collaboration with DNOs, optimize V2G operations by considering technical and economic factors, (iii) *EV Charger Adjustment*: The aggregator updates the settings of the EV charger to align with the owner's request and the overall system optimization.

Effective communication infrastructure requires the establishment of clear information-sharing protocols and operational limits. This involves defining the specific data to be exchanged between participants and setting technical standards for control mechanisms. While various international standards address communication among EVs, EVSEs, and the grid, a universal standard has yet to emerge. Notable standards include

- **EV-to-EVSE communication:** ISO 15118, IEC 61851, and SAE J3072;
- **Grid-to-EVSE communication:** IEEE 2030.5, OpenADR, and IEC 61850;
- **Charging station operator-to-EVSE communication**: OCPP and IEC 63110.

1.5   Conclusions and future outlook

V2G technology presents a significant opportunity to transform the energy landscape by leveraging the vast energy storage capacity of EV batteries. It offers numerous potential benefits, including providing a new source of revenue for EV owners through participation in electricity markets. By discharging energy back to the grid during peak demand periods, EVs can contribute to reducing the need for costly grid infrastructure upgrades. Furthermore, V2G can enhance grid resilience and reliability, particularly in regional and remote areas with weaker power networks, by providing DERs and backup power during outages. This decentralized approach can significantly improve energy security and reduce reliance on centralized power generation.

Implementing V2G on a large scale presents several technical and social challenges. Effective aggregation of numerous EVs to provide meaningful grid services requires sophisticated communication and control systems. These systems must seamlessly integrate with existing grid infrastructure and electricity market operations, allowing for real-time energy trading and ancillary service provision. Furthermore, clear regulatory frameworks and appropriate tariff structures are essential to incentivize EV owner participation and ensure fair compensation for their contributions to grid stability. Social acceptance and understanding of V2G technology are also crucial, as widespread adoption relies on overcoming concerns about battery degradation and ensuring user-friendly interfaces for participation in V2G programs. Establishing economically sustainable V2G programs requires careful consideration of factors such as infrastructure costs, operational expenses, and the value of grid services provided by EVs.

However, recent advancements in battery technology and research findings are alleviating some key concerns surrounding V2G implementation, particularly regarding battery degradation. A Geotab analysis of 10,000 EVs revealed an average battery degradation rate of just 1.8% per year, suggesting a potential lifespan of 20 years or more, with 80% capacity retention after 12 years – often exceeding the typical lifespan of a fleet vehicle. Furthermore, CATL's recent announcement of a new EV battery with a 15-year warranty and zero degradation across the first 1,000 cycles demonstrates the rapid pace of battery innovation. Research from NREL further supports this positive trend, indicating that participation in V2G programs can actually *extend* battery life by maintaining a lower SoC during parking periods. These developments suggest that the long-term viability and economic sustainability of V2G are becoming increasingly promising, paving the way for wider adoption and integration of this technology into future energy systems.

Looking ahead, various pilot projects and trials are demonstrating the real-world potential of V2G. For example, NUVVE has been conducting V2G programs using electric school buses in the US, demonstrating the feasibility of using fleet vehicles to provide grid services [11]. These trials provide valuable data and insights for refining V2G technologies and business models. The development of more sophisticated aggregation topologies within the power distribution grid's operational structure is also an emerging field of research. This involves designing hierarchical control systems that can effectively manage large numbers of EVs connected to the grid. Robust and secure communication protocols are also crucial, enabling seamless data exchange between EVs, aggregators, DNOs, and the network operator. Advanced optimization and control algorithms are being developed to manage V2G operations at scale, considering factors such as grid stability, energy market prices, and individual EV owner preferences. These algorithms will enable efficient dispatch of V2G services, maximizing benefits for both grid operators and EV owners. Further research and development are also focused on advanced battery chemistries that are more resilient to frequent charge/discharge cycles, as well as the development of standardized V2G communication protocols to ensure interoperability between different EV and EVSE manufacturers. The convergence of these advancements promises a future where EVs play a crucial role in balancing energy supply and demand, supporting the transition to a cleaner and more sustainable energy future.

References

[1] Australian Government – Department of Climate Change, Energy, the Environment and Water. (2023). *Reducing transport emissions*. https://www.dcceew.gov.au/energy/transport.

[2] Environmental Protection Agency (US). (2025). https://www.epa.gov/ghge-missions/inventory-us-greenhouse-gas-emissions-and-sinks.

[3] International Energy Agency. (2024). *Global EV Outlook: Moving towards increased affordability*. https://www.iea.org/reports/global-ev-outlook-2024.

[4] McKinsey Center for Future Mobility. (2024). Exploring consumer sentiment on electric-vehicle *charging*. https://www.mckinsey.com/features/mckinsey-center-for-future-mobility/our-insights/exploring-consumer-sentiment-on-electric-vehicle-charging.

[5]   Lu, J., and Hossain, J. (2015). *Vehicle-to-grid: Linking electric vehicles to the smart grid*. IET, UK. http://ndl.ethernet.edu.et/bitstream/123456789/17379/1/472.pdf.

[6]   Thompson, A.W., and Perez, Y.(2020). Vehicle-to-everything (V2X) energy services, value streams, and regulatory policy implications. *Energy Policy*, 137, 111136. https://doi.org/10.1016/j.enpol.2019.111136.

[7]   Cook, C. (2019). Octopus EV - *Tariffs for Charging and Using Smart Charging, (Ohme cable and V2G)*. presented at the Connected Automated Mobility, Milbrook, Bedfordshire. https://www.cenexcam.co.uk/seminars/session/2019-day-2-hall-5-v2g-business-case.

[8]   Datta, U., Saiprasad, N., Kalam, A., Shi, J., and Zayegh, A. (2019). A price-regulated electric vehicle charge-discharge strategy for G2V, V2H, and V2G. *International Journal of Energy Research*, 43(2), 1032–1042. https://doi.org/10.1002/er.4330.

[9]   Gschwendtner, C., Sinsel, S. R., and Stephan, A. (2021). Vehicle-to-X (V2X) implementation: An overview of predominate trial configurations and technical, social and regulatory challenges. *Renewable and Sustainable Energy Reviews*, 145, 110977. https://doi.org/10.1016/j.rser.2021.110977.

[10] Foundation for Traffic Safety. (2021, April). New American Driving Survey. https://aaafoundation.org/wp-content/uploads/2021/04/New-American-Driving-Survey-Report-April-2021-1.pdf.

[11] Jones, L., Lucas-Healey, K., Sturmberg, B., Temby, H., and Islam, M. (2021). The A to Z of V2G. REVS/ANU. https://apo.org.au/sites/default/files/resource-files/2021-02/apo-nid311127.pdf.

Chapter 2

Adoption of EV and V2G infrastructures in existing grid and renewable energy systems

Hui Li[1]

2.1 Introduction

The 26th United Nations Climate Change Conference was held in Glasgow, Scotland, in 2021, following the famous Paris Agreement of 2014. The Paris Agreement set a long-term goal to limit global temperature rise to well below 2 °C in order to avoid the potential damage caused by global warming. Countries around the world were asked to achieve net-zero emissions targets by the second half of the twenty-first century and reduce their emissions. On November 3, 2021, at the United Nations Climate Change Conference, more than 450 financial companies from 45 countries pledged to use the $130 trillion in assets under their management to achieve the Paris Agreement's climate change goals. On November 13, 2021, the conference concluded in Glasgow. The conference reached a resolution document and consensus on the implementation details of the Paris Agreement. To achieve these ambitious goals, countries must first accelerate the phase-out of coal, second, reduce deforestation, and third, accelerate the deployment of electric vehicles (EVs) and increase investment in renewable energy [1].

To achieve these goals, the world's attention has turned to clean energy. Sustainable energy, particularly solar power, is considered a reliable and promising solution to reduce reliance on non-renewable energy sources. However, the unreliability of unconventional or renewable energy sources, namely wind and solar photovoltaic (PV), poses a significant challenge to the operation of microgrids.

With growing concerns about atmospheric issues and energy demands, EVs have gained increasing interest from consumers, governments, and industries. EVs are considered one of the clean commodities that can reduce dependence on non-renewable resources. Not only can EVs act as variable loads on the power grid, but they can also provide electrical energy to the grid. In this way, EVs work like energy storage devices. However, as EV deployment increases, the electricity demand for adding load is also increasing. Compared to standard charging

[1]College of Automation Engineering, Shanghai University of Electric Power, China

methods, EV charging methods add costs due to factors such as uncertainty, fluidity, anxiety, and nonlinearity caused by the maximum input electrical energy, and most of the electrical energy is absorbed through a single charger. The randomness and uncontrollability of the effective load of EVs have a complex impact on the positive stable operation and economic allocation of the system and have an impact on the layout of the conventional power grid. At the same time, due to the positive aspect of low pollution from EVs, the number of EVs will inevitably increase significantly in the future to limit the growth of fuel consumption and greenhouse gas (GHG) emissions [1]. Therefore, during the development process, research on energy storage systems and control systems will inevitably become key and difficult points.

Vehicle-to-grid (V2G) technology is centered around the interaction between EVs and the power grid, as shown in Figure 2.1. The core idea of V2G is to enable bidirectional communication and energy flow between EVs and the grid. This technology can effectively manage the charging and discharging process of EVs, minimizing the impact of EV load on the grid. At the same time, it fully utilizes the battery resources of EVs to increase the flexibility and stability of the grid's energy management. Therefore, we must believe in the tremendous development potential of V2G technology and be well-prepared to face various challenges during its development.

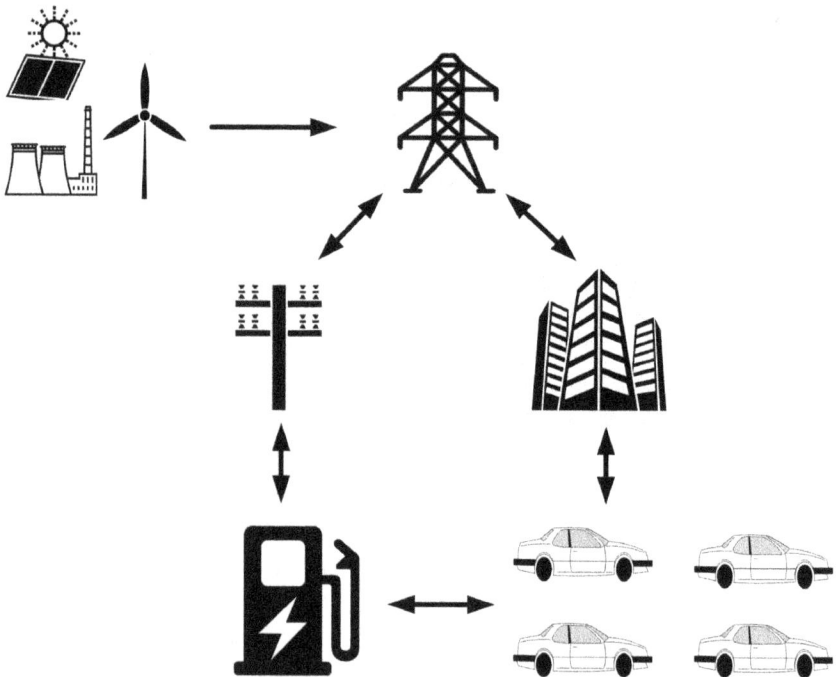

Figure 2.1   Schematic diagram of the V2G model

2.2   Overview of existing power grid and renewable energy

2.2.1   Structure and characteristics of traditional power grids

The traditional power grid consists of power generation, transmission, and distribution systems, as shown in Figure 2.2. In general, the power generation points are usually hydroelectric or thermal power plants, located far from consumers. Therefore, the generated electrical voltage is converted to high voltage by the primary substation transformer and transmitted over long distances to the secondary substation, which steps down the voltage and distributes it throughout the user area. The above description is the general cycle of electricity from generation to reaching consumers. Power and information flow only in one direction. Customers do not participate in decisions about when and how to receive electricity, and they also do not know the cost of the electricity they consume at a particular point in time [2].

The advantage of the traditional power grid lies in the fact that since power plants are usually located in remote areas, in order to deliver the electricity they generate to users in cities and reduce power losses during transmission, the voltage is stepped up at the generation end to reduce losses during long-distance transmission and gradually stepped down near the user end. At the same time, to improve the stability of the power supply quality, the transmission network operates multiple different types of power generation systems in parallel. The transmission network is basically an application structure of topology. Any node on the network must be designed to be able to join or disconnect from the network at any time without causing a significant impact on the power supply quality of the transmission network.

The traditional power grid has some shortcomings that need to be improved as it has developed to the present day. For example, the mechanical electric meters used in the traditional power grid can easily lead to inaccurate measurement of electricity after long-term operation due to mechanical wear and tear. Second, mechanical

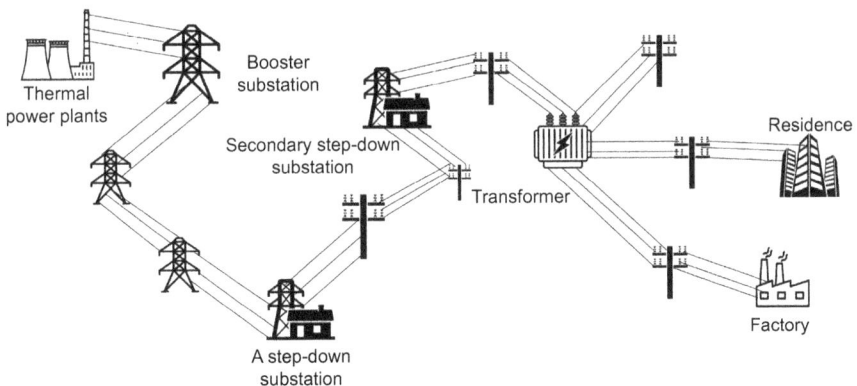

Figure 2.2   Schematic diagram of the traditional power grid structure

electric meters can only measure electricity consumption and cannot measure information such as voltage, current, and electrical load. In addition, mechanical power grids require manual transcription, so data cannot be obtained very frequently. From small to large, the traditional power grid also has many similar disadvantages. For example, when a fault occurs in the traditional power grid, monitoring and repair need to be completed manually, and local faults are likely to cause large-scale power outages. Moreover, the traditional power grid has a weak ability to integrate other energy sources. Small-scale and unstable renewable energy sources, such as wind and solar energy, are not suitable for integration into the traditional power grid [3].

2.2.2   Types and characteristics of renewable energy systems

Renewable energy refers to natural resources that can be regenerated in a short period of time, compared to fossil fuels that take over hundreds of millions of years to form and do not produce other pollutants during the process of energy conversion. Examples include solar energy, wind energy, geothermal energy, hydropower, tidal energy, biomass energy, etc., which are currently developing renewable energy sources worldwide [4].

The utilization of wind energy is mainly achieved through wind turbines using large rotor blades installed at high altitudes on land and sea to capture the kinetic energy generated by the wind. As the wind flows past the blades, the air pressure on one side of the blade decreases, pulling it down with a force called lift. The pressure difference between the two sides causes the blades to rotate, which in turn rotates the rotor. The rotor is connected to a turbine generator, which rotates to convert the kinetic energy of the wind into electrical energy.

Solar energy technology captures light or electromagnetic radiation from the sun and converts it into electrical energy. PV solar cells contain a semiconductor chip with a positive pole on one side and a negative pole on the other, forming an electric field. When light hits the cell, the semiconductor absorbs the sunlight and transfers energy in the form of electrons. These electrons are captured by the electric field in the form of current. The power generation capacity of a solar energy system depends on the semiconductor material, as well as environmental conditions such as heat, dirt, and shade.

Geothermal energy comes directly from the Earth's core. The heat from the core causes underground water reservoirs to boil, which are known as geothermal resources. Geothermal power plants typically use water wells to extract hot water from geothermal resources and convert it into steam for turbine generators. The extracted water and steam can then be reinjected, making it a renewable energy source.

Similar to wind turbines, hydroelectric power plants convert the kinetic energy of flowing water into electrical energy using turbine generators. Hydroelectric power plants are usually located near bodies of water and use flow-altering structures such as dams to change the flow of water. Power generation depends on the volume of flowing water and changes in height or head of water. Larger volumes of water and higher heads of water produce more energy and electricity, and vice versa.

Since our ancestors learned how to make fire, humans have likely been using energy from biomass or bioenergy for heating. Biomass, such as wood, dry leaves, and agricultural waste, is often burned but considered renewable because it can be regenerated or replenished. Burning biomass in boilers produces high-pressure steam, which rotates turbine generators to generate electricity. Biomass is also converted into liquid or gaseous fuels for transportation. However, emissions from biomass vary depending on the material being burned and are often higher than other clean sources.

Most countries are in the early stages of energy transition, with only a few countries having the majority of their electricity coming from clean energy sources. However, growth in the next decade may be greater compared to recent record-setting years. Therefore, regardless of when renewable energy takes over traditional energy, it is clear that the global energy economy will continue to change [5].

2.2.3 Challenges and limitations faced by existing systems

As an important means to reduce carbon emissions and promote the transformation of traditional energy structures, the construction of new power systems faces many challenges. Among them, the influx of a high proportion of renewable energy will have a huge impact on the existing power system. Since the various characteristics and operation mechanisms of the new power system are not yet clear, its development will not only face security and stability issues, but also encounter multiple challenges such as supply guarantee, dispatch optimization, transmission and consumption, and economic efficiency [6].

From the perspective of the power supply side, new energy power supplies mostly exhibit distributed characteristics, such as scattered wind power and distributed PVs. Compared with centralized power generation, distributed energy supply has the advantages of flexible operation and less pollution, but faces problems such as decentralized management and difficulty in large-scale centralized dispatching, which will weaken the regulation ability of the power grid and affect the security and stability of the power grid. Moreover, the development of new energy is constrained by geographical, environmental, and meteorological conditions and has large intermittency, fluctuation, and uncontrollability, and the energy density is also low, which increases the difficulty of ensuring power supply, resulting in new energy power sources only being able to provide electricity, but unable to participate in power balance. During peak electricity consumption periods, traditional conventional power sources are still required to achieve power and electricity balance. In the future, a higher proportion of new energy power sources will appear and be connected to the power grid as electricity providers, and the security and stability of guaranteed power supply still face huge challenges.

From the perspective of the power grid side, the traditional power grid form is mainly based on AC synchronization. The substitution of a large amount of new energy power weakens the fundamental factors of security and stability of the AC power grid and intensifies the complexity of the power grid system stability problem. The use of a high proportion of power electronic equipment has caused huge changes in the dynamic characteristics of the power grid system, and a series of problems such as reduced system inertia, weakened anti-interference ability, and enhanced wide-frequency oscillation have emerged, making traditional centralized control strategies increasingly difficult to adapt. The access of a high proportion of new energy power to the grid has led to an increase in the uncertainty of supply and demand balance regulation on both sides of the grid, and the system balance mechanism of the grid urgently needs to be optimized to meet the consumption demand of new energy with greater regulation capabilities. The utilization hours of new energy are low, the guaranteed output is low, and there is a seasonal deviation in the output. As an "intermediary" between power sources and loads, the power grid may experience reduced efficiency due to supply and demand mismatches, making it difficult to play its maximum role.

From the perspective of the load side, some countries have abundant resources but uneven distribution, which will lead to difficulties in local consumption of new energy and limitations on development. For example, in China, the electricity load centers are concentrated in the economically developed eastern regions where energy is relatively scarce, while new energy gathering areas are mostly in the western and three northern regions, which leads to difficulties in local consumption of new energy and limited development. Although China's new energy has developed rapidly in recent years, large-scale "wind abandonment" and "solar abandonment" phenomena have occurred in new energy-gathering areas. On one hand, there is high-speed growth of rigid demand on the load side, and on the other hand, there are difficulties in new energy consumption, with uncertainties on both sides, which leads to difficulties in responding to demand on the load side, requiring a new scheduling model to solve the problem [6].

In summary, building a new power system with new energy as the main body is an inevitable requirement. This is a complex large-scale system engineering that requires the strong investment of the entire industrial chain of generation, transmission, distribution, utilization, and storage, and the task is arduous and long.

2.3   EV and V2G infrastructures

2.3.1   Types and characteristics of EV charging infrastructures

Charging facilities connect EVs and the power grid and affect the concentration and dispersion of charging loads as well as the magnitude of charging power. According to the charging rate, charging facilities are mainly divided into two types: conventional charging and fast charging. The former has a more concentrated charging load, while the latter has a more dispersed charging load. Charging facilities are generally installed in residential areas, commercial centers, and centralized charging stations. The distribution of charging loads varies depending on the location where EVs are connected [7].

Based on the charging contact method, it can be further divided into EV charging and discharging based on electrical conductivity and wireless charging and discharging. Wireless charging and discharging is in the research and development stage and is gradually being applied, but currently, the main application is EV charging and discharging based on electrical conductivity. As shown in Figure 2.3, EV charging and discharging based on electrical conductivity is further divided into on-board chargers, off-board chargers, and integrated chargers, among which on-board chargers are widely used [8].

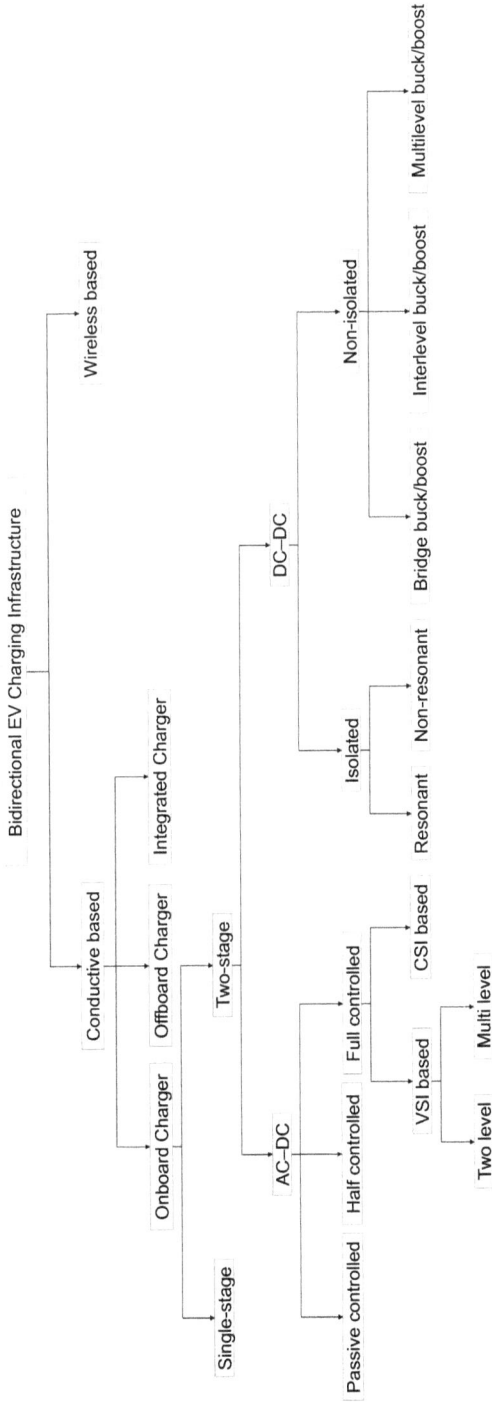

Figure 2.3   Classification of EV charging infrastructure

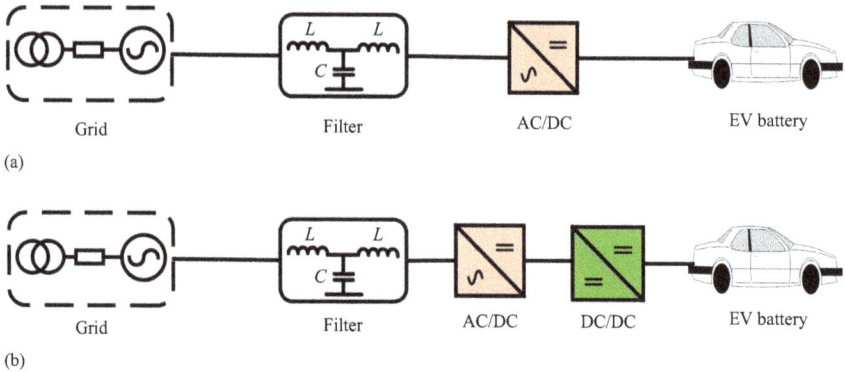

(a)

(b)

Figure 2.4   Diagram of V2G bidirectional converter topologies (a) Single-stage converter, (b) Two-stage converter

For on-board chargers, there are usually two main types of topologies for bidirectional AC–DC converters used for V2G, as shown in Figure 2.4, including single-stage and two-stage topologies. The single-stage topology is the simplest, with the least power components, which can reduce the volume and weight of the converter. The two-stage topology consists of an AC–DC converter on the grid side and a DC–DC converter on the battery side. The single-stage topology is more complex in control strategy and more difficult in electrical isolation, with high requirements for the circuit, while the two-stage topology has relatively independent front and rear stages, without coupling in control, and can also achieve more accurate control of the battery's charging and discharging power, making it the mainstream choice today. The grid-side is usually an AC–DC converter with power factor correction capability, while the battery-side DC–DC converter plays the role of controlling the charging and discharging power [9].

2.3.2   *Principle and key components of V2G technology*

2.3.2.1   Principle

From the perspective of participation structure, V2G can be divided into the following four layers: vehicle layer, intelligent charging and discharging equipment layer, control layer, and power grid layer.

The specific working principle of V2G is as follows:

1. The vehicle and the power system need to exchange information. For this purpose, the vehicle is connected to an intelligent bidirectional control device. The intelligent bidirectional control device can obtain the status information of the EV, and then the smart meter (SM) can send the acquired status information to the fleet and battery management system (FBMS).
2. The FBMS analyzes and integrates the collected information and then provides the integrated data to the supervisory control and data acquisition (SCADA) system of the power grid.

3. SCADA formulates dispatch strategies based on the obtained data and sends the dispatch strategies to the FBMS.
4. After receiving the dispatch instructions, the FBMS analyzes and calculates the corresponding control methods and controls the lower-level intelligent bidirectional control devices.
5. The intelligent bidirectional control devices control the EVs currently connected to the system according to the control instructions, realizing intelligent charging and discharging of the EVs.

The schematic diagram of the application principle of V2G technology in the power system is shown in Figure 2.5.

2.3.2.2 Key components

The intelligent charging and discharging equipment layer of V2G is the key layer of V2G technology, which mainly includes the following equipment:

1. Bidirectional charging pile
   The bidirectional charging pile is the core part of the V2G system. The EV bidirectional charging pile consists of a filtering circuit, a bidirectional AC–DC converter, a bidirectional DC–DC converter, a control module, and a PWM rectifier. The structure diagram is shown in Figure 2.6. Among them: AC/DC realizes the flow of electrical energy from the power grid to the EV; DC/AC realizes the flow of energy from the EV to the power grid, feeding electrical energy back to the grid; DC/DC realizes the regulation of the current magnitude during the charging and discharging process of the EV. In addition to charging and discharging control, the bidirectional charging pile also has functions such as overcurrent protection, overvoltage protection, and communication.

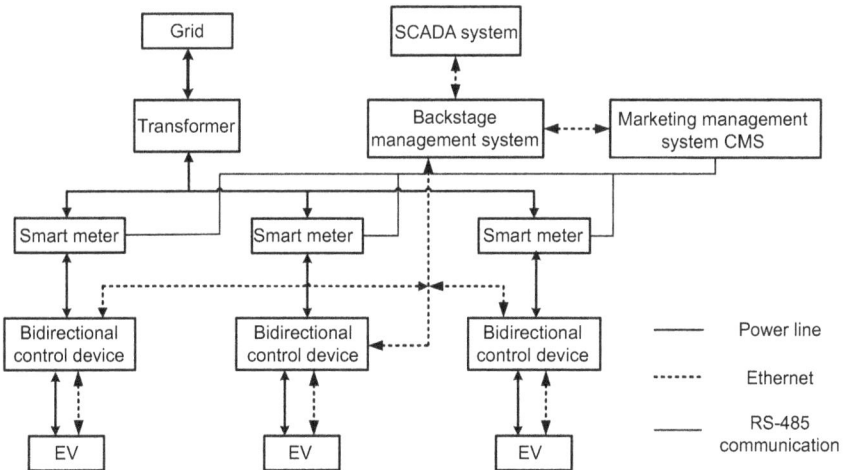

Figure 2.5 Schematic diagram of the application principle of V2G technology in the power system

2. Human–machine interaction system
    The main functions of the human–machine interaction system include identity recognition, operation prompts, ticket printing, data management, etc. The specific structure is shown in Figure 2.7.
3. Smart meter
    The SM is mainly responsible for calculation, data storage, information transmission, etc. The structure of the SM is shown in Figure 2.8.

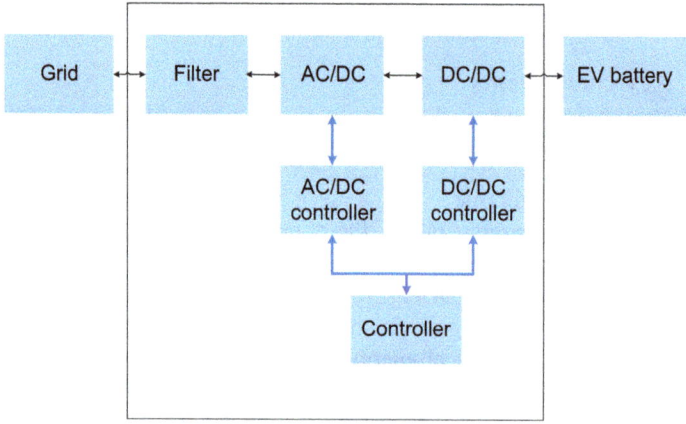

Figure 2.6   Structural diagram of EV bidirectional charging pile

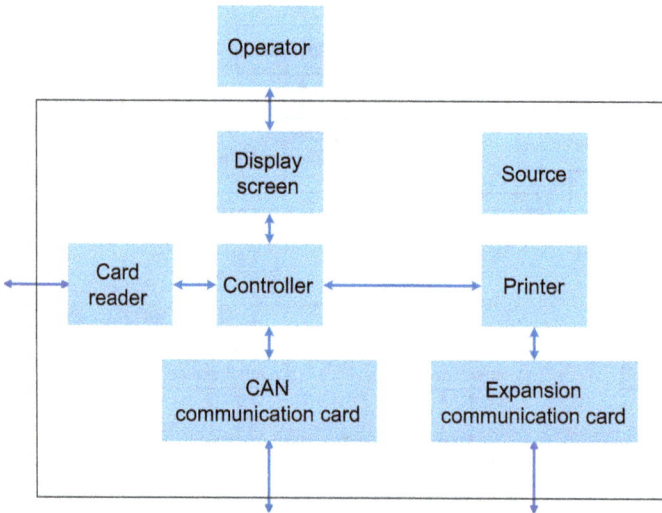

Figure 2.7   Structural diagram of the human–machine interaction system

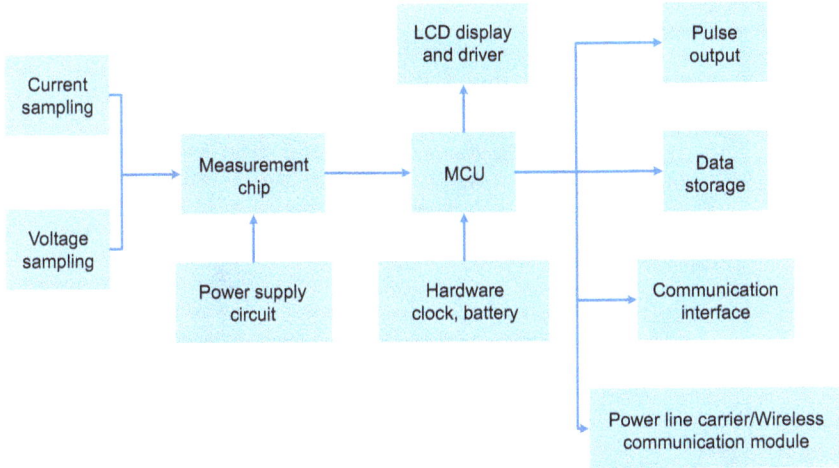

Figure 2.8   Structural diagram of the SM

2.3.3   Current status and development trends of EV and V2G infrastructure construction

The theoretical origin of V2G technology is in the United States, and it was further developed in Western Europe, the United States, and other regions, with related demonstrations being carried out [10].

As China's EV market continues to expand, V2G applications have drawn widespread attention across the country. Pilot demonstration projects are progressing rapidly, facilitating the extensive adoption of V2G technology in China. As the world leader in EV capacity, China possesses enormous market potential for V2G interaction among vehicle networks. Since 2020, national policies have been implemented to promote advancements in V2G technology, while major Chinese universities have also conducted in-depth research on V2G technology [11].

As the world's leading EV market, China has announced its goal to build sufficient charging infrastructure by 2025 to accommodate 20 million zero-emission vehicles, including fuel cell hybrid, plug-in hybrid, and pure EVs. This information is based on data provided by the International Energy Agency Global EV Policy Explorer [12].

According to incomplete statistics from the V2G Hub website, there are currently more than 100 V2G pilot demonstration projects worldwide, mainly concentrated in Europe, North America, and Japan, as shown in Figure 2.9. The "European Energy Storage Targets for 2030 and 2050 Research Report" released by the European Association for Storage of Energy shows that by 2030, the total energy storage demand of European countries is expected to be at least 187 GWh, of which V2G energy storage will be 33 GWh, accounting for approximately 17.6%. Europe is expected to become the world's largest V2G market.

Figure 2.9   Statistical results chart from the V2G Hub website

Among them, V2G-related vehicle charging standards such as ISO 15118-20, led by Europe and the United States, were officially released in April 2022. Currently, regions such as California in the United States are focusing on promoting the demonstration and popularization of V2G-related vehicle charging standards such as ISO 15118-20. For example, the California Energy Commission (CEC) has proposed a plan to implement ISO 15118-20 by 2025. As a vehicle–grid interaction standard led by the CharIN organization, ISO 15118 is very likely to become the technical route that Europe will focus on supporting [13–15].

Furthermore, in 2022, the US federal government announced its goal to increase the use of EVs in the transportation sector. Electric vehicles have gained significant attention from both the government and private sectors as a strategy to reduce GHG emissions. As a result, many governments and countries have recognized the need for EVs and have begun to focus on accelerating the electrification of vehicles while gradually reducing the sales of traditional internal combustion engine vehicles. Due to the environmental benefits and advancements in battery technology, EVs are also becoming increasingly popular among various stakeholders, including consumers, researchers, manufacturers, and regulatory agencies.

2.4 Integrating EV charging infrastructure into the existing power grid

2.4.1 The impact and challenges of EV charging on the power grid

With the increasing number of EVs, the integration of a large number of EV charging stations and power electronic devices has brought significant impacts on the power grid, power quality, and user terminals. This has caused issues such as power imbalance in the grid, voltage and frequency deviations, harmonic pollution, overheating of electrical equipment, equipment aging, and overloading of transformers and cables. The rapid development of EVs has both advantages and disadvantages for the power supply network of the power system. Disorderly charging can have a huge impact on the grid, which requires intelligent and orderly charging rules to optimize the operation of the grid and compensate for the grid's shortcomings in special situations. However, at the same time, EVs, as individual energy storage units, provide active power supplementation and reactive power compensation for regional loads and power sources, greatly optimizing the operation mode of the power grid. Another example is to utilize the peak-shaving and valley-filling capabilities of EVs to meet the operational needs of the power grid, which can fully leverage the social benefits of EVs.

The charging behavior of EVs has significant randomness and uncertainty. Connecting them to the grid for charging without distribution network regulation will inevitably have a non-negligible adverse impact on the distribution network. Specifically:

1. Increase the peak-to-valley difference of the distribution network load. In the disorderly charging mode without the participation of the distribution network regulation, the charging behavior of vehicle owners will be influenced by

living habits and remain consistent with the base load. If EVs are charged only according to the owners' electricity consumption habits, it is likely to occur during the peak period of the base load, increasing the peak-to-valley difference of the distribution network load.

2. Increase the difficulty of distribution network operation and regulation. In the disorderly charging mode without the participation of the distribution network regulation, EV owners will choose to charge at any time and any location according to their own preferences. This randomness will increase the difficulty of monitoring and dispatching the distribution network, making it impossible to accurately predict the real-time load changes of the distribution network in a timely manner.

3. Increase the losses of the distribution network. In the disorderly charging mode without the participation of the distribution network regulation, the time and location chosen by EVs for charging are uncertain. If a large number of EVs are connected to areas with local power supply tensions or during peak load periods, it will cause adverse situations such as a decrease in the node amplitude of the grid voltage and an increase in transformer losses.

4. Reduce the economic efficiency of the distribution network. In order to encourage vehicle owners to participate in unified scheduling, grid companies will provide charging compensation fees to vehicle owners, making them more willing to participate and stimulating their enthusiasm to respond to the call. Electric vehicles that do not participate in unified regulation often charge during peak electricity price periods, avoiding off-peak periods, resulting in higher charging costs and missing out on the grid's charging discounts, causing certain economic losses [16].

The impact and challenges of EV charging on the power grid cannot be ignored. To address these impacts and challenges, it is necessary to consider the layout and expansion of charging facilities in grid planning and operation, formulate reasonable charging strategies and scheduling mechanisms, and promote the development of intelligent management and technical support. Only through comprehensive measures and cooperation we can ensure the smooth interaction between EV charging systems and the power grid and provide reliable support for the development of a sustainable energy future.

2.4.2 Intelligent charging strategies and algorithms

Intelligent charging strategies and algorithms can be formulated based on the power grid load conditions, the charging needs of EV users, and the supply of renewable energy. The following are some common intelligent charging strategies and algorithms proposed to address the adverse effects of EV charging behavior on the distribution network mentioned in Section 2.4.1:

(1) Increase the peak-to-valley difference of the distribution network load. Literature [17] studies the travel characteristics of private cars, divides residential areas into detailed categories, and analyzes the capacity of residential areas to accommodate EV charging. Simulation results show that the concentrated charging

periods of EVs highly overlap with the peak periods of the base load in residential areas, causing the peak-to-valley difference of the load in different types of residential areas to continuously increase with the number of EVs, thus increasing the difficulty of peak regulation in the distribution network. After adopting the peak-valley electricity price guidance strategy, the peak-to-valley difference of the load is effectively reduced. Literature [18] uses a probabilistic statistical model to simulate the characteristics of EV users' electricity consumption behavior. Through orderly charging and discharging control, EVs are controlled to discharge during the peak period of the base load, alleviating the power supply pressure of the distribution network. The charging behavior of EVs is shifted to the valley period, taking full advantage of the low electricity consumption and low electricity prices during the valley period. While meeting their own electricity needs, it alleviates the power supply pressure of the distribution network.

(2) Increase the losses of the distribution network. Literature [19] analyzes the performance of distribution network nodes by constructing a network loss sensitivity index. It uses power flow calculation and convex optimization algorithms to obtain an orderly charging and discharging strategy for EVs. Simulation results show that the proposed method can effectively reduce charging costs and network loss costs under different penetration rates and proportions of responsive V2G modes. Literature [20] establishes a reactive power optimization model considering orderly charging and discharging to achieve active and reactive power regulation between the distribution network and EV users. Simulation results show that the proposed method can not only improve the voltage curve and reduce the network loss rate but also meet users' temporary travel needs.

(3) Reduce the economic efficiency of the distribution network. Literature [21] combines price-based and incentive-based demand response measures and proposes two EV load aggregator scheduling strategies: fixed contract strategy and flexible contract strategy. Simulation results show that both fixed and flexible contract strategies can effectively improve the benefits of load aggregators, while the flexible contract strategy can further reduce the charging costs of EVs. Literature [22] addresses the problem that traditional time-of-use electricity prices and real-time electricity price demand response mechanisms will generate new load peaks during low load periods. It proposes a time-of-use electricity price dynamic optimization method. The particle swarm algorithm is used to perform a two-stage optimization of the charging and discharging behavior of EVs, and simulation analysis is conducted on the charging demand under different charging and discharging strategies, different optimization weights, and different participation levels. The results show that compared with other strategies, the proposed optimization strategy can significantly reduce users' charging costs.

There are also some other potential issues and intelligent charging strategies and algorithms, which will not be elaborated on further here. In summary, the combination of these intelligent charging strategies and algorithms can achieve intelligent management of EV charging, improve the stability and reliability of the power grid, reduce energy waste, and promote the virtuous interaction between EVs and the power grid.

2.4.3   Grid upgrades and retrofits to accommodate EV charging needs

With the popularization of EVs, the demand on the power grid is also increasing, posing new challenges and opportunities for the grid. To meet the needs of EV charging, the power grid must undergo upgrades and retrofits to adapt to the growing electricity demand and ensure stable operation.

First, the infrastructure of the power grid needs to be expanded and retrofitted to increase the number and capacity of charging stations. This includes installing more charging stations in city centers, commercial areas, parking lots, and other locations to provide convenient charging options for EV owners. At the same time, to meet the charging needs along highways and major traffic arteries, more fast charging stations and super charging stations need to be built to provide quick and convenient charging services.

Second, the power grid needs to adopt intelligent technologies to achieve dynamic scheduling and management of charging demands. Through smart charging stations and intelligent charging networks, the grid can optimize the allocation of charging resources based on real-time power supply and demand conditions and charging needs, avoiding power shortages during peak charging periods and ensuring stable operation of the grid. There are still many scholars conducting research in this field, and more in-depth research is needed.

Furthermore, to improve the reliability and security of the power grid, it is necessary to strengthen the monitoring and fault detection capabilities of the grid. By real-time monitoring of the operating status of the power grid and the working conditions of charging equipment, potential faults and problems can be detected and handled in a timely manner, ensuring that the grid can continuously and stably provide services for EV charging.

In summary, upgrading and retrofitting the power grid is one of the key measures to adapt to the EV charging needs. By expanding charging facilities, adopting intelligent technologies, and strengthening monitoring and management, the power grid can better meet the growing demand for EV charging, promote the popularization of EVs, and drive the development of the electric power industry.

2.5   Integrating V2G technology into existing power grids and renewable energy systems

2.5.1   The role of V2G technology in balancing peak-valley differences in existing power grids and suppressing fluctuations in renewable energy

The sustainable development of power systems faces many challenges, one of which is the volatility of power grid loads and the instability of renewable energy. V2G technology, by connecting EVs to the grid and using their battery storage as adjustable power sources and loads, can effectively address the fluctuations in grid loads and renewable energy.

V2G technology is based on the bidirectional charging and discharging function of EVs, allowing them to supply power to the grid or draw power from the
grid. When the grid load is light, the V2G system can achieve load regulation by
connecting EVs to the grid and releasing their battery storage to the grid; when the
grid load is heavy, the V2G system can use the battery storage of EVs as a backup
power source to help balance the grid load.

Enhancing grid stability: The application of V2G technology can enhance the
stability of power networks to a certain extent. Traditional power networks often
struggle to meet demand during peak usage, leading to unstable power supply. By
connecting EVs to the grid, vehicle owners can release the energy stored in their
vehicles to the power network to meet the demand during peak periods. For
example, when the power network is unable to meet demand, power companies can
purchase the storage capacity in vehicle owners' vehicles to make up for the power
shortage and increase grid stability. This approach not only relieves pressure on the
power network but also improves energy utilization.

Promoting renewable energy development: As global demand for renewable
energy continues to grow, the application of V2G technology also helps promote
the development of renewable energy. Renewable energy sources generate electricity continuously, but due to the instability of energy demand, there are often
situations where it cannot be fully utilized. V2G technology allows EV owners to
use their vehicle's storage capacity to store and distribute renewable energy,
improving energy utilization. For example, if wind power and solar power generate
more electricity than the demand during a certain period, this excess power can be
stored in vehicles for use at night or on cloudy days. Through this approach, the
utilization of renewable energy is increased, while reducing the demand for traditional non-renewable energy sources such as coal-fired power generation.

2.5.2   V2G interfaces and communication with existing power grids and renewable energy systems

The smart grid will enhance the security, performance, and flexibility of the traditional power grid through advanced communication and control architectures.
Integrating EVs into the smart grid helps fill valleys, regulate peaks, and reduce EV
charging costs.

To control the large-scale integration of EVs in the smart grid, a reliable
communication network is required. The smart grid communication system plays a
crucial role in load forecasting, EV driving patterns, and cost transfer between
different entities in V2G.

Different entities in the smart grid infrastructure generate a large amount of data.
To control and analyze this data, various communication technologies are needed,
which can be mainly divided into two different types: wireless and wired technologies. Wireless communication technologies support connections to remote areas with
low infrastructure costs, but signal paths may cause interference, attenuation, and
errors, while wired technologies may not be affected by these issues.

The information flow of the V2G infrastructure in the smart grid scenario
occurs in two ways. The first way is to transmit data from EVs and sensors to SMs

in charging stations. The other way is to transmit data from SMs in charging stations to utilities or data centers. Based on the distance between entities in the V2G infrastructure, different combinations of wired and wireless communication technologies are considered to exchange electronic device data.

The communication system of the smart grid includes different networks, such as home area network (HAN), neighborhood area network (NAN), and wide area network (WAN). HAN connects SMs in charging stations and EVs. NAN collects data from different HANs and transmits it to aggregators with the support of WAN. By integrating EVs into the smart grid through charging stations, the SM in charging stations can obtain real-time information about the required or consumed power. In this way, SMs support load forecasting, while aggregators act as intermediaries for communication between utilities and EVs.

By integrating EVs into the smart grid through charging stations, the SM in charging stations can obtain real-time information about the required or consumed power. In this way, charging stations support load forecasting, while aggregators act as intermediaries for communication between utilities and EVs.

The overall overview of a V2G system with an integrated communication network is shown in Figure 2.10. EVs exchange energy with the grid through charging or discharging. With the support of different aggregators at the node and substation levels, fault information in the distribution system is transmitted to the data center using a combination of distance-based wired and wireless technologies. Necessary measures can be taken to resolve problems in the distribution system in a short time. Continuous monitoring of the V2G system in the smart grid is a significant advantage.

In the V2G infrastructure, various electrical parameters need to be communicated from the EV fleet to the power company and vice versa. This communication should take place at short intervals when EVs are parked at charging stations. If the transmitted data packets do not reach the destination electrical

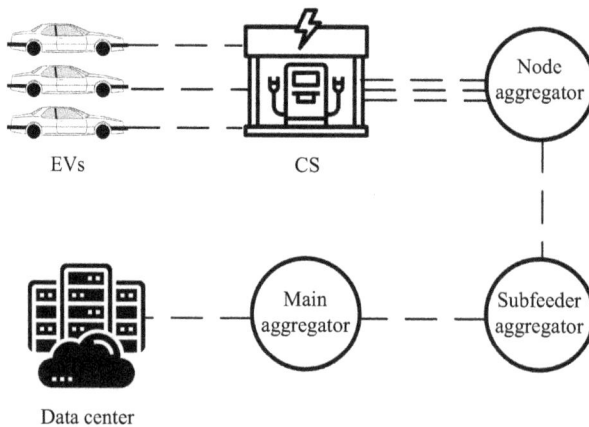

Figure 2.10   V2G system communication network

components within a fixed time interval, it will affect the state of charge (SOC) of the EV batteries and the grid voltage. Electric vehicles can only charge when the distributed node voltage is higher than the charging point voltage [23]. It is evident that the smart grid communication system is essential in V2G.

2.5.3 Optimized scheduling and control strategy for V2G integration

2.5.3.1 Optimized scheduling

With the expansion of EV scale and the development of V2G technology, the application purpose of V2G has gradually shifted from initially providing reliability and stability for the power system to improving the benefits and efficiency of multiple participants. Therefore, when conducting V2G scheduling, the dispatch center needs to fully balance the interests of all parties and provide optimal scheduling strategies for each participant. For users, the optimal scheduling strategy can maximize their revenue and usage efficiency while ensuring vehicle safety and comfort; for operators, the optimal scheduling strategy can reduce costs and risks, improve operating efficiency and competitiveness; for the power system, the optimal scheduling strategy can improve the reliability and stability of the power system, achieve a balance between power supply and demand, and thus promote the large-scale application of renewable energy. However, the optimal scheduling strategies under different scenarios and requirements will vary due to factors such as climate, policies, and social development. In order to find the most suitable scheduling method, many scholars have conducted research on this topic in recent years.

Currently, research in the international field of V2G optimization scheduling primarily focuses on the following aspects:

1. V2G optimal scheduling based on energy management: This type of research takes improving the reliability and stability of the power system as the main purpose, takes load fluctuation, node voltage deviation, network energy loss, etc., as optimization objectives, completes the optimal scheduling of V2G, and provides strategies for distribution network planning.
2. V2G optimal scheduling based on cost–benefit: With the main purpose of reducing operating costs and improving economic benefits, the scheduling strategy of V2G is optimized for different electricity markets and business models to achieve optimal cost–benefit. Relevant research mainly includes cost–benefit analysis and market mechanism construction. The emergence of V2G brings together a large number of EVs and improves their ancillary service capabilities. During peak power consumption periods, V2G can maximize ancillary service revenue by providing backup capacity and frequency regulation services.
3. V2G optimal scheduling based on user behavior: In order to improve user experience and satisfaction, reasonable V2G scheduling strategies are designed by deeply analyzing user behavior characteristics, charging needs, travel habits, and other factors to maximize user efficiency and revenue. Relevant

research mainly includes user behavior modeling, demand forecasting, scheduling strategy design, and other aspects.

4. V2G optimal scheduling based on environmental requirements: With the main purpose of achieving low-carbon and green development, the charging and discharging behavior of EVs is optimized through scheduling to achieve sustainable development and environmental protection of the power system. Relevant research mainly includes environmental demand analysis, scheduling strategy design, energy management, and other aspects.

5. V2G optimal scheduling based on battery capacity protection: Considering that the reduction of battery life will directly affect the economic benefits and sustainability of the V2G system, this type of research usually quantifies the battery aging generated in the V2G process through different means and introduces it into the objective function; some studies also limit battery aging to an acceptable range by adding constraints [24].

2.5.3.2   Control strategy

In addition to the selection and integration of topological structures, appropriate control strategies need to be researched. Current V2G control strategy studies are specifically designed for selected mature topologies or proposed new topologies, aiming to improve converter performance in certain aspects, as shown in Figure 2.11. Based on a comprehensive review of the literature, research on V2G converter control strategies focuses on the following aspects:

1. Utilizing decoupling and filtering techniques to reduce the values and sizes of capacitors and inductors in the circuit, thereby reducing the converter volume and lowering costs.

2. Applying novel control strategies to suppress harmonic distortion in grid-connected currents and improve the quality of grid-connected electrical energy.

3. Achieving a wide voltage range for input or output, optimizing charging and discharging efficiency, and enhancing response speed.

4. Combining grid scheduling control and battery energy management technologies to expand V2G functions such as reactive power control, allowing EVs to provide more services through converters.

5. In the field of digital control, optimizing control processes and further designing functions such as human–machine interaction and metering and billing for productization [25].

2.5.4   *Case study: Application projects of V2G technology in existing systems*

On February 7, 2024, China's first real-time settlement V2G new energy charging station was put into operation in the Zhengzhou Airport Economic Zone. The V2G integrated energy charging station in the Zhengzhou Airport Economic Zone was invested and constructed by Xinggang Electric Power Company, the owner of the

Forward charging

Bidirectional AC–DC module   Bidirectional DC–DC module

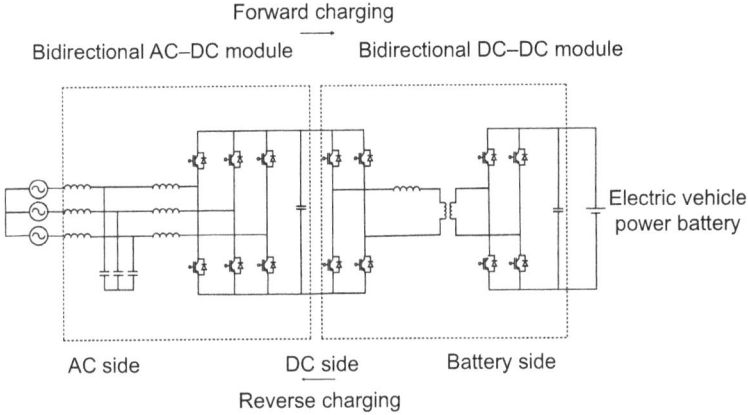

AC side   DC side   Battery side

Reverse charging

Figure 2.11   V2G converter system topology diagram

incremental power distribution reform pilot project in the zone. It is located on the south side of the company's Zhiyang Road Smart Energy Center. The station is not only equipped with a 1,000 kW super fast charging system but also integrates various application scenarios such as wind, solar, storage, charging, and discharging. By leveraging the institutional advantages of the incremental power distribution network, it fully realizes convenient functions such as bidirectional charging and discharging, convenient QR code scanning, instant charging and unplugging, and real-time settlement.

In December 2021, the third large-scale charging station invested and constructed by the State Grid Urumqi Power Supply Company, the Degang Wanda Charging Station, was put into operation. It is the first public charging station in Xinjiang that adopts V2G charging technology. It employs two advanced charging technologies: V2G charging and orderly charging. The station is equipped with eight 60 kW DC V2G charging piles and three 60 kW DC orderly charging piles, which can simultaneously meet the charging needs of 11 EVs. V2G charging significantly improves vehicle charging efficiency. It takes only 40 min for a vehicle to charge from low power to 80% and 1 h to fully charge. It is estimated that the station can provide 3,000 charging services annually, generating 80,000 kWh of charging electricity. Compared with fuel vehicles, it can reduce fuel consumption by 32,000 l and carbon dioxide (CO_2) emissions by 76,000 kg annually.

In 2019, the Renault Group launched a large-scale trial of V2G technology, introducing a fleet of 15 ZOE vehicles in Europe to achieve V2G charging. In June 2023, Renault announced that the upcoming Renault 5 EV will be the first to support bidirectional charging, with the ability to provide up to 11 kW of power to the home grid and save users 50% on charging costs.

It is evident that with the continuous development and promotion of V2G technology, EVs are poised to become a crucial component of future energy systems, surpassing traditional energy storage devices. However, it is important to note that this requires intensified efforts in technology research and development, standard formulation, market promotion, and other areas to ensure the safety, reliability, and efficiency of V2G technology.

2.6   Economic and environmental benefit analysis of EV and V2G infrastructure integration

2.6.1   Cost–benefit analysis of EV and V2G integration

Charging pricing generally consists of the sum of electricity charges and service fees, which vary among different operators. Taking the mainstream charging operators in the Chinese market, such as StarCharge and TELD, as examples, currently both derive their charging rates from electricity and service fees. For instance, during the same time period (15:00–18:00), at StarCharge (location: Yansha Friendship Mall, Liangmaqiao, Chaoyang District, Beijing), the charging rate (original price) is 1.4265 RMB/kWh, comprising an electricity charge of 0.8265 RMB/kWh and a service fee of 0.6 RMB/kWh. At TELD (location: North Area of Taikoo Li, Sanlitun, Chaoyang District, Beijing), the charging rate is 1.6469 RMB/kWh, with an electricity charge of 0.8469 RMB/kWh and a service fee of 0.8 RMB/kWh.

Based on calculations, the annual average income for a single 50 kW DC charging pile is approximately 17,500 RMB. Considering that consumers are sensitive to electricity prices, charging operators typically adopt a policy of charging electricity fees at cost, aligning with grid tariffs without arbitrage. For simplification, this aspect is not factored into the model. Thus, the revenue from charging piles primarily relies on service fees. Taking into account the current market conditions of both operators, assuming an efficiency utilization rate of 8% per charging pile, meaning it operates 8% of the year (700 h annually out of 365 days), a single 50 kW DC charging pile averages an annual charge of 35,000 kWh. Based on the service fee standard mentioned earlier of 0.5 RMB/kWh, the estimated annual service fee income for a single charging pile is approximately 17,500 RMB.

The initial investment cost for a single 50 kW DC charging pile is approximately 57,000–60,000 RMB. This corresponds to an initial investment of about 1.14–1.2 RMB per watt. The detailed calculation is as follows:

- **Equipment investment:** The cost of the DC charging equipment (including the monitoring system) is approximately 0.5 RMB per watt. Therefore, the equipment investment for a single 50 kW DC charging pile is around 25,000 RMB.
- **Civil and construction investment:** This includes site layout, purchase and assembly of cables and wires, and canopy assembly. The averaged cost per pile is approximately 10,000–13,000 RMB.
- **Power distribution equipment cost:** This includes relays, protection devices, low-voltage electrical appliances, and meters. The averaged cost per pile is about 22,000 RMB.

The annual total cost for a single charging pile typically consists of fixed costs and variable costs. Based on the above, the fixed cost for a single 50 kW DC charging pile is approximately 13,800 RMB. The detailed calculation is as follows:

- **Interest expense:** Assuming a loan ratio of 50% and an interest rate of 6%, the annual interest expense is approximately 1,800 RMB.
- **Depreciation expense:** Assuming a depreciation period of 10 years and an initial investment cost of approximately 57,000–60,000 RMB, the annual depreciation expense is approximately 6,000 RMB.
- **Equipment maintenance and labor costs:** Assuming it is 10% of the initial investment, the estimated annual cost is approximately 6,000 RMB.

The variable costs for a single charging pile generally include two parts:

1. **Electricity cost:** As mentioned earlier, adopting a flat-rate policy without engaging in price arbitrage. For the sake of simplicity in the model, this aspect is not considered.
2. **Sharing of parking lot service fees and operation and maintenance costs:** Some parking lots require a certain percentage of charging service fees to be shared, along with other ongoing operational and maintenance costs. Assuming these costs amount to 10% of the service fee, it totals approximately 1,750 RMB.

Based on the above, for a single 50 kW DC charging pile, the annual revenue is approximately 17,500 RMB, with annual total costs of around 15,500 RMB, resulting in an annual profit of about 2,000 RMB. Considering annual depreciation costs of approximately 6,000 RMB, the annual net cash flow is approximately 8,000 RMB. With initial investment costs estimated at around 57,000–60,000 RMB, the static payback period for the investment is approximately 8 years, as shown in Tables 2.1–2.3.

Based on the above calculation process, the key factors for increasing the profit of charging piles and shortening the investment payback period are the

Table 2.1   Calculation of the average annual cost of a single DC charging pile

Project	Cost (RMB)
Investment of pile equipment	25,000
Civil and construction investment	13,000
The investment of equipment on the distribution side	22,000
Initial investment (subtotal)	**60,000**
Annual interest	1800
Average annual depreciation expense	6000
Average annual equipment maintenance and labor costs	6000
Average annual fixed costs (subtotal)	**13,800**
Annual variable costs	**1750**
Average annual total cost	**15,550**

Table 2.2   Calculation of the average annual profit of a single DC charging pile

Project	Income and expenses (RMB)
Charging service fee income	+17,500
Average annual total cost	−15,550
Average annual profit	+2000

Table 2.3   Calculation of the static payback period

Project	Amount (RMB)
Initial investment	60,000
Annual net cash flow	8000
Static payback period	8 years

improvement of single-pile utilization efficiency and sensitivity analysis of electricity service fees. The results are as follows:

Assuming the electricity service fee remains at 0.5 yuan/kWh, increasing single-pile utilization efficiency from 8% to 10% results in a rise in single-pile profit from 20,000 to 59,000 RMB. On average, for every 1% increase in single-pile efficiency, profit roughly increases by 20,000 RMB. Assuming the electricity service fee remains at 0.5 yuan/kWh, increasing single-pile utilization efficiency from 8% to 10% reduces the investment payback period from 8 years to 5 years, significantly less than the equipment's usable life (approximately 10 years) [26]. This indicates that attracting a large number of consumers to use EVs through various means is a key factor in shortening the investment payback period.

Furthermore, continuous advancements in battery technology and cost reduction are prerequisites for V2G deployment. For EV users, the primary condition for participating in V2G is not compromising normal driving needs, which requires further improvement in battery energy density to achieve higher overall lifecycle range and enhance the available capacity and economics of power batteries participating in V2G. Only under the premise of rapid increase in battery energy and continuous decrease in battery costs can the willingness of EV users to participate in V2G be guaranteed, thus realizing the large-scale development of V2G [27].

2.6.2   The contribution of EV and V2G integration to the environment and sustainable development

V2G integration, as an emerging energy management technology, makes significant contributions to the environment and sustainable development. By incorporating EVs into the power system and enabling bidirectional energy flow and intelligent control, V2G integration provides strong support for building a clean and sustainable energy future.

V2G integration can significantly reduce carbon emissions. Traditional gasoline vehicles emit large amounts of CO_2 and other harmful gases, which negatively impact climate change and air quality. In contrast, EVs, as an essential component of V2G integration, have zero tailpipe emissions. Electric vehicles are powered by batteries and use electric energy to drive the vehicle, eliminating the issue of exhaust pollution. This zero-emission feature provides strong support for improving urban air quality and has a positive impact on reducing global carbon emissions.

V2G integration also promotes the utilization and promotion of renewable energy. Renewable energy sources such as solar and wind power are intermittent and fluctuating, which poses challenges to the stable supply of the power system. However, by connecting EVs with renewable energy systems, V2G integration can treat EVs as mobile energy storage devices. When renewable energy generates excess energy, it can be charged at charging piles, storing the surplus electrical energy. When energy demand peaks, this stored energy can be released to supply the grid, reducing the load pressure. By achieving energy balance, V2G integration improves the utilization of renewable energy, reduces the demand for high-emission energy sources such as traditional coal-fired power generation, and thus has a positive impact on the environment and sustainable development.

V2G integration also promotes energy balance. Electric vehicles, as energy storage devices, can intelligently regulate energy according to the needs of the power system. During off-peak periods of grid demand, EVs can absorb excess electrical energy for charging and store it. During peak periods of grid demand, EVs can release the stored energy to meet the energy needs of the power system. This intelligent energy scheduling mechanism can balance load demand, reduce imbalances between supply and demand, decrease energy waste, and improve energy utilization efficiency. By alleviating the load pressure on the power system, V2G integration provides strong support for energy balance and contributes to a sustainable energy system.

2.7   Challenges faced by the integration of EV and V2G infrastructure into the existing power grid

2.7.1   *Compatibility issues of technologies and standards*

Currently, there are relatively few smart charging standards, and there are international differences. In the process of building smart charging infrastructure, attention needs to be paid to hardware (such as chargers and compatible vehicles), software (such as orchestration platforms), and standards to ensure the compatibility of system components. At the international level, Europe has made efforts in the standardization of smart charging, aiming to achieve interoperable, seamless, and secure systems. To unleash the potential of smart charging, the European Environmental Citizens' Organisation for Standardisation (ECOS) recommends continuing to advance the standardization process of smart charging in line with the EU's regulatory requirements for smart charging [28].

To support "all smart charging scenarios," international and European standardization organizations are developing, revising, or still formulating new standards, such as the International Organization for Standardization (ISO), the International Electrotechnical Commission (IEC), and the European body CENELEC. IEC 61851 on "Electric vehicle conductive charging system" defines the safety rules for charging using plugs and cables (AC or DC) and the necessary low-level communication between the charging station and the EV, while ISO 15118 on "Road vehicles – Vehicle to grid communication interface" defines the high-level communication between the charging station and the EV for controlling charging services on top of IEC 61851. In addition, IEC 63119 "Charging service provider" will become the international standard for roaming and payment in the context of EV charging services.

Although the development activities of these technical standards are ongoing, more standardization work is still needed in terms of interoperability to ensure that smart chargers and infrastructure are compatible with all standardized hardware and equipment from energy suppliers.

In some countries, the standard system is still to be improved. For example, the current standards for new energy vehicles and charging facilities do not effectively standardize the V2G function, and the existing grid-connection and metering standards do not fully consider the needs of vehicles and equipment with V2G capabilities. In addition, the standard system for data interaction, operation control, and information security related to V2G aggregation participating in electricity trading also needs to be established and improved. The current V2G pilot demonstrations mainly rely on enterprises adopting private protocols for implementation, which cannot support commercial applications.

Therefore, to solve these problems, it is necessary to further improve the standard system for smart charging. This will help ensure interoperability between different devices and systems and provide support for commercial applications. At the same time, strengthening cooperation with international standardization organizations and learning from the experience of other countries and regions can promote the development and unification of standards, which can drive the widespread application of smart charging technology and the sustainable development of the EV market.

When integrating EVs and V2G infrastructure into the existing power grid, effective communication and coordination must be achieved between the various technologies and equipment involved, such as charging piles, EVs, and energy management systems. However, due to different standards and specifications among different manufacturers and technology providers, compatibility issues may arise, affecting the normal operation of equipment or the achievement of optimal performance.

The key to solving this problem is to establish unified technical standards and interoperability protocols. This requires close cooperation between manufacturers and standards-setting organizations to ensure seamless communication and collaboration between devices and systems. At the same time, it is necessary to develop unified testing and certification standards to ensure that all devices comply with the same technical specifications and performance requirements.

In addition, the updating and upgrading of devices and systems is also an important issue. As technology develops and updates, devices and systems may become outdated, leading to compatibility issues. Therefore, it is necessary to ensure that devices and systems have good upgradability to be updated and upgraded in a timely manner to adapt to new technical standards and requirements. Considering future technological evolution during the design and production stages and providing users with convenient upgrade paths is essential.

2.7.2 Improvement of policies, regulations, and market mechanisms

According to the policy document "Implementation Opinions on Strengthening the Integration and Interaction of New Energy Vehicles and Power Grid" released by the Chinese government on January 4, 2024, hereinafter referred to as the "Opinions," a series of measures have been proposed to promote the development of the integration and interaction of new energy vehicles and the power grid. These measures include accelerating the formulation and revision of national and industry standards for vehicle–grid interaction, optimizing and improving supporting electricity prices and market mechanisms, and strengthening the supporting and safeguarding capabilities of power grid enterprises. The goal of these initiatives is to achieve the construction of a vehicle–grid interaction technology standard system, promote the intelligent and orderly charging and large-scale application, and make new energy vehicles an important part of the electrochemical energy storage system, providing bidirectional flexibility regulation capabilities for the power system [29–31].

In the United States, the Biden administration has introduced policies to promote clean energy, driving the continued increase in the production, sales, and penetration rate of new energy vehicles. According to statistical data, in 2022, the sales of battery electric vehicles (BEVs) in the United States increased against the trend, exceeding 800,000 units, a year-on-year increase of 65%, and the penetration rate in the new car market reached 5.8%. In addition, the United States has also passed the "Build Back Better Act," which increases the upper limit of tax credits for new energy vehicles and is expected to further promote the sales and penetration rate of new energy vehicles.

Taken together, these policies are aimed at accelerating the formulation of standards, optimizing market mechanisms, strengthening the supporting and safeguarding capabilities of power grid enterprises, and other measures, which are expected to provide a favorable development environment for the integration and interaction of new energy vehicles and the power grid. These efforts may also further promote the sales and penetration rate of new energy vehicles and drive the realization of the goal of making new energy vehicles an important part of the electrochemical energy storage system and providing flexibility regulation capabilities for the power system.

Furthermore, with the continuous development of EV and V2G bidirectional power flow technology, the improvement of policies, regulations, and market mechanisms becomes crucial. The following are some suggestions regarding the improvement of policies, regulations, and market mechanisms:

1. Establish a clear policy and regulatory framework: The government should establish a clear policy and regulatory framework to ensure the widespread application of EV and V2G technologies. This includes formulating construction standards, layout requirements, access regulations, and regulatory mechanisms to provide a stable and predictable environment.
2. Optimize power market mechanisms: To adapt to the needs of EV and V2G, power market mechanisms need to be adjusted and optimized accordingly. Explore differentiated electricity pricing mechanisms to encourage users to charge during periods of lower prices and balance supply and demand. Establish a competitive mechanism for charging services to reduce users' charging costs.
3. Support technology research and development and innovation: The government and industry should strengthen support for research and development and innovation of EV and V2G technologies to promote technological breakthroughs and innovation. Provide funding and incentive measures to drive market development and popularization.
4. Strengthen international cooperation and standards unification: The development of EV and V2G technologies is a global trend, so strengthening international cooperation and standards unification is crucial. Jointly formulate standards and specifications to promote the interoperability and global sharing of technologies. Cooperate with other countries and regions to share experiences and best practices and jointly address challenges.

By establishing a clear policy and regulatory framework, optimizing power market mechanisms, supporting technology research and development and innovation, and strengthening international cooperation and standards unification, the widespread application of EV and V2G technologies can be promoted, achieving the development goals of sustainable energy and smart grids.

2.7.3   Consumer acceptance and willingness to participate

To further understand users' willingness to participate in orderly charging and V2G, with the support of the Energy Foundation, the China Society of Automotive Engineers, together with other research institutions, jointly carried out the project research on "Policy Research on Promoting the Coordinated Interaction between Electric Vehicles and Power Grid in China: Investigation and Suggestions on Vehicle–Grid Interaction Based on User Willingness." From May to July 2023, the project team conducted surveys on users in 19 cities, including Beijing, Shanghai, Tianjin, and Chongqing, through online and offline methods, and made detailed analyses of users' willingness to participate in orderly charging and V2G, and formed a research report.

From the survey data, it can be seen that:

1. The impact on battery life is the main concern for users to participate in V2G. The top four concerns for car owners to participate in V2G are the impact on

battery life, the impact of V2G on warranty and second-hand car resale, range anxiety when the car is not fully charged, and potential safety risks.

2. Users have a low tolerance for battery life loss caused by participating in V2G. More than 70% of car owners can only accept battery life loss within 5%. Battery life loss is an important concern for car owners to participate in V2G.

3. When the income from charging and discharging (V2G) reaches above 1 RMB per kWh, users are more willing to participate in the interaction. Users need to spend time, have a certain level of driving flexibility, and bear the cost of vehicle battery life loss to participate in charging and discharging (V2G). Through research on car owners with a certain experience in vehicle–grid interaction, it is found that more than 60% of users hope to obtain an income of more than 20 RMB (discharging 20 kWh, with an income of more than 1 RMB per kWh) for each participation. Among them, the income of 1–1.5 RMB per kWh accounts for 20.31%, 1.5–2 RMB accounts for 16.48%, and more than 2 RMB accounts for 30.46%.

In addition, a Dutch scholar also conducted a corresponding study and established a V2G charging point test at a solar car park at Delft University of Technology. Seventeen research participants received V2G-compatible Nissan LEAFs and built V2G charging facilities, and then they were interviewed.

Through the research, it was found that various forms of range anxiety, insight and control over charging through the user interface, and compensation are the most important determinants of consumer acceptance of V2G charging. Clear communication of the impact of V2G charging cycles on electric vehicle batteries (EVBs), economic compensation covering these impacts, real-time insight into battery SOC, and the ability to set operating parameters through a user-friendly interface are all considered factors that promote acceptance. The main obstacles affecting acceptance include uncertainty related to battery SOC, increased need for planning charging and trips, increased concerns about the vehicle's ability to reach its destination, economic and performance-related impacts on EVBs, and restrictions on user freedom associated with personal vehicles.

Users consider the most important conditions for accepting V2G to be:

1. Clear communication about the economic benefits to the user, the impact of V2G cycles on vehicle batteries, and the social/environmental value of V2G charging.

2. Economic compensation that at least covers battery degradation caused by V2G cycles.

3. Transparent real-time display of battery charging and status information.

4. Ability to set charging parameters and opt out of V2G charging.

Their research found that compared to previous studies, compensation is a more important factor. Control over V2G charging cycles was popular among respondents; however, their preference was to set parameters around minimum battery capacity, minimum driving range, etc., rather than actively controlling the charging itself, whereas in earlier experiments, participants preferred charging.

With the further popularization of EVs, the authors indicate that it is uncertain whether their research results can predict the acceptance of V2G charging among future EV users. However, among current users, out of the 17 study participants, 4 consumers had low acceptance, 6 consumers had high acceptance, and 7 consumers had conditional acceptance. Consumer acceptance is very high, and they are early adopters. They accept V2G charging in its current form and are likely to continue doing so in the future. Most drivers need to meet certain conditions before using V2G charging. The authors expect that as the EV driving population evolves, the proportion of high acceptance will decrease over time, but the total number of future high-acceptance drivers will increase [32,33].

2.7.4   Data security and privacy protection

In the field of modern intelligent vehicles, various sensors such as cameras, millimeter-wave radars, and LiDARs have become common equipment. To achieve comprehensive data collection, intelligent vehicles need to rely on these sensors to collect broader and larger-scale vehicle data. Comparative statistics show that an intelligent connected vehicle generates about 10TB of data per day, which is five to ten times that of a traditional fuel vehicle. At the same time, since 2020, many Chinese enterprises related to vehicle manufacturing and connected vehicle information services have suffered more than 2.8 million malicious data attacks. In addition, according to incomplete statistics, since 2023, there have been more than 20 data leakage incidents involving vehicle enterprises in China. According to a survey report released by J.D. Power, a US market research firm, 47.8% of the interviewed vehicle owners said that they had never received any notification from car brands or dealers about the collection of personal information. This indicates that many vehicle enterprises have collected data without the knowledge of users. Once this data is leaked or illegally shared, it will bring risks such as eavesdropping and identity theft, affecting the personal and property safety of enterprises and individuals [34,35].

Facing the wave of intelligent connected technology development, the global automotive industry has always attached great importance to vehicle data security issues, and regulatory agencies in various countries have successively introduced a series of regulations and standards based on actual situations. For example, in January 2021, the United Nations Economic Commission for Europe implemented Regulation No. 156 (UN R156). This regulation requires vehicle manufacturers to establish a software update management system and fulfill obligations such as information recording, preservation, update evaluation, vehicle information security, and user notification.

Data security and privacy protection should also be highly valued when EVs collect and store charging data in V2G scenarios. The following are some key measures to ensure that personal identities are not compromised:

Anonymization measures: To prevent personal identity leakage, anonymization measures should be taken when collecting and storing charging data, separating the collected personal identity information from the charging data to ensure that the data cannot be associated with specific users.

Encrypted communication: In the communication between EVs and charging infrastructure, encryption technology should be used to ensure the secure transmission of charging data. Through encrypted communication protocols, data interception or tampering by unauthorized third parties can be prevented.

Strict access control: To protect the privacy of charging data, only authorized users, service providers, and regulatory agencies should be able to access and use this data. A robust access control and review mechanism should be established to ensure that only legally authorized personnel can obtain charging data.

Secure data storage: The storage of charging data should use secure storage media, such as encrypted hard drives or secure cloud storage, to prevent unauthorized access or data leakage. At the same time, data should be regularly backed up, and a recovery mechanism should be established to deal with potential data loss or damage situations.

User authorization and informed consent: Before collecting charging data, the purpose of the data must be clearly explained to users, and their informed consent must be obtained. Users have the right to decide the scope and sharing method of personal data usage, ensuring that the use of data meets users' expectations.

Regulation and compliance: Governments and regulatory agencies should formulate relevant policies and regulations to clarify the privacy protection requirements for EV charging data and supervise the compliance of relevant enterprises and institutions. A clear regulatory framework can ensure effective supervision and enforcement of data security and privacy protection.

Education and awareness raising: Users and relevant participants need to understand the importance of charging data privacy protection and receive education on data security and privacy protection to raise awareness of personal data protection. Regular publicity and education activities can help users better understand and maintain their data security.

By implementing these measures, the security and privacy protection of EV charging data can be ensured, providing users with reliable charging services and promoting the sustainable development of the intelligent connected vehicle industry.

2.8   Future outlook

2.8.1   *Exploration of innovative technologies and business models for EV and V2G*

After the research by the Vehicle–Grid Interaction Team at the Institute of Energy Internet of Tsinghua University, the team believes that by around 2025, the first countries or regions to commercialize vehicle–grid interaction will emerge globally. Electric vehicles will no longer be merely a means of transportation but will be upgraded to mobile energy storage power stations. This will reshape the automotive industry value chain and competitive landscape, possibly impacting the industry competition pattern earlier and more intensely than intelligent technologies.

Currently, there are two main bottlenecks in the trend of vehicle electrification. The first is the installation and power connection bottleneck of charging facilities, and the second is the cost bottleneck of pure EVs. Vehicle–grid interaction has a direct role in promoting the resolution of these "two major bottlenecks" and is expected to become a "catalyst" for the comprehensive electrification transformation of new energy vehicles.

The difficulty of private installation in communities. The traditional model of individual charging pile power access in communities relies on high power redundancy to ensure the safety of residents' electricity use. This model has the disadvantages of large construction scale, large occupation of public space, and high investment cost, which is the fundamental reason for the current difficulty in accessing individual charging piles in communities. Excessive occupation of public space is likely to provoke opposition from property management and community owners, making it difficult to change. High investment costs make it difficult to carry out large-scale renovation in advance, making it unaffordable. With the support of vehicle–grid interaction technology, the access capacity of a power distribution branch box can be increased to more than twice the original capacity, effectively solving the problem of difficult private installation in communities.

The difficulty of accessing high-power fast charging stations. After the promotion of high-power super charging piles, it will become common for a fast charging station to have a power level of MW in the future. The increase in power level means that the difficulty of nearby access will also increase significantly. Not only will the grid connection cost increase significantly, but in some cases, there will also be problems such as insufficient urban power distribution channels and corridor resources, making it impossible to carry out construction. Therefore, it is necessary to find a low-cost grid connection solution for future MW-level fast charging stations, which also requires the support of vehicle–grid interaction technology. This station–grid interaction model is also known as flexible grid connection or active load management mode. Specifically, it allows charging stations to apply for larger-capacity transformers according to the supply margin during off-peak periods of the line, and sign a flexible grid connection agreement with the power grid, promising to reduce the available capacity upper limit of the transformer during peak periods of the line or according to the line operation status. At present, under the guidance of the peak-valley electricity price mechanism, the charging load of fast charging stations during the peak period of the power grid is not high, while the charging load of fast charging stations during the off-peak and flat periods of the power grid is the highest. Therefore, for fast charging stations, although the flexible grid connection mode limits their maximum electricity consumption capacity during the peak period of the power grid, it does not affect their actual ability to provide charging services. Allowing stations to install larger-capacity transformers to provide high-power fast charging services during off-peak and flat periods of the power grid is the real demand of charging operators. In summary, the flexible grid connection mode not only meets the real needs of charging operators but also better utilizes the existing power distribution network to meet the demand of fast charging stations for larger-capacity transformers. The support of vehicle–grid interaction technology can alleviate the difficulty of accessing high-power fast charging stations.

Regarding the cost bottleneck problem of pure EVs, taking China's basic national conditions as an example, the taxi and ride-hailing vehicles with the highest driving intensity and mileage have basically achieved comprehensive electrification. Currently, in the new vehicle market for taxis and ride-hailing in first- and second-tier cities, the penetration rate of EVs has exceeded 90%. Even in third-tier and lower cities with very few fast charging stations, the penetration rate of EVs has exceeded 50%. Taxis and ride-hailing vehicles will definitely become another market segment to achieve comprehensive electrification after public transportation. The reason is simple: the price difference between oil and electricity is achieved through driving mileage, and the economic advantage of electric taxis and ride-hailing vehicles is too great. According to calculations by industry organizations, an annual driving mileage exceeding 10,000 km is a prerequisite for the total cost of pure electric passenger vehicles to break even with fuel vehicles. However, according to statistics from the "Beijing Transportation Development Annual Report," in 2019, private cars in Beijing with an annual driving mileage of less than 10,000 km accounted for about 40%, which means that for about 40% of the private car market users, choosing EVs is not more economical than fuel vehicles. To encourage this 40% of low-driving intensity users to purchase EVs, it is necessary to realize the value of their mobile energy storage.

Currently, the unit investment cost of fixed energy storage power stations in China is about 2000 RMB/kWh. For a mainstream electric private car with a 70 kWh battery, if it is equivalent to an energy storage power station at a 1:1 ratio, the corresponding energy storage value can reach up to 140,000 RMB. However, since electric private cars need to be used for travel and cannot be on standby at any time like fixed energy storage power stations, their "equivalent energy storage" value is difficult to reach a 1:1 level with fixed energy storage. According to relevant statistics from the "Beijing Transportation Development Annual Report," currently, about 20%–25% of private cars in Beijing stay in the garage all day without any travel, and more than 50% of the vehicles return to their fixed parking spaces at home before 7 pm. This means that after large-scale EVs are aggregated in the future, in a statistical sense, the average "equivalent energy storage" potential of each vehicle is expected to exceed about 20% of fixed power stations, and the "equivalent energy storage" realization value is around 30,000 RMB. A realization value of 30,000 RMB for "equivalent energy storage" means that mainstream EVs priced at around 150,000 RMB can reduce costs by 20%. For vehicle companies that regard cost reduction as their lifeline, a 20% cost reduction space is an irresistible.

In view of the huge potential in breaking through the "two major bottlenecks" of charging access and vehicle cost, vehicle–grid interaction technology will inevitably become a new competitive high ground for the global new energy vehicle industry in the next 2–3 years, forming a competitive situation of "mid-game decisive battle" in the new energy vehicle industry.

Around 2025, with the global vehicle–grid interaction, especially V2G, beginning to enter the commercial introduction period, the competitive landscape of the new energy vehicle industry will undergo drastic changes.

Overall, leading regions in Europe and the United States generally regard vehicle-grid interaction as an important pillar of the zero-carbon transportation transformation strategy, especially in California and the United Kingdom, which are the most representative. For example, the United Kingdom has proposed goals to achieve zero-carbon new sales of light vehicles by 2035 and zero-carbon new sales of medium and heavy commercial vehicles by 2040. California aims to achieve zero-carbon new sales of passenger cars and trucks by 2035 and zero-carbon medium and heavy commercial vehicles (including existing ones) by 2045. Both of these regions consider vehicle–grid interaction as a key pillar of their zero-carbon transportation strategy and have incorporated it into the overall design of their policy systems. As a leading region, California has taken the lead in introducing vehicle–grid interaction bills and has proposed a vehicle–grid interaction policy roadmap through joint research by multiple departments. Currently, it appears that California is using 2025 as a key node for commercial introduction and aims to achieve full application by 2030. The United Kingdom has also proposed 2025 as the target node for the large-scale commercial use of smart charging in the process of formulating smart charging policies and has officially launched the policy consultation process for V2X (Vehicle-to-Everything). In addition to top-level design, leading regions such as California and the United Kingdom have also made several substantive breakthroughs in the preparatory work for the commercial introduction of vehicle–grid interaction.

2.8.2   International cooperation and experience exchange

The concept of V2G was first proposed in 1995, but due to the battery technology at that time and the low penetration rate of EVs, it was not implemented. In recent years, with the explosive growth of the EV market, V2G technology has been put on the agenda again.

In 1995, Amory Lovins, the chief scientist of the Rocky Mountain Institute in the United States, proposed the V2G concept prototype. Subsequently, Professor William Kempton of the University of Delaware further developed it and carried out related demonstrations. However, in the following decade or so, due to the rapid development of electronic fuel injection technology for internal combustion engines, the emission problem was greatly alleviated, and the development of EVs slowed down.

It was not until 2017 that Nissan launched a new generation of Leaf models that supported V2G technology. In 2018, they cooperated with the energy supplier Fermata Energy to start trying the feasibility of EVs charging civil buildings.

In 2019, the Renault Group announced the launch of a fleet of 15 Zoe cars to conduct V2G technology pilot projects in 7 European countries. The action was aimed at verifying the feasibility of large-scale pilots and estimating potential benefits.

In 2020, Nissan, together with the power transmission system operator Tennet and the technology company The Mobility House, launched a V2G project in

Germany with the theme of absorbing renewable energy. Northern Germany is rich in wind power resources, but it is difficult to allocate them to the south, directly leading to northern grid operators having to shut down some wind power equipment, while southern operators are forced to increase conventional power generation capacity. After the introduction of V2G technology, the excess wind power in the north is absorbed by EVs, and during peak power consumption in the south, it can absorb the "back-feeding" from EVs, achieving the effect of peak shaving and valley filling.

China's V2G attempts began in 2020. In this year, Weltmeister (WM Motor) reached a cooperation agreement with the State Grid and successfully passed the vehicle and charging pile full-item V2G technology test and road test, becoming the first new energy vehicle company in China to commercialize V2G. In November of the same year, Dongfeng Group signed a series of strategic cooperation agreements with the State Grid, including the "V2G Pilot Cooperation Agreement," marking the official pilot of V2G technology by Dongfeng Group.

In 2021, BYD UK announced a partnership with Alexander Dennis Ltd (ADL) and provided 28 BYD ADL Enviro400EV double-decker electric buses that support high-power V2G technology, as shown in Figure 2.12. In October of the same year, BYD will deliver up to 5000 BYD medium and heavy-duty pure EVs supporting V2G technology to Levo Mobility LLC over the next 5 years. At this point, BYD became the first company in China's commercial vehicle field to commercialize V2G technology.

Figure 2.12   BYD double-decker electric bus

2.9 Conclusion

With the advancement of EV technology, charging infrastructure, and grid-connected facilities, the popularity of EVs is expected to increase significantly in the next decade. Therefore, further technological advancements, such as appropriate smart charging infrastructure, reliable communication systems, and coordinated charging systems to quantify the impact on the grid, are crucial to ensure the maximum benefits of distributed generator EVs. Furthermore, the Energy Internet may be a future grid technology that will fully automate the power system with advanced energy management systems. This article discusses various aspects of EV charging and grid-connected infrastructure. Establishing unified global standards for EVs and their charging infrastructure is the primary condition for the widespread adoption of EVs in the market. The popular standards related to EV charging and grid integration are discussed so that future researchers can better understand the specifications that need to be met. In addition, a rigorous review of the existing charging and grid-connected infrastructure, including power, communication, control, coordination, and other aspects, has been conducted, analyzing their respective advantages and disadvantages. This article also provides recommendations for future research to overcome current challenges. The discussion on the future development prospects of EVs proves the necessity of further exploration in this research field.

References

[1]   S. Kirmani, I. Akhtar and M. Umar Rehman (2023). Electric vehicle infrastructure management with maximum utilization of renewable energy: A review. *2023 IEEE 3rd International Conference on Sustainable Energy and Future Electric Transportation (SEFET)*. IEEE, Piscataway, NJ.

[2]   B. Tetteh and K. Awodele (2019). Power system protection evolutions from traditional to smart grid protection. *2019 IEEE 7th International Conference on Smart Energy Grid Engineering (SEGE)*. IEEE, Piscataway, NJ.

[3]   W. Ce. (2021).What are the advantages and disadvantages of traditional power grids. Retrieved from https://www.ligongku.com/question/308.

[4]   Yang Lan Science. (2023). What are the renewable energy sources? Solar energy, wind power, water power, geothermal, biomass, 5 major types to understand at once! Retrieved from https://zhuanlan.zhihu.com/p/643721682.

[5]   J. Chen. (2022). What are the five main types of renewable energy? Retrieved from https://zhuanlan.zhihu.com/p/534723081.

[6]   H. Liu. (2022). Challenges facing the development of new power systems. *Electric Age*. 12(1):14–16.

[7]   Y. Wu. (2021). Research on the charging and discharging scheduling strategy of electric vehicles in V2G application scenarios. *MA Thesis*. China University of Mining and Technology.

[8]   R. P. Upputuri and B. Subudhi. (2023). A comprehensive review and performance evaluation of bidirectional charger topologies for V2G/G2V

operations in EV applications. *IEEE Transactions on Transportation Electrification* 10(1):583–595.

[9]   Y. Zhang (2022). Research on control strategy of two-stage bidirectional AC–DC converter. *MA Thesis.* Nanjing University of Information Science and Technology.

[10] X. Luo. (2022).V2G: The important fulcrum for promoting China's energy transformation. Chengdu Automotive Industry Research Institute. Retrieved from https://mp.weixin.qq.com/s/Xtb3brt8toOI-U0tmFpsEg.

[11] Recaino New Energy. (2023). V2G technology and present situation at home and abroad. Retrieved from https://mp.weixin.qq.com/s/Uv_0vj7S1Dhz6 GXSgE3tOA.

[12] S. Mlindelwa, S. P. Daniel Chowdhury and M. J. Lencwe. (2024). Overview of the challenges, developments, and solutions for electric vehicle charging infrastructure. *2024 32nd Southern African Universities Power Engineering Conference (SAUPEC).* IEEE, Piscataway, NJ.

[13] Energy Talk Center. (2024). Talk about the new "vehicle network interaction" (V2G) for electric vehicles. Retrieved from https://zhuanlan.zhihu.com/p/682228787.

[14] CT Carbon Ring. (2021). Impact and technical requirements of electric vehicle charging access on power grid. Retrieved from https://mp.weixin.qq.com/s/mEzX_cL3dlIgx9yElNZ7fg.

[15] eTran Electrification of transportation. (2021). Electric vehicle charging infrastructureand its impact on the grid. Retrieved from https://mp.weixin.qq.com/s/xD87-Iko6FZVKMu-M2fylw.

[16] Y. Cao. (2022). Ordered charging and discharging strategy for electric vehicles based on complex network in V2G mode. *MA Thesis.* Nanjing University of Posts and Telecommunications.

[17] L. Chen and S. Lu. (2018). Research on the impact of electric vehicle charging on residential community distribution networks. *Southern Energy Construction.* 5(1):51–58.

[18] J. Sun, Y. Wan, P. Zheng, *et al.* (2014). Ordered charging and discharging strategy for electric vehicles based on demand side management. *Transactions of China Electrotechnical Society.* 29(8):64–69.

[19] K. Chen and Y. Niu. (2019). Real-time scheduling strategy for electric vehicles based on V2G technology. *Power System Protection and Control.* 47(14):1–9.

[20] L. Yao, Q. Li, J. Yang, *et al.* (2022). Comprehensive reactive power optimization of distribution systems considering support from electric vehicle charging and discharging. *Automation of Electric Power Systems.* 46(6):39–47.

[21] H. Hou, Y. Wang, B. Zhao, *et al.* (2022). Dispatch strategy for aggregated electric vehicle loads under price and incentive demand response. *Power System Technology.* 46(4):1259–1269.

[22] L. Zhang, C. Sun, G. Cai, *et al.* (2022). Two-stage optimization strategy for ordered charging and discharging of electric vehicles based on PSO algorithm. *Proceedings of the CSEE.* 42(5):1837–1852.

[23] K. Pavan Inala and K. Thirugnanam. (2022). Role of communication networks on vehicle-to-grid (V2G) system in a smart grid environment. *2022 4th International Conference on Energy, Power and Environment (ICEPE)*. IEEE, Piscataway, NJ.

[24] M. Li. (2023). Research on optimization control of ordered charging and discharging for electric vehicles towards V2G. *MA Thesis*. Harbin University of Science and Technology.

[25] B. Xu. (2018). Research on V2G converter and control technology for electric vehicles. *MA Thesis*. Shandong University.

[26] Citic Securities. (2023). New energy vehicle charging operation industry research: Meet the dawn of the post-"horse racing" era. Retrieved from https://baijiahao.baidu.com/s?id=1765284514965925369&wfr=spider&for=pc.

[27] J. Yang and Z. Cao. (2020). Cost and benefit analysis of energy storage V2G mode for electric vehicles. *Energy Storage Science and Technology*. 9(S1): 45–51.

[28] S. Tirunagari, M. Gu and L. Meegahapola. (2022). Reaping the benefits of smart electric vehicle charging and vehicle-to-grid technologies: Regulatory, policy and technical aspects. *IEEE Access*. 10:114657–114672.

[29] Central People's Government of the People's Republic of China. (2023). Implementation opinions of the National Development and Reform Commission and other departments on strengthening the integration and interaction between new energy vehicles and the power grid. *National Development and Reform Commission website*. National Development and Reform Commission. December 13, 2023. Retrieved from https://www.gov.cn/zhengce/zhengceku/202401/content_6924347.htm.

[30] Sohu. (2023). Overcharge, V2G, liquid cooling module new energy technology research and development project feasibility study report. Retrieved from https://www.sohu.com/a/651661652_120244537.

[31] Q. Shi, W. Z. Kong and J. Hu. (2023). Analysis of the impact of electric vehicle users' participation in vehicle network interaction on battery life. Retrieved from https://www.escn.com.cn/20231123/a2521f590c71415da55c800287133580/c.html.

[32] R. Ghotge, K. P. Nijssen, J. A. Annema, *et al.* (2022). Use before you choose: What do EV drivers think about V2G after experiencing it? *Energies*. 15(13):4907.

[33] Z. Zhang. (2024). Strictly observe the security of car data, so that personal privacy is no longer "naked". Retrieved from https://mp.weixin.qq.com/s/B4keecIVT4HthSjROliXkg.

[34] Amarantine. (2023). What is the electric vehicle access technology (V2G), and what is the current research or application? Retrieved from https://www.zhihu.com/question/28866374/answer/3309226878.

[35] H. S. Das, M. M. Rahman, S. Li, *et al.* (2020). Electric vehicles standards, charging infrastructure, and impact on grid integration: A technological review. *Renewable and Sustainable Energy Reviews*. 120:109618.

Chapter 3

Transactive EV management with V2G capability: network-aware approaches

Md. Murshadul Hoque[1,2], M. Imran Azim[1,3], Mohsen Khorasany[1] and Reza Razzaghi[1]

3.1 Introduction

3.1.1 Background

Renewable energy sources (RESs) coupled with energy storage systems play a vital role in the energy transition and net zero future. Green transports, such as electric vehicles (EVs) [1], with vehicle-to-grid (V2G) capabilities can be used to manage the demand for RESs and balance the power systems. V2G technology allows EVs to return energy from their batteries to the grids, incorporating bi-directional charging and discharging [2,3]. EVs can assist with managing power networks and can participate in local energy markets (LEMs) to provide ancillary services through their managed integration and charging/discharging [4].

In practice with V2G, it is important that EVs are managed with sufficient energy levels for their batteries to meet EV owners' needs. Besides, increasing integration of EVs with V2G in power distribution networks poses challenges [5]. Uncontrolled EV integration and charging with V2G pose issues with managing voltage and line loading, power losses, peak load demand, overload, thermal effects and energy imbalance, and battery degradation [6,7]. Therefore, their effective coordination within distribution networks is being emphasized with the deliberate deployment of smart infrastructure and technologies as well as coordination strategies for EV integration with V2G [7].

3.1.2 Various schemes of EV management with V2G

Various techniques have been explored in the literature for EV coordination in distribution networks. Figure 3.1 shows EV coordination with V2G approaches [8]. These are primarily divided into direct control and indirect control schemes. The

[1]Department of Electrical and Computer Systems Engineering, Monash University, Australia
[2]Department of Electrical and Electronic Engineering, University of Chittagong, Bangladesh
[3]School of Engineering, RMIT University, Australia

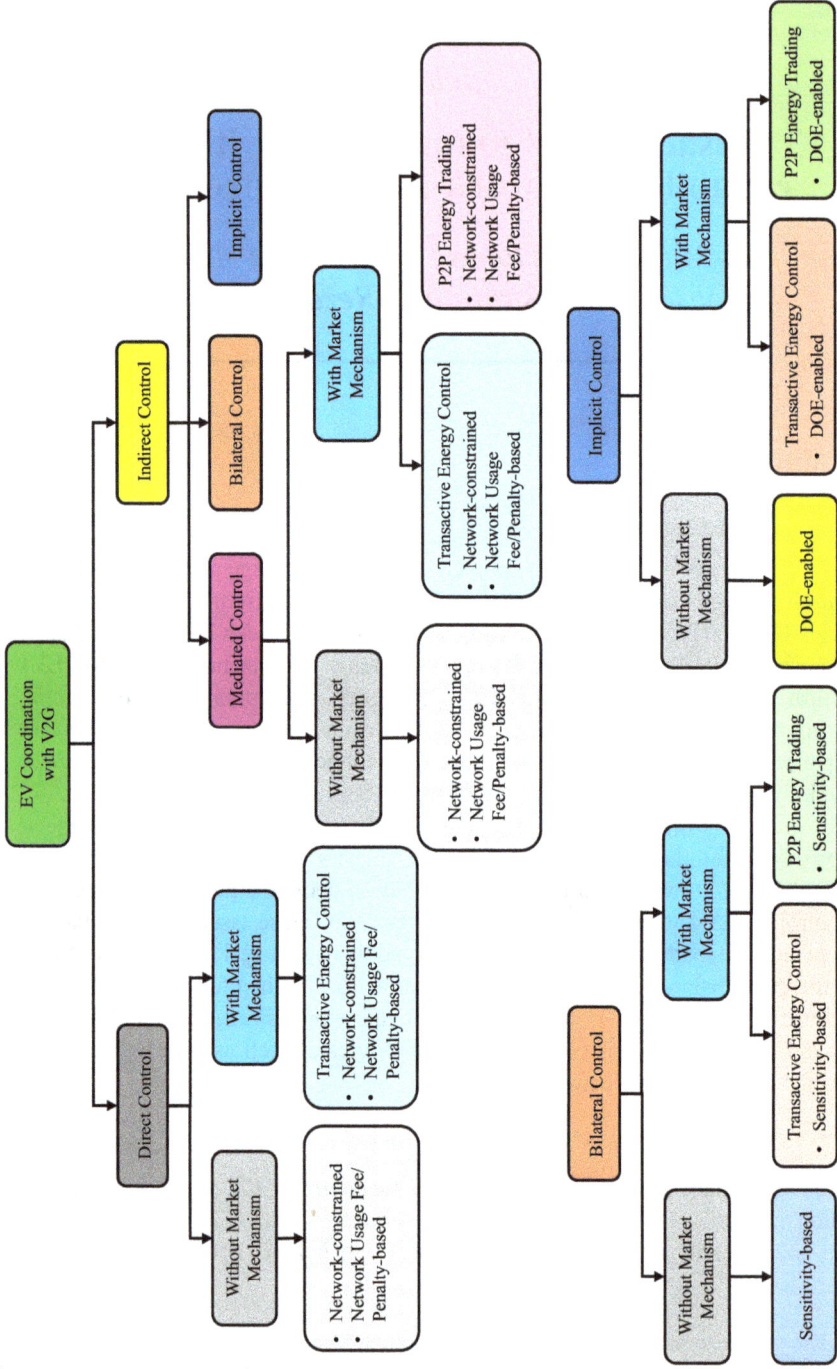

Figure 3.1 EVs coordination schemes

scheme for coordinating EVs with direct control involves a central coordinator, like the distribution system operator (DSO), who has complete access to EV data and control of EVs in order to optimize the coordination for EV charging and V2G and ensure secure system management. Direct control processes can be successfully implemented in small settings without or with market mechanisms that involve fewer facilities and communication support with access permits for the central coordinator to EV data and control. Direct control coordination schemes are studied to integrate EVs into distribution networks while taking network constraints into account and making optimal use of their resources and flexibility [9]. Moreover, it can be implemented by minimizing energy losses and charging time [10], based on the priority of charge management [11], and using a real-time event-driven dual control [12].

EV management in power networks has been investigated by incorporating the transactive energy concept. Transactive energy control is proposed to provide economic as well as control mechanisms in order to manage distributed energy resources (DERs), lower energy costs, enhance energy balance, and support grids. In the process of transactive EV coordination, DSO centrally determines energy prices, coordinates EV charges and V2G [13,14], and simultaneously maintains network operations without causing any constraint breaches [15–17]. For instance, transactive energy-based EV management frameworks have been developed to control energy supply and demand in the LEM while adhering to voltage constraints [15], to control the transformer capacity limit by reducing the charging cost of EVs and maximizing the profit of the DSO [16], and to respect network constraints by maximizing social welfare [17]. However, in the direct control EV with V2G management frameworks, the central coordinator has direct access to the EV data, hence the privacy-related issues.

On the contrary, the management of EV charge and V2G can be implemented in indirect control mechanisms in which a local agent – an EV, an EV charging station, or an EV aggregator, for example – accomplishes local optimization problems to integrate EVs with managed charge and V2G. These are decentralized control approaches that can be efficient both economically and computationally for large EV coordination and enhance the privacy of EVs and system reliability. In contrast to direct control, which requires EVs to divulge their personal information to both the central organization and their peers, decentralized EV coordination is not facilitated by the central entity. In these approaches, an EV owner uses local data to make an independent control decision and to carry out optimization problems by the local controller at each EV charger point or their home energy management system. As depicted in Figure 3.1, the indirect controls are classified into mediated control, bilateral control, and implicit control [8]. In mediated control, a central body interacts with EVs to obtain their data and sends signals to them without directly participating in the EV charge and V2G management. In the bilateral control method for EV coordination without a centralized operation, EVs, aggregators, and the DSO communicate bilaterally for exchanging data and signals. The EV charge and V2G can also be managed without disclosing their data to the central entity or their peers, where the local management systems of EVs carry out the implicit control to coordinate them.

Indirect controls for the management of EV charging and V2G can be applied without or with market mechanisms. For example, in [6], while managing EVs in the power network without network issues, EV aggregators run optimal problems in a decentralized manner to minimize EV charging costs and protect EV owners' autonomy and privacy. Optimal EV charge and V2G management are implemented by the photovoltaic (PV)-integrated EV charging station in [18] to minimize voltage deviation in the network. These approaches are not established based on market mechanisms that let EVs gain economic advantages. Indirect control schemes for EV coordination can be implemented, considering market mechanisms. For instance, decentralized transactive energy-based coordination of EVs is implemented via aggregators to improve network and market efficiencies [19] and to mitigate computational complexities and power imbalance [20]. A network-constrained transactive EV coordination framework is developed for aggregated EVs to minimize transformer overloading [21], lower system peak load [22], and minimize system operator costs [23].

EVs can also trade with their peers in an LEM, where the negotiation and peer matching for the trading can be accomplished locally without the involvement of a central entity [24,25]. For example, in [26], a peer-to-peer (P2P) energy trading system is presented that derives trading prices and energy amounts for EVs by maximizing their social welfare. P2P energy trading can be regulated to control network constraints by either imposing a network usage fee in addition to the market clearing price or restricting the power of EV charging/V2G [27]. These methods, however, may charge prosumers more for using the network while trading in the LEM and are inefficient at utilizing the grid capacity for DERs. Market operations can be indirectly managed for the energy trading of EVs while the DSO secures the network operation. The dynamic operating envelope (DOE) concept is currently being trialed in Australia to secure the integrity of DER-rich networks by allocating dynamic limits for the import/export power of prosumers [28]. In the context of coordinating EV charge and V2G, EV owner prosumers can be managed to trade in the LEM within the allocated DOEs so that EV integrations into networks do not create any network constraint breaches. Using DOE, a game-based P2P trading framework [29] and prosumers' asset management frameworks [30,31] are developed to manage prosumer exports within DOEs that improve network hosting capacity and avoid network constraint issues.

This chapter presents several network-aware strategies for transactive EV management with V2G, in which EVs can be involved in the LEM willingly through coordinated charge and discharge for effective management of the network and for participation in the prosumer-centric energy market. These strategies incorporate sensitivity-based dynamic charge and V2G (discharge) management, EVs' charge–discharge flexibility, the DOE concept, bidding strategies, and double-auction mechanisms to effectively manage EV integration into networks and ensure secured network operation for energy trading. The management approaches for EV integration with V2G presented in this chapter can provide real-time solutions for network and market management and are compatible with real-world applications. Case studies are carried out to analyze the performance of the presented strategies in managing EV integration with V2G without causing any anomalies in the network operation. The simulation results show that the presented EV management strategies can provide network security and services through managed EV charging and V2G, as well as financial benefits to participating EVs in the LEM.

3.2   Models of EV management with V2G

3.2.1   EV model

The EV is modeled in terms of the state-of-charge (SOC) level ($S_{b,t}^{EV}$) of the EV battery as (3.1). In (3.1), $P_{b,t}^{EVCh}$ and $P_{b,t}^{EVDch}$ are the EV charge power and the EV discharge power for V2G, $\bar{E}_b^{EV}$ is the maximum capacity of the EV battery, Δt is the duration of a time step, and η^{EVCh} and η^{EVDch} are the charging and discharging efficiencies of the EV, respectively. Each EV is connected to a node $b \Rightarrow b \in \mathcal{B} \triangleq \{1,\ldots,b,\ldots,B\}$, linked through lines denoted by a set of $d \in \mathcal{D} \triangleq \{1,\ldots,d,\ldots,D\}$ in a distribution network.

$$S_{b,t+\Delta t}^{EV} = S_{b,t}^{EV} + \frac{P_{b,t}^{EVCh}\eta^{EVCh}\Delta t}{\bar{E}_b^{EV}} - \frac{P_{b,t}^{EVDch}\Delta t}{\eta^{EVDch}\bar{E}_b^{EV}} ; \forall b \in \mathcal{B}, \forall t \in \mathcal{T} \tag{3.1}$$

where t is a set of time steps defined as $t \in \mathcal{T} \triangleq \{1,\ldots,t,\ldots,T\}$. $S_{b,t}^{EV}$ is constrained by (3.2), where $\bar{S}_b^{EV}$ and $\bar{S}_b^{EV}$ are the maximum SOC and the minimum SOC of the bth EV battery, respectively

$$\bar{S}_b^{EV} \leq S_{b,t}^{EV} \leq \bar{S}_b^{EV} ; \forall b \in \mathcal{B}, \forall t \in \mathcal{T} \tag{3.2}$$

EVs can start charging or discharging at their arrival time t_b^{Ar} with the initial SOC S_b^{In} and end at their departure time t_b^{Dp} with the desired SOC S_b^{Ds} as

$$S_{b,t_b^{Ar}}^{EV} = S_b^{In}, \ S_{b,t_b^{Dp}}^{EV} = S_b^{Ds} ; \forall b \in \mathcal{B} \tag{3.3}$$

EVs charge power $P_{b,t}^{EVCh}$ and discharge power $P_{b,t}^{EVDch}$ can be determined by (3.4) and (3.5), respectively, where $\bar{P}_b^{EV}$ is the maximum EV charge or discharge rating and $\rho_{b,t}^{EV}$ is an integer variable

$$0 \leq P_{b,t}^{EVCh} \leq \min\left(\bar{P}_b^{EV}, \frac{\left(S_b^{Ds} - S_{b,t}^{EV}\right)\bar{E}_b^{EV}}{\eta^{EVCh}\Delta t}\right) \frac{1}{2}\rho_{b,t}^{EV}\left(1 + \rho_{b,t}^{EV}\right) \tag{3.4}$$

$$0 \leq P_{b,t}^{EVDch} \leq \min\left(\bar{P}_b^{EV}, \frac{\left(S_{b,t}^{EV} - \bar{S}_b^{EV}\right)\bar{E}_b^{EV}\eta^{EVDch}}{\Delta t}\right) \frac{1}{2}\rho_{b,t}^{EV}\left(1 - \rho_{b,t}^{EV}\right) \tag{3.5}$$

Based on the preference of the EV owner and their mobility pattern, EVs can be configured either for charging ($\rho_{b,t}^{EV} = 1$) when $S_{b,t}^{EV} < S_b^{Ds} < \bar{S}_b^{EV}$, the available time $t_{b,t}^{Av}$ before EV departs is less than the required time $t_{b,t}^{Rq}$ to reach S_b^{Ds} and the local PV system generation surplus $P_{b,t}^{Sr}$ is available after consuming household load demand $P_{b,t}^{Load}$, or discharging $\left(\rho_{b,t}^{EV} = -1 \right)$ when $S_{b,t}^{EV} > \bar{S}_b^{EV}$ and $t_{b,t}^{Av} > t_{b,t}^{Rq}$, or idle $\left(\rho_{b,t}^{EV} = 0 \right)$. $t_{b,t}^{Av}$ is estimated by subtracting the action time $t_{b,t}^{Ac}$ from the departure time t_b^{dep} and $t_{b,t}^{Rq}$ is defined as

$$t_{b,t}^{Rq} = \left(S_b^{Ds} - S_{b,t}^{EV} \right) \frac{\bar{E}_b^{EV}}{\bar{P}_b^{EV}} \; ; \forall b \in \mathcal{B}, \forall t \in \mathcal{T} \tag{3.6}$$

3.2.2 Prosumer model

Prosumers ($b \in \mathcal{B}$), who are assigned to be connected to corresponding nodes ($b \in \mathcal{B}$) of the distribution network, can be equipped with an EV, a PV system, and a household load. According to the definition, prosumer b can behave as a consumer, which can be defined as $\mathcal{B}_t^{Cs} := b \in \mathcal{B}$ when $P_{b,t} > 0$ and as a producer, which can be defined as $\mathcal{B}_t^{Pd} := b \in \mathcal{B}$ when $P_{b,t} \leq 0$, where $\mathcal{B}_t^{Cs} \cup \mathcal{B}_t^{Pd} = \mathcal{B}, \mathcal{B}_t^{Cs} \cap \mathcal{B}_t^{Pd} = \varnothing$. The net power $P_{b,t}^{Net}$ absorbed as import $P_{b,t}^{Ip}$ when $P_{b,t}^{Net} \geq 0$ or injected as export $P_{b,t}^{Ep}$ when $P_{b,t}^{Net} < 0$ at prosumer node b is calculated by

$$P_{b,t}^{Net} = P_{b,t}^{EVCh} + P_{b,t}^{Load} - P_{b,t}^{EVDch} - P_{b,t}^{PV} \; ; \forall b \in \mathcal{B}, \forall t \in \mathcal{T} \tag{3.7}$$

where $P_{b,t}^{PV}$ is the local PV system generation of a residential prosumer at node b, which is limited by its maximum generation $\bar{P}_b^{PV}$ as

$$0 \leq P_{b,t}^{PV} \leq \bar{P}_b^{PV} \; ; \forall b \in \mathcal{B}, \forall t \in \mathcal{T} \tag{3.8}$$

3.2.3 Price model

The bid price model for EV charging $\lambda_{b,t}^{Bid}$ is formulated based on the comfort of achieving the desired charge level of EV batteries during departure and the economic aspect of reducing the charging cost of EVs [5,32,33] as

$$\lambda_{b,t}^{Bid} = \bar{\lambda}_{b,t}^{DLM} + \left(\kappa_b^{CF} \frac{t_b^{Rq}}{t_b^{Av}} + \left(1 - \kappa_b^{CF} \right) \left(1 - 2\kappa_h^{EF} \right) \right) \hat{\lambda}_{b,t}^{DLM} \tag{3.9}$$

where $\bar{\lambda}_{b,t}^{DLM}$ and $\hat{\lambda}_{b,t}^{DLM}$ are the mean and the standard deviation of the DLMP $\lambda_{b,t}^{DLM}$ price, respectively. The comfort factor κ_b^{CF} and the economy factor κ_b^{EF} of an EV can be assigned between 0 and 1, depending on the preferences of the EV owner. κ_b^{EF} is a dimensionless ratio of the actual deviation of $\lambda_{b,t}^{DLM}$ from its minimum to the maximum deviation.

The offer price model for EV discharging $\lambda_{b,t}^{Offer}$ can be formulated depending on the costs related to EV battery discharge, including the average recharge cost $RC_{b,t}^{EV}$ and the depreciation cost DC_b^{EV} for the charge–discharge round-trip efficiency loss of the EV battery as follows:

$$\lambda_{b,t}^{Offer} = \frac{RC_{b,t}^{EV} + DC_b^{EV}}{E_b^{EVOut}} = \frac{\bar{\lambda}_{b,t}^{DLM}}{\eta^{EVCh}\eta^{EVDch}} + \frac{2\omega^{DR}}{\eta^{EVDch}} \tag{3.10}$$

where E_b^{EVOut} is the output of the EV battery at t, $\bar{\lambda}_{b,t}^{DLM}$ is the average of $\lambda_{b,t}^{DLM}$ over the remaining time until departure, and ω^{DR} is the rate of the depreciation cost.

3.2.4 Network model

The distribution network is configured with a set of EV/prosumer connection nodes $b \in \mathcal{B}$, a set of lines $d \in \mathcal{D}$, and a set of phases defined as $\phi \in \Phi = \{a, b, c\}$. The import/export current $I_{b,t}$ at prosumer node b and phase ϕ are calculated by

$$I_{b,t} = I_{b,\phi,t}^{Re} + jI_{b,\phi,t}^{Im}; \ \forall b \in \mathcal{B}, \forall \phi \in \Phi, \forall t \in \mathcal{T} \tag{3.11}$$

where $I_{b,t}^{Re}$ and $I_{b,t}^{Im}$ are the real part and the imaginary part of $I_{b,t}$, which are calculated by (3.12) and (3.13), respectively, as

$$I_{b,t}^{Re} = \frac{1}{\tilde{V}_{b,t}}\left(P_{b,t}^{Net}\cos\left(\tilde{\vartheta}_{b,\phi}\right) + \tilde{Q}_{b,t}^{Net}\sin\left(\tilde{\vartheta}_{b,\phi}\right)\right); \ \forall b \in \mathcal{B}, \forall t \in \mathcal{T}, \phi \in \Phi \quad (3.12)$$

$$I_{b,t}^{Im} = \frac{1}{\tilde{V}_{b,t}}\left(P_{b,t}^{Net}\sin\left(\tilde{\vartheta}_{b,\phi}\right) - \tilde{Q}_{b,t}^{Net}\cos\left(\tilde{\vartheta}_{b,\phi}\right)\right); \ \forall b \in \mathcal{B}, \forall t \in \mathcal{T}, \phi \in \Phi \quad (3.13)$$

where $\tilde{V}_{b,t}$ is the voltage magnitude (i.e., $\approx$1 p.u.), $\tilde{\vartheta}_{b,\phi}$ is the voltage angle at phase ϕ, and $\tilde{Q}_{b,t}^{Net}$ is the net reactive power of prosumer's household load. The line current $I_{d,t}$, as in (3.14), flows through a line segment d between the upstream node $(b-1)$ and node b

$$I_{d,t} = I_{b,t} + \sum_{h\in\mathcal{H}} I_{h,t}; \ \forall b \in \mathcal{B}, \forall d \in \mathcal{D}, \forall h \in \mathcal{H}, \forall t \in \mathcal{T} \tag{3.14}$$

where $I_{h,t}$ is the current on line h, which is a set of downstream lines connected to node b defined as $h \in \mathcal{H} \triangleq \{1, \dots, h, \dots, H\}$. The node voltage $V_{b,t}$ at node b of the network is obtained by (3.15) as

$$V_{b,t} = V_{b-1,t} - (R_d + jX_d)I_{d,t}; \ \forall b \in \mathcal{B}, \forall d \in \mathcal{D}, \forall t \in \mathcal{T} \tag{3.15}$$

where $V_{b-1,t}$ is the voltage at the upstream node $(b-1)$ and R_d and X_d are the resistance and reactance matrices of the distribution network. Equation (3.16) presents the voltage constraint of nodes for both the over-voltage and under-voltage of the network, where $\bar{V}$ and $\underline{V}$ are the upper and lower voltage limits, respectively,

$$\underline{V} \le |V_{b,t}| \le \bar{V} \ ; \ \forall b \in \mathcal{B}, \forall t \in \mathcal{T} \tag{3.16}$$

The line loading corresponding to line d of the network is constrained by its maximum line loading limit $\bar{I}_d$ as

$$0 \le |I_{d,t}| \le \bar{I}_d \; ; \forall d \in \mathcal{D}, \forall t \in \mathcal{T} \tag{3.17}$$

3.2.5   DOE assignment

DOEs are assigned in a three-phase unbalanced distribution network by employing an optimal iterative power flow process considering voltage and line loading constraints to limit prosumers' intended import/export power to secure the integrity of the network. In this chapter, an objective function is presented to determine DOEs by minimizing the absolute difference between the intended import/export power $P_{b,t}^{Net}$ of prosumers and the DOE $P_{b,t}^{Doe}$ by

$$\min \sum\nolimits_{b \in \mathcal{B}} \left| P_{b,t}^{Net} - P_{b,t}^{Doe} \right| \; ; \forall t \in \mathcal{T} \tag{3.18}$$

subject to : $(3.11)-(3.17)$, (3.19), (3.20)

where $P_{b,t}^{Doe}$ is constrained by (3.19) for import power and (3.20) for export power of the prosumer

$$0 \le P_{b,t}^{Doe} \le P_{b,t}^{Net}, \text{ if } P_{b,t}^{Net} \ge 0 \; ; \forall b \in \mathcal{B}, \forall t \in \mathcal{T} \tag{3.19}$$

$$P_{b,t}^{Net} \le P_{b,t}^{Doe} \le 0, \text{ if } P_{b,t}^{Net} < 0 \; ; \forall b \in \mathcal{B}, \forall t \in \mathcal{T} \tag{3.20}$$

3.3   Frameworks of network-aware EV management with V2G

3.3.1   Sensitivity-based transactive EV management

3.3.1.1   Overview

This framework can manage EV charges and V2G in the distribution network, considering EV owners' preferences, through the participation of EVs in a real-time LEM to mitigate voltage problems under sensitivity-based transactive energy control. In the framework, the retail market operator (e.g., DSO) determines the retail prices (e.g., distribution locational marginal price, DLMP) depending on the wholesale prices and the estimated net load demands and communicates the price signals to EVs for the transactive feedback bid/offer prices. EV owners estimate their bid/offer prices by their local controllers, considering market prices, EV mobility and battery data, preferences, and battery-wearing costs. After receiving the feedback price signals from EVs, the DSO performs a market clearing optimization problem to maximize the social welfare of all participating agents and determines the market clearing prices for EVs to get their action (i.e., EV charge and V2G) on a real-time basis. In the proposed framework, a sensitivity-based voltage management algorithm is executed on the actions of EVs through an incentivized dynamic power curtailment of EV charge and V2G to control node voltages within the operating ranges and fulfill the EV owners' desired energy levels of EV batteries.

3.3.1.2   Methodology

The sensitivity-based transactive EV management is executed by employing the market clearing optimization problem modeled by maximizing the social welfare from the buying and selling of energy in the LEM in (3.21) as follows:

$$
\max \sum_{b \in B} \left(\left(\lambda_{b,t}^{Bid} P_{b,t}^{EVCh} \right) - \lambda_{b,t}^{Offer} P_{b,t}^{EVDch} - \lambda_{b,t}^{DLM} P_{b,t}^{Imp} + \lambda_t^{FiT} P_{b,t}^{Exp} \right)
$$
$$
- \lambda_{b,t}^{Clear} \left(P_{b,t}^{EVCh} - P_{b,t}^{EVDch} - P_{b,t}^{Imp} + P_{b,t}^{Exp} \right) \right) \Delta t \tag{3.21}
$$
$$
\text{subject to : } (3.4), (3.5), (3.22)–(3.24)
$$

The objective function (3.21) has a unique optimal solution using Karush–Kuhn–Tucker (KKT) conditions, and the relaxation of constraints is carried out through the Lagrange multiplier (λ_{clear}), where the power balance equation (3.22) corresponds to λ_{clear} in (3.21). Constraints (3.23) and (3.24) limit the import power $P_{b,t}^{Imp}$ from the upstream grid and the export power $P_{b,t}^{Exp}$ to the upstream grid. A binary variable (α) is used to state the power import from the grid ($\alpha = 1$) or export to the grid ($\alpha = 0$), where $\alpha + \alpha' = 1$

$$
P_{b,t}^{EVCh} + P_{b,t}^{Exp} = P_{b,t}^{EVDch} + P_{b,t}^{Imp} \forall b \in B, \forall t \in T \tag{3.22}
$$

$$
0 \leq P_{b,t}^{Imp} \leq \bar{P}^{Grid} \alpha; \forall b \in B, \forall t \in T \tag{3.23}
$$

$$
0 \leq P_{b,t}^{Exp} \leq \bar{P}^{Grid} \alpha'; \forall b \in B, \forall t \in T \tag{3.24}
$$

With the actions of EVs of buying (charge) if $\lambda_{b,t}^{Clear} < \lambda_{b,t}^{Bid}$ and selling (V2G) if $\lambda_{b,t}^{Clear} > \lambda_{b,t}^{Offer}$ energy in the LEM, the voltage problem in the network and EV owners' satisfaction are investigated. If there is a voltage issue, the framework employs the proposed sensitivity-based dynamic power management through an iterative process to permit EV charging and discharging (i.e., V2G) at a dynamic rate so that the related voltage problem can be resolved and the EV owners' satisfaction can be achieved [5]. If a voltage problem occurs in node b at time step t of the network, $P_{b,t}^{EVCh}/P_{b,t}^{EVDch}$ is configured by curtailing $P_{b,t}^{EVCh}/P_{b,t}^{EVDch}$ one-by-one at a rate β in (3.25) until the network operates within constraint (3.16), where k is the total iteration number. The curtailment depends on the highest value of voltage sensitivity coefficients χ^P, which corresponds to the highest voltage deviation ΔV for the highest power change ΔP, as in (3.26). χ^P are computed by using (3.27), where R_{mn} is the arithmetic sum of resistances (R_d) of lines that make a path path $h_{m,n}$ in which power is absorbed between nodes m and n [5,34]. According to the updated values of $P_{b,t}^{EVCh}$ and $P_{b,t}^{EVDch}$, $S_{b,t}^{EV}$ is updated at the end of each t using the EV model demonstrated in (3.1)

$$
P_{b,t}^{EVCh} = P_{b,t}^{EVCh} - k\beta P_{b,t}^{EVCh}
$$
$$
P_{b,t}^{EVDch} = P_{b,t}^{EVDch} - k\beta P_{b,t}^{EVDch} \tag{3.25}
$$

$$[\Delta V] = [\chi^P][\Delta P] \tag{3.26}$$

$$[\chi^P] = -\frac{1}{\tilde{V}_b}\left[\sum_{mn\in\text{path}_{m,n}} R_{mn}\right] \tag{3.27}$$

3.3.2   Aggregated EV management using flexibility of EV charge and discharge

3.3.2.1   Overview

An aggregated EV management is presented in this framework, considering the flexibility of EV charge and discharge. The management framework facilitates EVs to trade energy in the LEM with their preferred bid/offer prices so that they can benefit from reduced charging costs and contribute to the network services with V2G provisions. At the same time, it provides decentralized parallel control via EV aggregators to offer enhanced control and computational efficiency for large-scale EV integration into networks. The flexibility of EV charge and discharge is utilized to control line loading and node voltage in the network within constraints. In the proposed framework, EVs are connected to several sub-distribution areas where an independent EV aggregator manages EVs in its area for their market participation and network services through EV charging and V2G. EV aggregators inter-communicate with the DSO for market prices and with EVs for their bid/offer prices and obtain the charge and discharge requirements of EVs. On top of that, the DSO executes the market-clearing problem with the EV charge and discharge requirements and determines the market-clearing prices to operate energy trans-actions for EVs. At each time step, the EVs' charge and discharge profiles are updated while managing line loading and node voltages in the network by imple-menting the charge and discharge flexibility of EVs.

3.3.2.2   Methodology

Charge–discharge flexibility-based aggregated EV management is developed by executing two-level optimization problems. First, at the bottom, EV aggregators determine EVs' charge/discharge requirements by minimizing the operating cost of EVs as in (3.28), subject to constraints (3.4) and (3.5)

$$\min\sum_{b\in B_a}\left(P_{b,t}^{EVCh}\lambda_{b,t}^{Bid} - P_{b,t}^{EVDch}\lambda_{b,t}^{Offer}\right)\Delta t \tag{3.28}$$

subject to: (3.4), (3.5)

where b is the set of nodes of the distribution network under an EV aggregator defined as $b \in \mathcal{B}_a \triangleq \{1,\ldots,b,\ldots,B_a\}$. Second, at the top, the DSO performs the market clearing problem and generates the market clearing prices by maximizing the social welfare from the buying and selling of energy as in (3.29), subject to the power balance constraint (3.30), the aggregated EVs charge/discharge requirements $\left(P_{a,t}^{EVChAg}/P_{a,t}^{ECDchAg}\right)$ (3.31) and (3.32), and grid power constraints (3.33) and

(3.34). Let a be a set of EV aggregators defined as $a \in \mathcal{A} \triangleq \{1, \ldots, a, \ldots, A\}$

$$\max \sum_{b \in \mathcal{B}} \left(\left(P_{a,t}^{ImpAg} \lambda_{a,t}^{DLM} - P_{a,t}^{ExpAg} \lambda_t^{FiT} \right) \right.$$
$$\left. - \lambda_{a,t}^{Clear\ Ag} \left(P_{a,t}^{ImpAg} - P_{a,t}^{ExpAg} - P_{a,t}^{EVChAg} + P_{a,t}^{ECDchAg} \right) \right) \Delta t \tag{3.29}$$

subject to: $(3.30)-(3.34)$

where $P_{a,t}^{ImpAg} / P_{a,t}^{ExpAg}$ are the import/export power to/from the upstream grid for aggregator a and $\lambda_{a,t}^{ClearAg}$ are the market clearing prices. The objective function (3.29) has a unique optimal solution using KKT conditions, and constraints (3.30)–(3.34) are relaxed by the Lagrange multiplier ($\lambda_{a,t}^{ClearAg}$), where the power balance constraint (3.30) corresponds to $\lambda_{a,t}^{ClearAg}$ in (3.29). In (3.33) and (3.34), a binary variable (γ) is used to define the aggregator import power from the grid ($\gamma = 1$) or aggregator export power to the grid ($\gamma = 0$), where $\gamma + \gamma' = 1$

$$P_{a,t}^{ImpAg} + P_{a,t}^{ECDchAg} = P_{a,t}^{ExpAg} + P_{a,t}^{EVChAg}; \forall a \in \mathcal{A}, \forall t \in \mathcal{T} \tag{3.30}$$

$$P_{a,t}^{ImpAg} = P_{a,t}^{EVChAg} = \sum_{b \in \mathcal{B}_a} P_{b,t}^{EVCh}; \forall a \in \mathcal{A}, \forall t \in \mathcal{T} \tag{3.31}$$

$$P_{a,t}^{ExpAg} = P_{a,t}^{ECDchAg} = \sum_{b \in \mathcal{B}_a} P_{b,t}^{ECDch}; \forall a \in \mathcal{A}, \forall t \in \mathcal{T} \tag{3.32}$$

$$0 \leq P_{a,t}^{ImpAg} \leq \bar{P}^{GridAg} \gamma; \forall a \in \mathcal{A}, \forall t \in \mathcal{T} \tag{3.33}$$

$$0 \leq P_{a,t}^{ExpAg} \leq \bar{P}^{GridAg} \gamma'; \forall a \in \mathcal{A}, \forall t \in \mathcal{T} \tag{3.34}$$

Depending on $\lambda_{a,t}^{ClearAg}$, EVs decide their actions for buying (charging) if $\lambda_{b,t}^{Clear} < \lambda_{b,t}^{Bid}$ if $\lambda_{a,t}^{ClearAg} < \lambda_{b,t}^{Bid}$ and selling (V2G) if $\lambda_{a,t}^{ClearAg} > \lambda_{b,t}^{Bid}$ in the LEM. The network operation with EV actions for any violations of line loading and node voltage constraints is investigated by employing the proposed charge/discharge flexibility-based line loading and voltage control in the distribution network within constraints (3.16) and (3.17). For the line loading control, the flexibility of EV's charge/discharge power is managed in node b by curtailing the power with a quantity $\left(\Delta P_{b,t}^{CC} \right)$ using power transfer distribution factors Φ for both import $\left(P_{b,t}^{EVCh} \right)$ and export $\left(P_{b,t}^{EVDch} \right)$, as in (3.35) within the constraint defined in (3.17)

$$P_{b,t}^{EVCh} = P_{b,t}^{EVCh} - \Delta P_{b,t}^{CC}$$
$$P_{b,t}^{EVDch} = P_{b,t}^{EVDch} - \Delta P_{b,t}^{CC} \tag{3.35}$$

where $\Delta P_{b,t}^{CC}$ is the fractionally distributed line loading to be curtailed for line loading management in the network from the charge/discharge power of EVs in node b at time step t as calculated by (3.36) and (3.37). The factor $\Phi_{d,b}$ defines the power flow change through the network line d to the import/export power change at node b in the network, as in (3.36) and (3.37) [35]. Φ is computed by (3.38), where C^{DM} is a D by D diagonal of line susceptance matrix, E^{IM} is a D by B line incidence matrix containing 1 for the "from bus" and -1 for the "to bus," and F^{IM} is an inverse matrix of B by B line susceptance matrix of the network with a zero entry to the row and column corresponding to the slack bus

$$\Delta P_{d,t} = \Delta I_{d,t} V_{d,t}^+ = \sum_{b \in \mathcal{B} | b \neq 1} {}_{d,b} \Delta P_{b,t}^{CC}; \ \forall d \in \mathcal{D}, \forall t \in \mathcal{T} \tag{3.36}$$

$$\Delta I_{d,t} = I_{d,t} - \bar{I}_d \ ; \ \forall l \in \mathcal{L}, \forall t \in \mathcal{T} \tag{3.37}$$

$$[] \overset{\Delta}{=} [C^{DM}][E^{IM}][F^{IM}] \tag{3.38}$$

For the voltage control, the flexibility of EV's charge/discharge power for both import $\left(P_{b,t}^{EVCh} \right)$ and export $\left(P_{b,t}^{EVDch} \right)$ is managed in node b by the sensitivity-based dynamic power curtailment approach through an iterative process as described in section 3.3.1 using (3.25)–(3.27). Based on the updated values of $P_{b,t}^{EVCh}$ and $P_{b,t}^{EVDch}$ with the EVs' charge/discharge flexibility, $S_{b,t}^{EV}$ is updated at the end of each time step t using the EV model demonstrated in (3.1).

3.3.3   DOE-enabled market-based EV management

3.3.3.1   Overview
This framework can manage EV owner prosumers in the LEM within the assigned DOEs for both import and export power to operate the network without any violations of network constraints. In the framework, prosumers send their intended import and export power at each time step to the DSO to check whether or not the intended imports and exports can cause any breaches of network constraints. The DSO assigns DOEs for prosumers based on the intended imports and exports and network constraints so that prosumers can update their assets (e.g., EV charged and discharged power) using DOEs and trade in the LEM. A DOE-constrained bidding strategy is developed for prosumers, where prosumers can trade with their peers within generated bids by employing a peer matching mechanism to get economic benefits. The participation of EVs in the proposed LEM provides economic and network services while achieving the desired charge levels of EV batteries for driving before departure.

3.3.3.2   Methodology
DOE-enabled market-based EV management is accomplished by managing prosumers' import/export power within the assigned DOEs to trade with their peers in the LEM without any network constraint breaches. In the proposed framework, considering the intended import/export power of prosumers and network voltage and line loading constraints, DOEs $\left(P_{b,t}^{Doe} \right)$ are determined using (3.18) to communicate with prosumers and limit their import/export power so that prosumers can manage their assets, such as EVs' charge/discharge power and PV system generations, within DOEs to trade in the LEM with increased benefits and secured network operation. A DOE-constrained bidding strategy is proposed by

minimizing the total trading cost for prosumers to decide on the trading in each time step t as

$$\min \sum_{b \in \mathcal{B}} \left(P_{b,t}^{PeerBuy} \lambda_{b,t}^{Lem} + P_{b,t}^{GridBuy} \lambda_{b,t}^{ToU} - P_{b,t}^{PeerSell} \lambda_{b,t}^{Lem} - P_{b,t}^{GridSell} \lambda_{t}^{FiT} \right) \Delta t$$

subject to: $(3.40) - (3.42)$

$$(3.39)$$

where $P_{b,t}^{PeerBuy} / P_{b,t}^{PeerSell}$ are the buying/selling power with the peers, $P_{b,t}^{GridBuy} / P_{b,t}^{GridSell}$ are the buying/selling power with the grid, $\lambda_{b,t}^{Lem}, \lambda_{b,t}^{ToU}$, and λ_{t}^{FiT} are the LEM prices, the time-of-use (TOU) prices, and the feed-in-tariff (FIT) rates, respectively. The sum of $P_{b,t}^{PeerBuy}$ and $P_{b,t}^{GridBuy}$ of prosumer b at time step t should be equal to or less than $P_{b,t}^{Doe}$, and the sum of $P_{b,t}^{PeerSell}$ and $P_{b,t}^{GridSell}$ should be equal to or less than $\left| P_{b,t}^{Doe} \right| \left(P_{b,t}^{Doe} \leq 0 \right)$ as in (3.40) and (3.41), respectively

$$P_{b,t}^{PeerBuy} + P_{b,t}^{GridBuy} \leq P_{b,t}^{Doe}, \text{ if } P_{b,t}^{Doe} > 0; \forall b \in \mathcal{B}, \forall t \in \mathcal{T} \tag{3.40}$$

$$P_{b,t}^{PeerSell} + P_{b,t}^{GridSell} \leq \left| P_{b,t}^{Doe} \right|, \text{ if } P_{b,t}^{Doe} \leq 0; \forall b \in \mathcal{B}, \forall t \in \mathcal{T} \tag{3.41}$$

The power balance of the total buying and the total selling of the P2P trading in each t must be the same as in (3.42), so the remaining amounts of import and export power of prosumers within $P_{b,t}^{Doe}$ can be traded with the grid in the LEM

$$\sum_{b \in \mathcal{B}} P_{b,t}^{PeerBuy} = \sum_{b \in \mathcal{B}} P_{b,t}^{PeerSell}; \forall t \in \mathcal{T} \tag{3.42}$$

To generate the P2P trading quantities and prices, an optimal P2P trading mechanism is executed by minimizing the mismatches between individual P2P trades of peers of prosumers, as in (3.43) so that prosumers can independently decide on an optimal buy/sell energy from/to their matched peers. Let P2P trading be considered for buying (selling) energy $P_{b,n,t}^{PeerBuy} \left(P_{n,b,t}^{PeerSell} \right)$ by prosumer $b \in \mathcal{B}$ from (to) prosumer $n \in \mathcal{B} | n \neq b$

$$\min \sum_{b \in \mathcal{B}} \sum_{n \in \mathcal{B} | n \neq b} \left(\left(P_{b,n,t}^{PeerBuy} \lambda_{b,t}^{PeerBuy} - P_{n,b,t}^{PeerSell} \lambda_{n,t}^{PeerSell} \right) \right.$$

$$\left. - \lambda_{b,n,t}^{ClearP2P} \left(P_{b,n,t}^{PeerBuy} - P_{n,b,t}^{PeerSell} \right) \right) \Delta t \tag{3.43}$$

subject to: $(3.44)-(3.46)$

The objective function (3.43) has a unique optimal solution using KKT conditions, and constraints (3.44)–(3.46) are relaxed by the Lagrange multiplier $\left(\lambda_{b,n,t}^{ClearP2P} \right)$, where the power balance constraint (3.44) corresponds to $\lambda_{b,n,t}^{ClearP2P}$ in (3.43). In (3.45) and (3.46), a binary variable $\left(v_{b,n,t} \right)$ is used to represent the P2P trading status such that if $v_{b,n,t} = 1$, prosumer b can buy energy from prosumer n, or if $v_{b,n,t} = 0$, prosumer b can sell energy to prosumer n at t, where $v_{b,n,t} + v'_{b,n,t} = 1$

$$P_{b,n,t}^{PeerBuy} = P_{n,b,t}^{PeerSell}; \ \forall (b,n) \in \mathcal{B}|n \neq b, \forall t \in \mathcal{T} \tag{3.44}$$

$$0 \leq P_{b,n,t}^{PeerBuy} \leq P_{b,t}^{PeerBuy} v_{b,n,t}; \ \forall (b,n) \in \mathcal{B}|b \neq n, \forall t \in \mathcal{T} \tag{3.45}$$

$$0 \leq P_{n,b,t}^{PeerSell} \leq P_{b,t}^{PeerSell} v'_{b,n,t}; \ \forall (b,n) \in \mathcal{B}|b \neq n, \forall t \in \mathcal{T} \tag{3.46}$$

Prosumers manage their assets based on the assigned DOEs by updating EV charged and discharged power $\left(P_{b,t}^{EVCh} \text{ and } P_{b,t}^{EVDch} \right)$ and utilization of the PV system generation $\left(P_{b,t}^{PV} \right)$ with a curtailed amount $P_{b,t}^{Cur}$, which is the difference between $P_{b,t}^{Net}$ and $P_{b,t}^{Doe}$. If $p_{b,t}^{cur} \leq 0$, prosumer updates EV charge power using (3.47); if $p_{b,t}^{cur} < 0$, prosumer updates EV discharge power using (3.48)

$$P_{b,t}^{EVCh} = P_{b,t}^{EVCh} - P_{b,t}^{Cur}, \ \text{if } P_{b,t}^{Cur} \geq 0; \ \forall b \in \mathcal{B}, \forall t \in \mathcal{T} \tag{3.47}$$

$$P_{b,t}^{EVDch} = P_{b,t}^{EVDch} - \left| P_{b,t}^{Cur} \right|, \ \text{if } P_{b,t}^{Cur} < 0; \ \forall b \in \mathcal{B}, \forall t \in \mathcal{T} \tag{3.48}$$

If there is no EV charge/discharge, the PV system utilization power is updated using (3.49). Otherwise, no updates on EV charge/discharge and PV power are done. Based on the updated values of $P_{b,t}^{EVCh}$ and $P_{b,t}^{EVDch}$, the EV battery charge level $S_{b,t}^{EV}$ is updated at the end of each time step t using the EV model demonstrated in (3.1)

$$P_{b,t}^{PV} = P_{b,t}^{PV} - \left| P_{b,t}^{Cur} \right|; \ \forall b \in \mathcal{B}, \forall t \in \mathcal{T} \tag{3.49}$$

3.3.4 Network-secure EV management via P2P trading

3.3.4.1 Overview

A network-secure EV management framework is presented by employing a prosumer-centric P2P energy trading mechanism, which is developed based on the hybrid DER marketplace model in [36]. In this framework, the DSO keeps an eye out for any network issues that may arise from prosumers' trading for their intended power imports and exports and oversees the network within network operational constraints by allocating DOEs to prosumers. Conversely, the distribution market operator (DMO) oversees the distribution market to facilitate prosumers' local energy trading while taking into account their preferences within the DOEs that the DSO has assigned. Prosumers' trading offers and bids are updated within DOEs in order to manage the EV charged and discharged power. The DMO then initiates an auction mechanism to determine the winning peers based on the suggested price and quantity matching algorithm. In the P2P energy market, prosumers who win an auction can transact with other prosumers, and losers can transact with the upstream grid. This framework gives priority to prosumers charging their EVs from PV system surplus, satisfies the preferences of EV owners in terms of achieving the desired charge levels of EV batteries when they depart, allows them to participate equally in the market with their preferred bids and offers for both energy and prices, and ensures fairness by giving equal motivation to traders in terms of access to trading and payoff.

3.3.4.2 Methodology

The network-secure EV management via P2P trading is developed based on the double auction approach [37], where network-secure prosumers' trading quantities for both import and export (i.e., DOEs) are defined by the DSO for secured network operation. The DMO then performs the double auction-based P2P energy trading mechanism and determines the matched peers, prices, and quantities within the allocated DOEs. For price and quantity matching under a proposed double auction mechanism, a request vector is set for each buyer $(x \in \mathcal{B}^{Cs})$ and seller ($y \in \mathcal{B}^{Pd}$) prosumer consisting of the identification number (ID_u^B and ID_s^S), the net import and export power ($P_{x,t}^{Net}$ and $P_{y,t}^{Net}$), respectively, and the offer and bid prices $\left(\lambda_{x,t}^{Bid} \text{ and } \lambda_{y,t}^{Offer}\right)$, respectively, where the vectors are updated with assigned import and export power $\left(P_{x,t}^{Doe} \text{ and } P_{y,t}^{Doe}\right)$ based on the allocated DOEs. To provide equal opportunities for all prosumers to win the auction, the DMO finds out whether each buyer satisfies $\lambda_{x,t}^{Bid} \geq \lambda_t^{Avg}$ and each seller satisfies $\lambda_{y,t}^{Offer} \leq \lambda_t^{Avg}$. Otherwise, unsatisfied prosumers are listed in the respective loser matrix to trade with the gird at $\lambda_{b,t}^{ToU} / \lambda_t^{FiT}$ prices, where λ_t^{Avg} is the average price of all prosumers b in each t.

To start the auction process, the DMO arranges the request vectors from buyers in descending order (i.e., $\lambda_{1,t}^{Bid} > \lambda_{2,t}^{Bid} > \lambda_{3,t}^{Bid} > \ldots > \lambda_{x,t}^{Bid}$) and from sellers in ascending order (i.e., $\lambda_{1,t}^{Offer} < \lambda_{2,t}^{Offer} < \lambda_{3,t}^{Offer} < \ldots < \lambda_{y,t}^{Offer}$) to their bid and offer prices in each t corresponding to their assigned matrices, so that the buyer of the highest price bid will trade with the seller of the lowest offer price at the matched price and quantity from the respective winner buyer (seller) matrix and so on. The buyers (sellers) with the remaining import (export) in each auction trade are transferred to the end of the remaining buyer (seller) matrix. For P2P energy trading, the matched quantity $P_{x,y,t}^{P2P}$ is calculated by the minimum of the buyer's import and seller's export as in (3.50), and the matched price $\lambda_{x,y,t}^{P2P}$ is computed by the average of the buyer's bid and seller's offer prices as in (3.51)

$$P_{x,y,t}^{P2P} = \min\left(P_{x,t}^{Doe}, P_{y,t}^{Doe}\right); \forall x \in \mathcal{B}_t^{Cs}, \forall y \in \mathcal{B}_t^{Pd}, \forall t \in \mathcal{T} \tag{3.50}$$

$$\lambda_{x,y,t}^{P2P} = \frac{\lambda_{x,t}^{Bid} + \lambda_{y,t}^{Offer}}{2}; \forall x \in \mathcal{B}_t^{Cs}, \forall y \in \mathcal{B}_t^{Pd}, \forall t \in \mathcal{T} \tag{3.51}$$

The auction process continues for the remaining buyers and sellers until no buyer or seller with a matched peer exists in the remaining buyer and seller matrices. The remaining buyers (sellers) in the remaining buyer (seller) matrix without a matched opposite peer are treated as losers and placed in the corresponding loser matrices to trade with the grid at $\lambda_t^{ToU} \left(\lambda_t^{FiT}\right)$ prices. The proposed

P2P trading mechanism is prosumer-centric and facilitates all prosumers with equal market access and benefits.

Depending on the assigned import/export power $\left(P_{b,t}^{Doe}\right)$ for the P2P trading, prosumers update EV charged and discharged power ($P_{b,t}^{EVCh}$ and $P_{b,t}^{EVDch}$) and utilization of the PV system generation ($P_{b,t}^{PV}$) by (3.47), (3.48), and (3.49), respectively. According to the updates of $P_{b,t}^{EVCh}$ and $P_{b,t}^{EVDch}$, the SOC value $S_{b,t}^{EV}$ of the EV battery is updated at the end of each time step t using the EV model demonstrated in (3.1).

3.4   Case studies

This section provides case studies of the proposed EV management frameworks. The case studies in sections 3.4.1, 3.4.2, and 3.4.3 are conducted on an IEEE 55-node three-phase European low-voltage (LV) distribution test system (Figure 3.2) [38], and the case study in section 3.4.4 uses a 78-node three-phase Denish LV distribution test system (Figure 3.3) [37]. The network is configured for each connection node to be connected to any asset, such as an EV, a PV system, or household loads. All household data are taken from the IEEE European LV test feeder network [5,38]. The lithium-ion battery-based Tesla Model 3 with a 55 kWh capacity and 6.6 kW charger rating is used in the first three case studies, whereas the Hyundai Kona with a 64 kWh capacity and 7 kW charger rating is used in the last case study with 0.95 charge–discharge efficiencies. The market prices (e.g., the LMP price, the ToU price, the FiT rate) and EV mobility data (e.g., EV arrival and departure time and initial and desired SOC of EV) are taken from [4,5,37]. The DLMP price deviation is considered to be 0.1 in case studies. The time step resolution for the simulation analysis is set at 15 min. The upper voltage limit of 1.1 p. u. and the lower voltage limit of 0.94 p.u., and the 100% maximum line loading of the corresponding line capacity of the network are considered for the performance analysis of the networks.

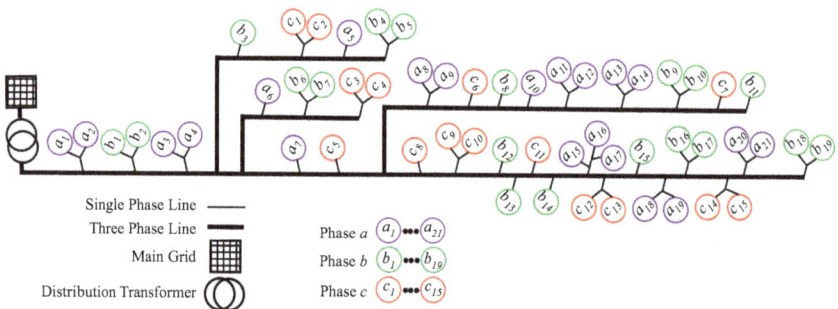

Figure 3.2   IEEE 55-node three-phase European LV distribution test system

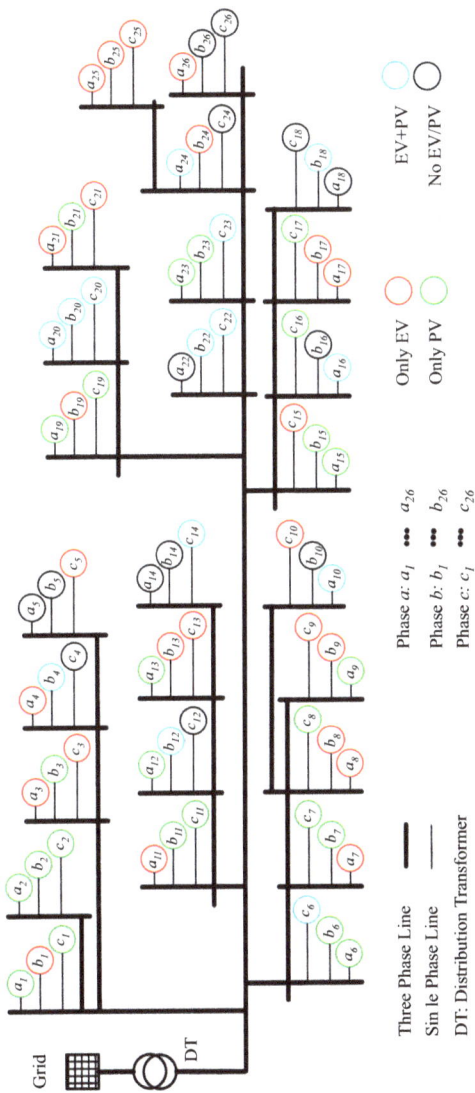

Figure 3.3 78-node three-phase Denish LV distribution test system

3.4.1   Case study 1: sensitivity-based transactive EV management

The performance of this framework is analyzed to observe the safe network operation for EV charge and V2G coordination into the distribution network considering 100% penetration level of EV. The results in Figure 3.4 depict for over 24 h that while EV management with No Control creates voltage problems in many nodes of the network, this framework can manage EVs' participation in the LEM by controlling the voltages in nodes of different phases within the safe operating ranges of the network.

As the proposed framework can manage EVs for both charging and discharging (i.e., V2G), the challenge is to manage EVs considering their preferences and desires while providing network services through managed EV charging and V2G. Figure 3.5 shows the EV charging and discharging (i.e., V2G) actions at sample nodes of different phases. As shown in Figure 3.5, the EV with V2G management is accomplished to achieve the desired SOC levels of EV batteries at the time of departure without any violations of network voltage constraints.

3.4.2   Case study 2: aggregated EV management using flexibility of EV charge and discharge

Aggregated EV management using the flexibility of EV charging and discharging is accomplished to manage the integration of large-scale EVs in the network in an

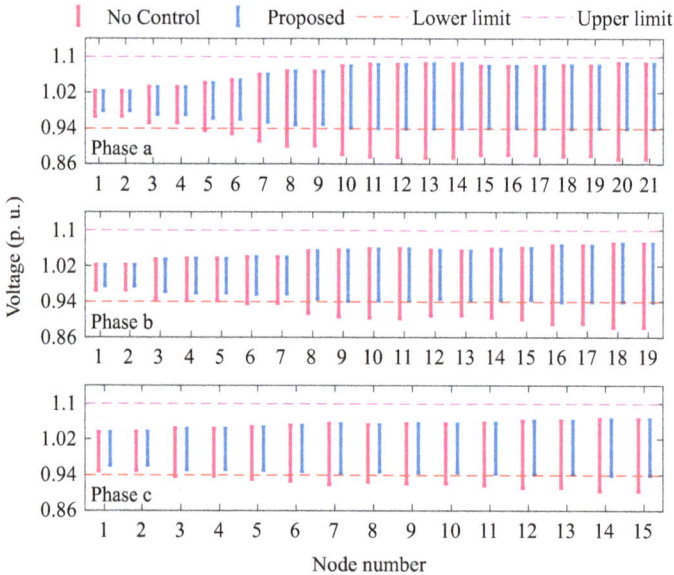

Figure 3.4   Node voltages comparison between the No Control and the proposed management at various nodes of different phases of the distribution network over 24 h (Case study 1)

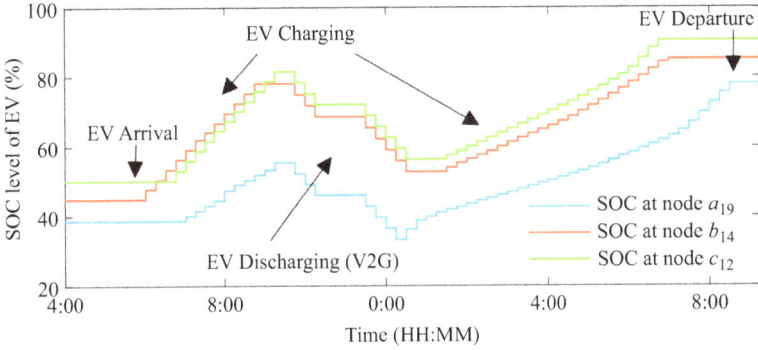

Figure 3.5   EV charging and discharging (i.e., V2G) actions at sample nodes of different phases in the proposed EV management (Case study 1)

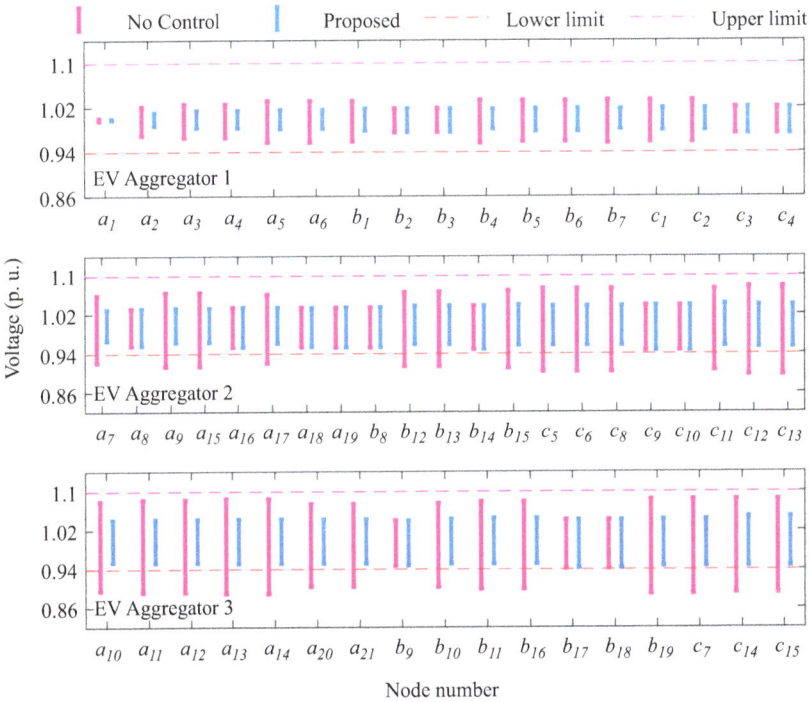

Figure 3.6   Node voltages comparison between the No Control and the proposed management at various nodes of the distribution network at different EV aggregators over 24 h (Case study 2)

aggregator-based distributed parallel control. The performances are analyzed to investigate the node voltages and line loadings of the network for EVs' buying (charge) and selling (V2G) power in the LEM in order to provide market and network services. Figure 3.6 presents the performance comparison regarding node voltages between the No Control and the proposed management at various nodes of the distribution network at different EV aggregators over 24 h. The proposed EV management framework does not have any breaches of node voltages in the network operation while considering EVs' transactions in the LEM with 100% EV penetration.

It also investigates the line-loading performance of the network for EV charging and V2G. The results in Figure 3.7 show the line loading values for the No Control and the proposed management method in various lines of the distribution network for different EV aggregators over 24 h. The proposed framework is effective for managing EVs with V2G within the maximum loading capacities of all network distribution lines.

EVs' participation in the LEM can be seen in Figure 3.8 at sample nodes for different EV aggregators, where EVs' charging and discharging (i.e., V2G) are managed for safe network operation and to fulfill the desires of EVs by achieving expected charge levels of EV batteries during departure.

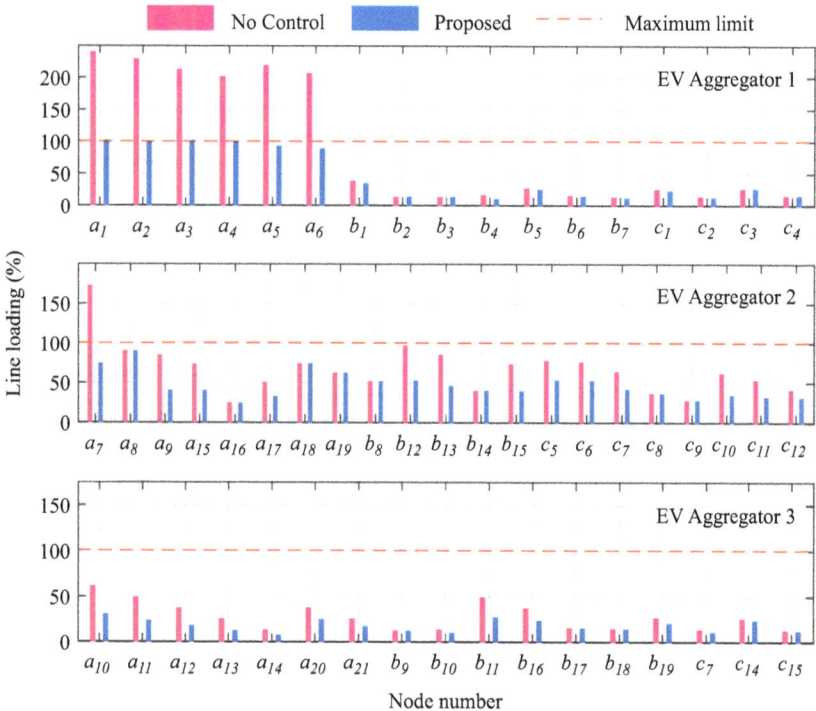

Figure 3.7   Comparison of line loading values in various lines of the distribution network for different EV aggregators over 24 h between the No Control and the proposed management framework (Case study 2)

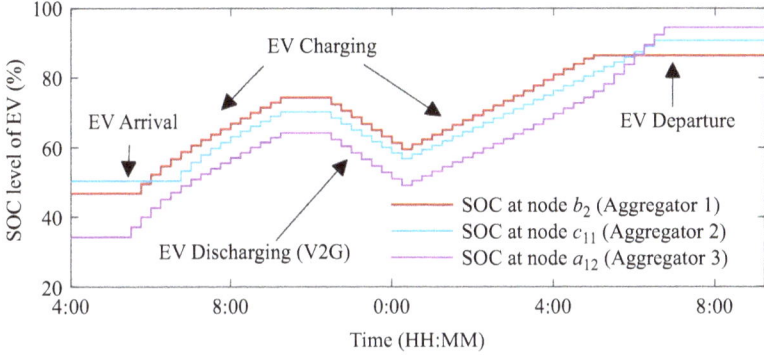

Figure 3.8   EV charging and discharging (i.e., V2G) actions at sample nodes at different EV aggregators in the proposed EV management (Case study 2)

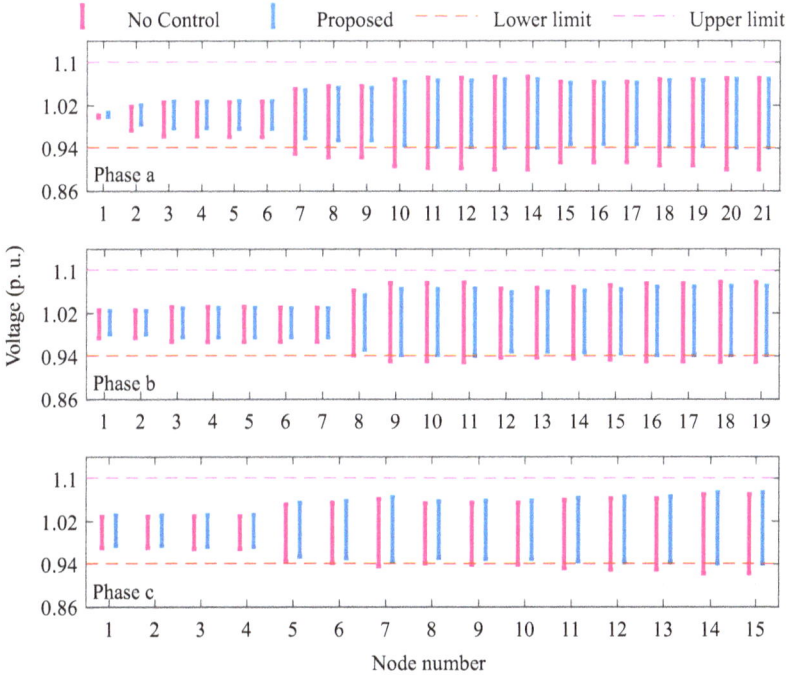

Figure 3.9   Node voltages comparison between the No Control and the proposed management at various nodes of different phases of the distribution network over 24 h (Case study 3)

3.4.3   Case study 3: DOE-enabled market-based EV management

In this framework, import/export DOEs are determined to secure the network operation for EV-based market management, considering the voltage and line loading constraints of the network. The results are illustrated in Figures 3.9 and 3.10

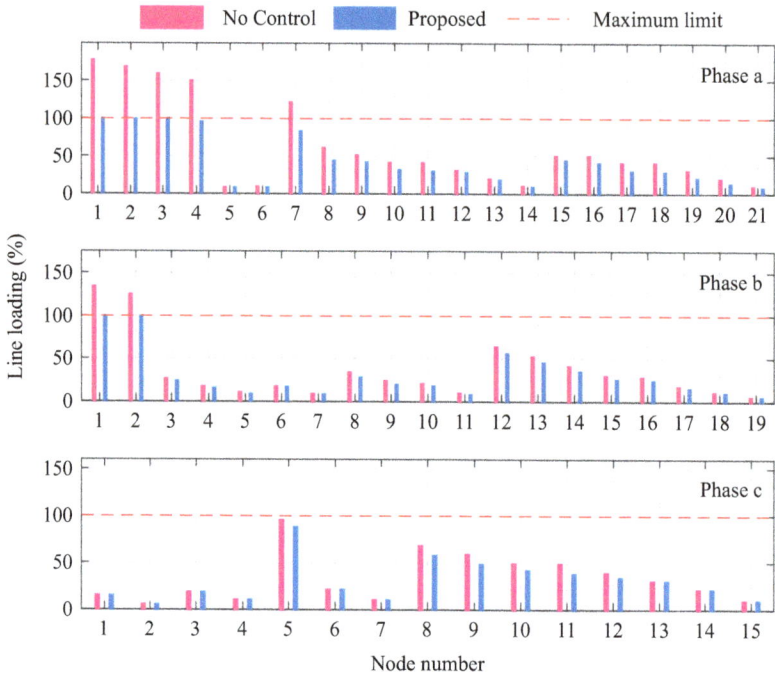

Figure 3.10   Line loading values comparison between the No Control and the proposed management in various lines of different phases of the distribution network over 24 h (Case study 3)

to investigate the voltage and line loading characteristics of the network while managing EVs with V2G in the network. Figure 3.9 depicts node voltages at various nodes of different phases of the distribution network, and Figure 3.10 shows line loading values in various lines of different phases of the distribution network using the proposed management over 24 h and a comparison with the No Control EV management. According to the results, the proposed management framework is able to secure the network operation regarding node voltages and line loadings in the network, which are operated in safe regions.

As DOEs are assigned to prosumers equipped with EVs and PV systems to trade in the LEM with the secured bids within DOEs for power import and export, it is necessary to manage the assets of prosumers considering their expectations. Accordingly, as shown in Figure 3.11, prioritizing EV charge from the surplus PV system generation, EV charged and discharged power are managed by the allowable trading bids within DOEs that secure the network integrity, and the SOC levels of EVs are maintained at their desired levels before departure.

The prosumers' transactions are accomplished by the proposed P2P energy trading mechanism, where prosumers can benefit from the LEM by saving energy buying costs and gaining more revenue from energy selling. Figure 3.12 illustrates the impact of the proposed framework in the market by receiving increased profit as

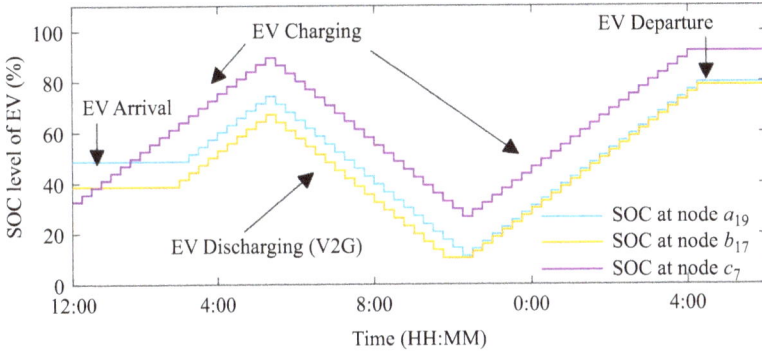

Figure 3.11 EV charging and discharging (i.e., V2G) actions at sample nodes of different phases in the proposed EV management (Case study 3)

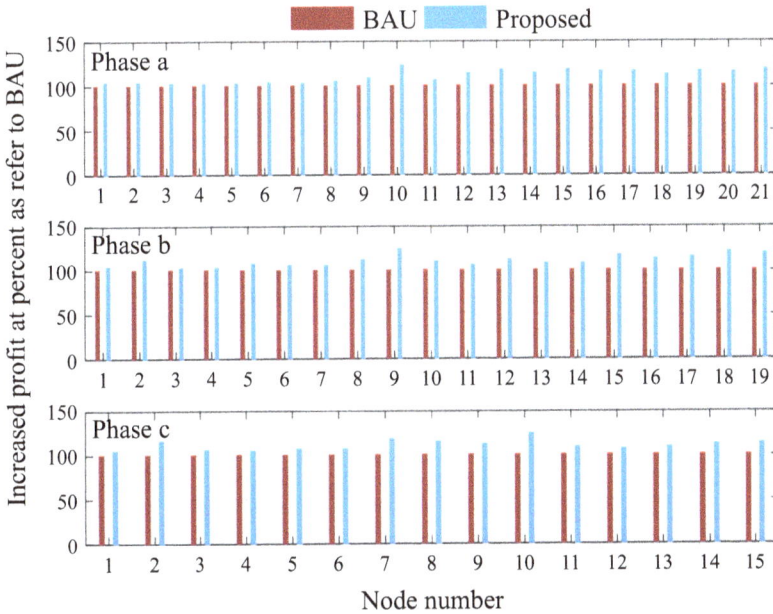

Figure 3.12 Increased profit comparison between the BAU and the proposed management as referred to the BAU at percent over 24 h

compared to the trade at business-as-usual (BAU) at percent value over 24 h. As displayed in Figure 3.12, every prosumer gets an increased amount of profit through participating in the proposed LEM. For example, both prosumers at node b_9 at phase b and node c_{10} at phase c of the distribution network receive around 25% increased profit as referred to the BAU from trading in the proposed market. BAU is defined as trading energy with the upstream grid at the buying TOU prices and at the selling FIT rates.

3.4.4   Case study 4: network secure EV management via P2P trading

This framework can manage EVs in the distribution network by allocating network-secure, flexible import/export limits for P2P energy trading. The operational performance of the distribution network can be seen in Figures 3.13 and 3.14, where Figure 3.13 shows node voltage results at various nodes of different phases of the distribution network and Figure 3.14 presents line loading results in various lines of different phases of the distribution network over 24 h using the proposed management and their comparisons with the No Control EV management. The network operates for prosumers' P2P trade without causing any voltage and line loading constraints violations.

With flexible network-secure import/export limits, prosumers independently manage their assets by updating the power of EV charge and V2G and achieving the desired SOC levels of EV batteries at the departure time. Figure 3.15 illustrates the EV charging and discharging (i.e., V2G) actions at sample nodes of different phases in the proposed EV management. In Figure 3.15, it is seen that after arrival, EVs take part in trading through EV charge and V2G, maintaining their mobility

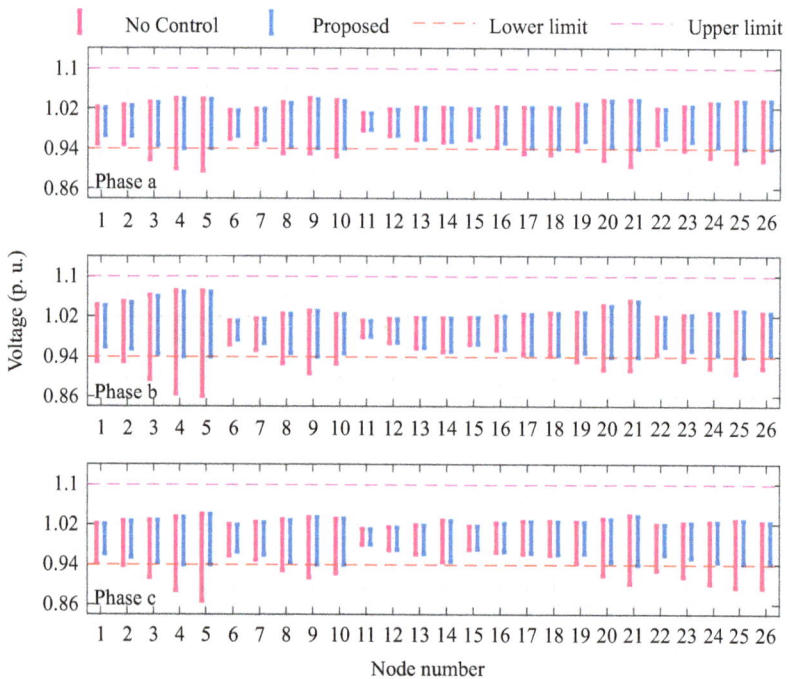

Figure 3.13   Node voltages comparison between the No Control and the proposed management at various nodes of different phases of the distribution network over 24 h (Case study 4)

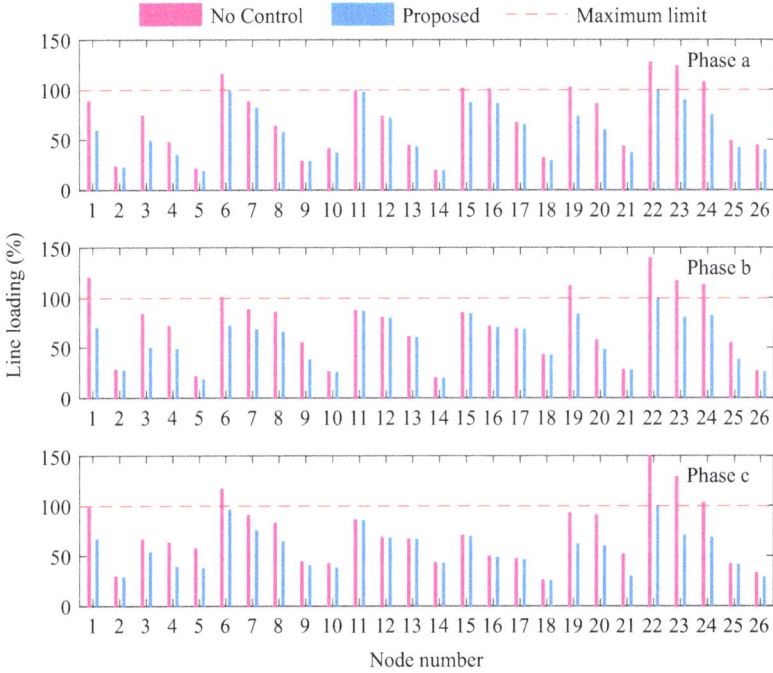

Figure 3.14 Line loading values comparison between the No Control and the proposed management in various lines of different phases of the distribution network over 24 h (Case study 4)

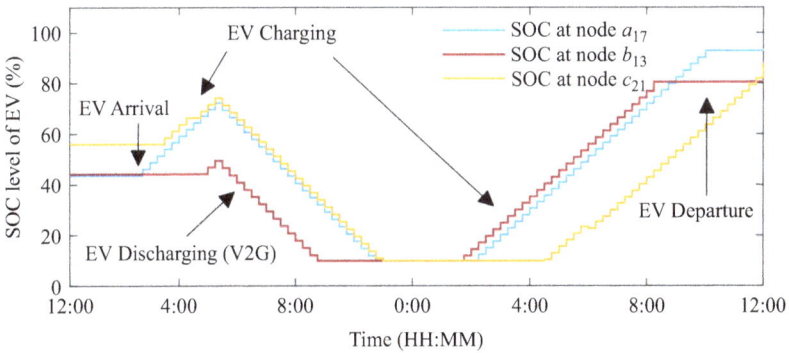

Figure 3.15 EV charging and discharging (i.e., V2G) actions at sample nodes of different phases in the proposed EV management (Case study 4)

and charge constraints and managing the desired charge levels of their batteries before departure.

The proposed framework manages prosumers' trading in the LEM by employing a prosumer-centric double auction mechanism, in which prosumers buy

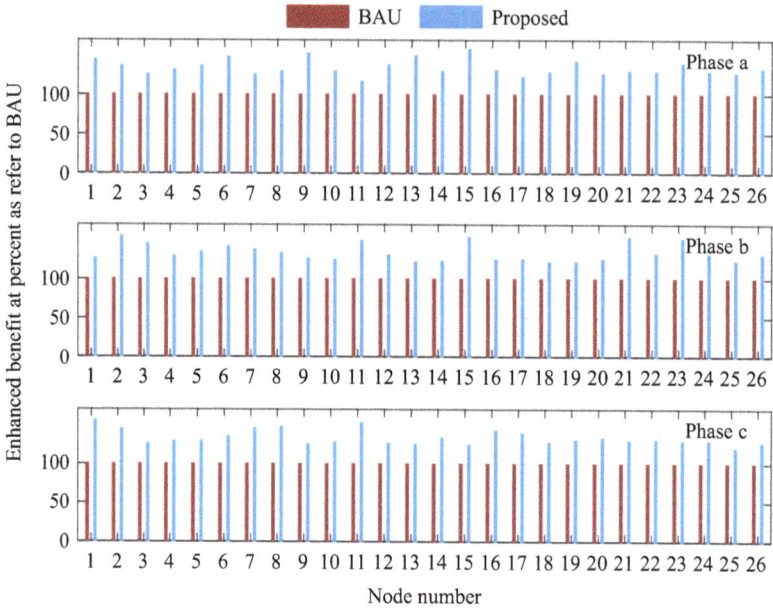

Figure 3.16   Enhanced benefit comparison between the BAU and the proposed management as referred to the BAU at percent over 24 h (Case study 4)

energy from their matched peers at reduced prices and sell energy to the matched peers at increased prices as compared to the BAU. Therefore, prosumers receive more benefits from the proposed market, as shown in Figure 3.16. As in Figure 3.16, each prosumer gets an enhanced benefit in the proposed management from buying and selling energy with their peers as referred and compared to the BAU at percent over 24 h. For instance, prosumers at node a_{15} at phase a and node c_1 at phase c of the distribution network receive around 58% and 55% enhanced benefit as refer to the BAU from P2P trading in the market.

3.5   Conclusion

Several management frameworks for EVs with V2G have been presented in this chapter. This chapter has begun with illustrations on the background of EV management, various types of EV management schemes, and related state-of-the-art reviews. The models of EV, prosumer, network, and DOE assignment have been presented for the development of EV management with V2G frameworks. Four different frameworks for network-aware EV management with V2G have been presented, including (1) *Sensitivity-based transactive EV management* that can transactively coordinate EV charging and V2G and resolve the voltage problems in the network by the sensitivity-based voltage management algorithm, (2) *Aggregated EV management using flexibility of EV charging and discharging* that can manage a large number of EVs in distributed parallel manner via aggregators and maintain the network operation without violating node voltage and line loading constraints by employing EVs' charge and discharge flexibility, (3) *DOE-enabled market-based EV management* which provides DOEs to manage prosumers import/export power to secure the network integrity and enables prosumers to trade with their peers in the LEM within allocated DOEs to get increased profit, and (4) *Network-secure EV management via P2P trading* that allows prosumers to trade in a double auction-based P2P energy market and manage their assets within the assigned flexible trade limits to maintain secured network operations. Further, case studies for all EV management frameworks have been presented to investigate the effectiveness of the proposed frameworks. From the results and performance analyses, all frameworks have been shown to be able to maintain the network operation within the network constraints for EVs with V2G management, manage the power for EV charging and V2G to achieve the desired SOC level of EV batteries, and financially benefit EV owner prosumers. Overall, these frameworks are effective enough to be applied to EVs with V2G management in the real world.

References

[1] Azim M.I., Khorasany M., and Razzaghi R. 'Techno-economic-environmental assessment of green transportation systems'. *Interconnected Modern Multi Energy Networks and Intelligent Transportation Systems*. Wiley, New York, 2024. ch. 3, pp. 39–58.

[2] Panossian N., Muratori M., Palmintier B., Meintz A., Lipman T., and Moffat K. 'Challenges and opportunities of integrating electric vehicles in electricity distribution systems'. *Curr. Sustain. Renew. Energy Rep.* 2022, vol. 9(2), pp. 27–40.

[3] Mouli G.R.C., Kefayati M., Baldick R., and Bauer P. 'Integrated PV charging of EV fleet based on energy prices, V2G, and offer of reserves'. *IEEE Trans. Smart Grid.* 2019, vol. 10(2), pp. 1313–1325.

[4] Hoque M.M., Khorasany M., Azim M.I., Razzaghi R., and Jalili M. 'Dynamic operating envelope-based local energy market for prosumers with electric vehicles'. *IEEE Trans. Smart Grid.* 2024, vol. 15(2), pp. 1712–1724.

[5] Hoque M.M., Khorasany M., Razzaghi R., Wang H., and Jalili M. 'Transactive coordination of electric vehicles with voltage control in distribution networks'. *IEEE Trans. Sustain. Energy.* 2022, vol. 13(1), pp. 391–402.

[6] Nizami M.S.H., Hossain M., and Mahmud K. 'A coordinated electric vehicle management system for grid-support services in residential networks'. *IEEE Syst. J.* 2021, vol. 15(2), pp. 2066–2077.

[7] Quiros-Tortos J., Ochoa L., and Butler T. 'How electric vehicles and the grid work together: Lessons learned from one of the largest electric vehicle trials in the world'. *IEEE Power Energy Mag.* 2018, vol. 16(6), pp. 64–76.

[8]   Yong J.Y., Tan W.S., Khorasany M., and Razzaghi R. 'Electric vehicles desti-
      nation charging: An overview of charging tariffs, business models and coordi-
      nation strategies'. *Renew. Sustain. Energy Rev.* 2023, vol. 184, pp. 113534.
[9]   Rabiee A., Keane A., and Soroudi A. 'Enhanced transmission and distribu-
      tion network coordination to host more electric vehicles and PV'. *IEEE Syst.
      J.* 2022, vol. 16(2), pp. 2705–2716.
[10] Sun G., Zhang F., Liao D., Yu H., Du X., and Guizani M. 'Optimal energy
      trading for plug-in hybrid electric vehicles based on fog computing'. *IEEE
      Internet Things J.* 2019, vol. 6(2), pp. 2309–2324.
[11] Hoque M.M., Khorasany M., Razzaghi R., Jalili M., and Wang H.
      'Network-aware coordination of aggregated electric vehicles considering
      charge–discharge flexibility'. *IEEE Trans. Smart Grid.* 2023, vol. 14(3),
      pp. 2125–2139.
[12] De Craemer K., Vandael S., Claessens B., and Deconinck G. 'An event-
      driven dual coordination mechanism for demand side management of
      PHEVs'. *IEEE Trans. Smart Grid.* 2014, vol. 5(2), pp. 751–760.
[13] Mohammad A., Zamora R., and Lie T.T. 'Transactive energy management
      of PV-based EV integrated parking lots'. *IEEE Syst. J.* 2021, vol. 15(4),
      pp. 5674–5682.
[14] Liu Z., Wu Q., Shahidehpour M., Li C., Huang S., and Wei W. 'Transactive
      real-time electric vehicle charging management for commercial buildings
      with PV on-site generation'. *IEEE Trans. Smart Grid.* 2019, vol. 10(5),
      pp. 4939–4950.
[15] Astero P., Choi B.J., and Liang H. 'Multi-agent transactive energy
      management system considering high levels of renewable energy source
      and electric vehicles'. *IET Gen. Trans. Dist.* 2017, vol. 11(15), pp. 3713–
      3721.
[16] Kapoor A., Patel V.S., Sharma A., and Mohapatra A. 'Centralized and
      decentralized pricing strategies for optimal scheduling of electric vehicles'.
      IEEE Trans. Smart Grid. 2022, vol. 13(3), pp. 2234–2244.
[17] Saber H., Ehsan M., Aghtaie M.M., Firuzabad M.F., and Lehtonen M.
      'Network-constrained transactive coordination for plug-in electric vehicles
      participation in real-time retail electricity markets'. *IEEE Trans. Sustain.
      Energy.* 2021, vol. 12(2), pp. 1439–1448.
[18] Kikusato H., Mori K., Yoshizawa S., *et al.* 'Electric vehicle charge–dis-
      charge management for utilization of photovoltaic by coordination between
      home and grid energy management systems'. *IEEE Trans. Smart Grid.*
      2019, vol. 10(3), pp. 3186–3197.
[19] Liu Z., Wu Q., Ma K., Shahidehpour M., Xue Y., and Huang S. 'Two-stage
      optimal scheduling of electric vehicle charging based on transactive control'.
      '*IEEE Trans. Smart Grid.* 2019, vol. 10(3), pp. 2948–2958.
[20] Pan Z., Yu T., Li J., *et al.* 'Stochastic transactive control for electric vehicle
      aggregators coordination: A decentralized approximate dynamic program-
      ming approach'. *IEEE Trans. Smart Grid.* 2020, vol. 11(5), pp. 4261–4277.

[21] Nizami M.S.H., Hossain M.J., and Mahmud K. 'A nested transactive energy market model to trade demand-side flexibility of residential consumers'. *IEEE Trans. Smart Grid.* 2021, vol. 12(1), pp. 479–490.

[22] Masood A., Hu J., Xin A., Sayed A.R., and Yang G. 'Transactive energy for aggregated electric vehicles to reduce system peak load considering network constraints'. *IEEE Access.* 2020, vol. 8, pp. 31519–31529.

[23] Hou P., Yang G., Hu J., and Douglass P.J. 'A network–constrained rolling transactive energy model for EV aggregators participating in balancing market'. *IEEE Access.* 2020, vol. 8, pp. 47720–47729.

[24] Azim M.I., Tushar W., Saha T.K., Yuen C., and Smith D. 'Peer-to-peer kilowatt and negawatt trading: A review of challenges and recent advances in distribution networks'. *Renew. Sustain. Energy Rev.* 2022, vol. 169, pp. 112908: 1–23.

[25] Azim M.I., Gazafroudi A.S., and Khorasany M. 'Generation-side and demand-side player-centric tradings in the LEM: Rule-empowered models and case studies'. *Trading in Local Energy Markets and Energy Communities: Concepts, Structures and Technologies.* Springer International Publishing, Cham, 2023, ch. 3, pp. 221–239.

[26] Zhao K., Zhang M., Lu R., and Shen C. 'A secure intra-regional-inter-regional peer-to-peer electricity trading system for electric vehicles'. *IEEE Trans. Veh. Technol.* 2022, vol. 71(12), pp. 12576–12587.

[27] Guerrero J., Chapman A.C., and Verbic G. 'Decentralized P2P energy trading under network constraints in a low-voltage network'. *IEEE Trans. Smart Grid.* 2019, vol. 10(5), pp. 5163–5173.

[28] *Australian Energy Market Operator.* Project EDGE, Australia, 2023.

[29] Azim M.I., Lankeshwara G., Tushar W., *et al.* 'Dynamic operating envelope-enabled P2P trading to maximise financial returns of prosumers'. *IEEE Trans. Smart Grid.* 2024, vol. 15(2), pp. 1978–1990.

[30] Gerdroodbari Y.Z., Khorasany M., Razzaghi R., and Heidari R. 'Management of prosumers using dynamic export limits and shared community energy storage'. *Appl. Energy.* 2024, vol. 355, p. 122222.

[31] Mahmoodi M., Blackhall L., Noori R.A.S.M., Attarha A., Weise B., and Bhardwaj A. 'DER capacity assessment of active distribution systems using dynamic operating envelopes'. *IEEE Trans. Smart Grid.* 2024, vol. 15(2), pp. 1778–1791. doi:10.1109/TSG.2023.3313550.

[32] Saber H., Ehsan M., Moeini-Aghtaie M., Ranjbar H., and Lehtonen M. 'A user-friendly transactive coordination model for residential prosumers considering voltage unbalance in distribution networks'. *IEEE Trans. Ind. Informat.* 2022, vol. 18(9), pp. 5748–5759.

[33] Hammerstrom D., Ambrosio J.R., Carlon T.A., *et al. Pacific Northwest Gridwise Testbed Demonstration Projects, Part I: Olympic Peninsula Project.* Pacific Northwest National Lab (PNNL), Richland, WA, 2008.

[34] Brenna M., Berardinis E.D., Carpini L.D., *et al.* 'Automatic distributed voltage control algorithm in smart grids applications'. *IEEE Trans. Smart Grid.* 2013, vol. 4(2), pp. 877–885.

[35] Song C.S., Park C.H., Yoon M., and Jang G. 'Implementation of PTDFs and LODFs for power system security'. *J. Int. Council Electr. Eng.* 2011, vol. 1 (1), pp. 49–53.

[36] Energy Networks Australia. *Open Energy Networks Project: Energy Networks Australia Position Paper.* Australia, 2020.

[37] Hoque M.M., Khorasany M., Azim M.I., Razzaghi R., and Jalili M. 'A framework for prosumer-centric peer-to-peer energy trading using network-secure export–import limits'. *Appl. Energy.* 2024, vol. 361, p. 122906.

[38] IEEE 55-node LV European test feeder system [online]. 2021. Available from https://site.ieee.org/pes-testfeeders/ [Accessed 8 Feb 2022].

Chapter 4

Electric vehicle charging technologies: topologies, modulation and control

Inam Nutkani[1], Carlos Alberto Teixeira[1], Richardt Howard Wilkinson[1], Nil Rajeshkumar Patel[1] and Brendan Peter McGrath[1]

4.1 Overview

Electric vehicle (EV) chargers are the primary systems that enable the exchange of energy between the vehicle battery and the AC electricity grid. Charging systems are based on power electronic converter platforms that are operated using sophisticated control strategies that enable specific performance objectives to be realised. This chapter provides a comprehensive review of the fundamental elements that form the basis of EV chargers. The chapter begins with an overview of the critical charger classifications, specifications and standards. Next, a review of the fundamental knowledge underpinning power electronic topologies and their operating principles is provided. Single-phase on-board chargers (OBCs) are then described, including uni-directional and bidirectional variants. The chapter then provides a review of three-phase chargers, including two-stage, integrated and modular variations, before briefly discussing wireless charging systems. The control of EV chargers is more complex and less standardised compared to distributed generation since the operating require-ments for EV chargers are still being refined in response to fast-changing standards and regulations. Similarly, conversion technologies are continuously advancing, resulting in a wide variety of solutions. To address this evolving scenario, this chapter outlines the essential control features required for modern chargers to operate in dif-ferent modes, the control architectures for dual-stage chargers and single-stage char-gers and the principles that underpin grid-side and battery-side control.

4.1.1 On-board and off-board chargers

EV charging systems are categorised into OBCs and off-board chargers, depending on whether the power conversion takes place inside or outside the vehicle [1].

[1]Department of Electrical and Electronic Engineering, School of Engineering, RMIT University, Australia

These chargers play a crucial role in determining the charging speed, infrastructure requirements and overall convenience for EV users.

OBCs are integrated within the EV and are responsible for converting AC power from an external source, such as a wall outlet or a public charging station, into DC power to charge the battery. These chargers typically support Level 1 (3.3 kW, slow charging) and Level 2 (3.7–22 kW, moderate charging). Level 3 on-board charging (22–43.5 kW) is supported in some advanced EVs with a three-phase AC input, allowing for much faster charging, typically taking around 1–3 h for a full charge [2]. The key advantage of an OBC is its convenience, as it allows an EV to be charged anywhere with a standard power outlet, making home and workplace charging easy and cost-effective. However, since OBCs are constrained by space, weight and heat dissipation limitations, their power output is usually limited to 3–22 kW, leading to longer charging times compared to fast-charging solutions.

Off-board chargers are external charging stations that provide DC power directly to the EV battery, bypassing the vehicle's OBC [3]. These chargers typically support Level 3 (DC fast charging), with power outputs ranging from 50 to 350 kW or more, significantly reducing charging times. Off-board chargers are commonly found at highway charging stations, fleet depots and commercial locations where rapid charging is essential. Since the power conversion takes place externally, these chargers are not limited by size or weight constraints, allowing them to handle higher power levels. However, it requires expensive infrastructure and high-power grid connections, making them less accessible for everyday charging. Additionally, frequent use of DC fast charging can accelerate battery degradation due to increased heat generation and stress on the battery cells.

4.1.2   G2V, V2G and V2X

The increasing adoption of EVs has led to the development of advanced energy interaction frameworks between EVs and the power grid. These interactions are broadly categorised into grid-to-vehicle (G2V), vehicle-to-grid (V2G) and vehicle-to-everything (V2X) [4]. G2V refers to the conventional charging process where EVs draw power from the electrical grid. V2G extends this concept by enabling bidirectional power flow, allowing EVs to return energy to the grid when needed. V2X further expands these capabilities by integrating EVs with various energy systems, including homes, buildings and microgrids. These technologies play a crucial role in improving grid stability, optimising energy usage and promoting the integration of renewable energy sources into modern power networks.

The G2V operation represents the fundamental charging process where EVs receive power from the electrical grid. As EV adoption increases, efficient G2V strategies are essential to manage electricity demand and prevent grid congestion [5]. Smart charging mechanisms can optimise energy consumption by scheduling charging during off-peak hours, utilising surplus renewable energy and responding to real-time electricity prices. By incorporating demand response techniques, G2V helps balance the overall power system, reducing stress on the grid while ensuring a reliable energy supply for EV users. The integration of renewable energy sources further enhances the sustainability of G2V, making it a key element in the transition towards cleaner transportation and energy systems.

The V2G concept enables EVs to operate as distributed energy resources, allowing bidirectional energy exchange with the power grid. This capability provides multiple benefits, including peak demand reduction, frequency regulation and improved grid resilience [6]. By discharging stored energy during periods of high electricity demand, EVs can support grid stability while reducing the need for additional power generation from conventional sources. V2G also enhances the economic value of EV ownership by allowing users to participate in energy markets, potentially earning financial incentives for supplying power back to the grid [7]. However, the effective implementation of V2G requires coordination among EV owners, utilities and grid operators to ensure optimal energy flow while considering battery life and charging priorities [8].

The V2X paradigm extends beyond V2G by enabling EVs to interact with multiple energy systems, including homes (V2H), buildings (V2B), microgrids (V2M) and other vehicles (V2V) [4]. This approach allows EVs to serve as mobile energy storage units that can provide backup power during outages, enhance local energy management and facilitate peer-to-peer energy sharing. V2X integration supports the development of smart cities by enabling seamless energy exchange between EVs and intelligent infrastructure, improving overall energy efficiency and sustainability. Additionally, V2X enhances grid resilience by decentralising power distribution and integrating renewable energy more effectively. However, widespread V2X adoption requires advancements in energy management strategies, communication networks and regulatory policies to ensure seamless coordination between EVs and various energy systems.

4.2 Charger power level classification and standards

EV charging systems are classified according to whether the external power source is connected to the battery through AC or DC [2]. The AC charging system uses an OBC installed in the vehicle, drawing power from the grid. In contrast, the DC charging system has an off-board charger at fixed locations, which delivers DC power directly to the battery. EV charging systems are classified into three main levels according to their power levels, charging times and applications, each serving different needs in the evolving EV ecosystem. Table 4.1 summarises the classification of EV charging based on AC and DC power levels.

Table 4.1   EV charging power levels [1]

Expected power level	Power rating (kW)	Charging time (h)	Type of charger
Level 1 (AC)	3.3	4–36	On-board
Level 2 (AC)	3.7–22	1–6	On-board
Level 3 (AC)	22–43.5	0.5–1	On-board
Level 3 (DC) Fast charger	50–350	0.2–1	Off-board

4.2.1 Level 1 (AC) charging system

Level 1 charging is the most basic and accessible form, typically used in residential or household settings. It operates at a power rating of up to 3.3 kW with charging times ranging from 4 to 36 h [1,2]. The primary advantage of Level 1 charging is its simplicity, as it can be plugged into standard household outlets without requiring additional infrastructure. However, its slow charging speed limits its practicality for users needing frequent and rapid charging, making it more feasible for early adopters or overnight charging.

4.2.2 Level 2 (AC) charging system

Level 2 charging offers a significant improvement in charging speed and is widely used in both residential and public settings. With power ratings ranging from 3.7 to 22 kW, charging times can be reduced to 1–6 h [1,2]. This level of charging requires a larger investment in higher installation costs but provides a balanced solution between charging speed and infrastructure investment. Level 2 charging is versatile, suitable for home charging, workplace charging and public parking facilities. The integration of smart charging capabilities, such as load management and demand response, further enhances its efficiency and reduces grid strain, making it the standard for most residential and public charging needs.

4.2.3 Level 3 (AC) charging system

Level 3 AC charging is designed for commercial applications, operating at power ratings ranging from 22 to 43.5 kW. It requires charging times ranging from 0.5 to 1 h [1,2]. Level 3 AC charging requires dedicated electric vehicle supply equipment (EVSE) and utilises OBCs, making it ideal for commercial settings where faster charging is needed compared to Level 2 systems. However, it is less common than Level 3 DC fast charging due to its lower power output and limited application in high-speed charging scenarios.

4.2.4 Level 3 (DC) charging system

Level 3 DC fast charging is designed for commercial purposes in places like parking lots, shopping centres and restaurants. Operating at power ratings from 50 to 350 kW, it drastically reduces charging times to 0.2–1 h. This makes it ideal for EVs used for long-distance travel and in commercial fleets [2,3]. Level 3 DC charging requires significant infrastructure investment, including dedicated EVSE and off-board chargers, as well as potential upgrades to local electrical grids to handle the high power demand. Despite its high cost and complexity, Level 3 DC charging is critical for supporting the growing adoption of EVs, especially for applications requiring rapid charging. Current advancements in ultra-fast charging technologies, such as those exceeding 350 kW, promise even shorter charging times but necessitate careful consideration of grid capacity, infrastructure and cost-effectiveness.

4.2.5   Standards and specifications

The design, power ratings and regulatory standards for EV chargers vary across different regions. However, to enhance compatibility, regulatory bodies and manufacturers are working towards establishing international standards, protocols and couplers for both slow- and fast-charging systems, minimising potential conflicts and challenges [9]. Key components of EVSE, including connectors, power outlets, cords and plugs, play a cardinal role in ensuring efficient charging, discharging and system protection. Tables 4.2 and 4.3 present various commercially available AC and DC connectors along with their respective plugs, standards and specifications based on the power levels [10].

4.2.5.1   Specifications of AC charging connectors

AC charging connectors are essential for daily usage EV charging, particularly at residential and public locations due to their affordability and ease of installation.

Table 4.2   EV charger power level and plug standards [10]

Charging level	North America	Europe	China	Japan
Level 1 (AC)	SAE J1772 T1[a]	N/A	N/A	N/A
Level 2 (AC)	SAE J1772 T1[a]	IEC 62196-2 T2[b]	GB/T 20234 AC	SAE J1772 T1[a]
Level 3 (AC)	SAE J3068	IEC 62196-2 T2[b]	GB/T 20234 AC	N/A
Level 3 (DC fast charging)	CCS Combo 1 (SAE J1772 T1[a]/IEC 62196-3)	CCS Combo 2 (IEC 62196-3)	GB/T 20234 DC	CHAdeMO

[a]T1 denotes Type 1.
[b]T2 denotes Type 2.

Table 4.3   EV charger plug standard specifications [11]

EV charger plug standard	Power rating (kW)	Nominal supply Voltage (V)	Maximum Current (A)
SAE J1772 T1[a]	1.92/7.68	1-phase 120/240	≤16/32
IEC 62196-2 T2[b]	7.36/43.5	1-phase 230/3-phase 400	≤32
CCS Combo 1	75	600	≤125
CCS Combo 2	200	1000	≤200
CHAdeMO	200	500	≤400
GB/T 20234 AC	7.04/21.12	1-phase 220/3-phase 380	≤32
GB/T 20234 DC	187.5	750	≤250
Tesla Supercharger	140	480	≤300

[a]T1 denotes Type 1.
[b]T2 denotes Type 2.

The SAE J1772 (Type 1) connector, used predominantly in North America, operates at single-phase 120 V (Level 1) and 240 V (Level 2), delivering power from 1.4 to 19.2 kW [10]. Its five-pin design includes power, ground and communication pins for safe charging operations, ensuring interoperability with various EV models. In Europe, the IEC 62196-2 (Type 2) connector supports single-phase (up to 7.36 kW) and three-phase (up to 43.5 kW at 400 V) charging. Its seven-pin layout provides better current handling, higher power capacity and advanced communication protocols, making it suitable for both residential and commercial charging stations. China's GB/T 20234 AC connector supports single-phase charging at approximately 7 kW and three-phase charging up to approximately 43.5 kW, aligning with national standards for broad EV compatibility, while also incorporating safety mechanisms such as insulation monitoring and ground fault detection. AC connectors are preferred for their wide availability, cost-effectiveness and ease of integration with existing electrical infrastructure. They are particularly suitable for overnight charging at homes, workplaces and public parking spaces, ensuring that EVs are sufficiently charged for daily use.

4.2.5.2   Specifications of DC charging connectors

DC charging connectors facilitate rapid charging by directly feeding power to the EV battery, bypassing the OBC. The Combined Charging System (CCS) is widely adopted, with CCS Combo 1 used in North America and CCS Combo 2 in Europe. Both support up to 1000 V, delivering power levels from 50 to over 350 kW, thanks to their dual-pin DC layout combined with AC pins for versatility [9]. CCS connectors integrate advanced communication protocols, enabling features like load management, billing and remote diagnostics, making them ideal for high-traffic commercial charging stations.

Tesla's Supercharger connector in North America delivers up to 250 kW, with its proprietary design optimised for seamless integration with Tesla vehicles, while European Tesla vehicles utilise CCS Combo 2 for higher power capabilities and broader compatibility. The GB/T 20234 DC connector in China supports up to 750 V and currents exceeding 250 A, delivering more than 200 kW, with built-in safety features such as overcurrent protection and thermal management. Japan's CHAdeMO connector provides up to 500 V and 125 A, delivering around 50 kW, with newer versions aiming for up to 400 kW, and is renowned for its bidirectional charging capability, enabling V2G applications [12]. DC connectors are crucial for long-distance travel and high-demand charging scenarios, significantly reducing charging times from hours to minutes. Their continuous development focuses on increasing power delivery, improving safety and enhancing interoperability across different EV models and regions, ensuring a robust and future-ready charging infrastructure for the growing EV market.

4.2.5.3   Standards of EV charging and grid codes

EV charging infrastructure is governed by various international standards that ensure safety, interoperability, power quality and efficiency. Table 4.4 lists the various standards of EV charging and grid codes with their descriptions [13,14].

Table 4.4 Standards of EV charging and power quality standards [13,14]

Body	Standard	Description
IEEE	IEEE 519-1992	Harmonic distortion control in power systems
	IEEE 1159-1995	Power quality monitoring methodologies
	IEEE 1100-1999	Grounding recommendations for sensitive equipment
	IEEE 1366-2012	Reliability indices for power distribution
	IEEE 1547	DER interconnection standards
	P1547, P2100.1	DER grid connection and EV charging standardisation
SAE	SAE J2293	On-board/off-board EV charging standards
	SAE J1772	AC/DC charging levels 1 and 2
	SAE J1773	Inductive charging standards
	SAE J2847	Communication protocols for EV charging
	SAE J2953	EV and charger interoperability
	SAE J2847/1	EV-grid communication for demand response
	SAE J3068	Three-phase AC charging for heavy-duty EVs
	SAE J2931/7	Security protocols for EV communication
IEC	IEC 61851	Conductive charging operations
	IEC 61980	Wireless power transfer standards
	IEC 62196	Charging connectors, plugs and sockets
	IEC 61000-2,3,4	EMC standards for low-frequency conduction
	IEC 60038	Voltage levels for EV charging applications
	IEC 60664-1	Insulation coordination for low-voltage supply for EV charging
	IEC 62752	Cable control and protection requirements
NEC	NEC 625, NEC 626	Safety measures for off-board charging
UL	UL 2231, UL 2251	Safety standards for EV circuits
	UL 2202	
	UL 2594, UL 1741	EV power supply equipment standards
ISO	ISO 15118	V2G communication protocols
	ISO 17409	EV external energy connection requirements

The Institute of Electrical and Electronics Engineers (IEEE) provides key standards such as IEEE 519-1992 for harmonic control in power systems, IEEE 1159-1995 for monitoring electric power quality, IEEE 1100-1999 for grounding sensitive electronic equipment and IEEE 1366-2012 for power distribution reliability indices. Additionally, IEEE 1547 outlines the interconnection of distributed resources with electric power systems, while P1547 and P2100.1 define grid connection and charging system standardisation.

The Society of Automotive Engineers (SAE) has also established essential standards for EV charging. SAE J2293 covers on-board and off-board charging equipment for conductive AC, DC and inductive charging. SAE J1772 specifies voltage and current ratings for AC and DC charging, and SAE J1773 addresses inductively coupled charging systems. SAE J2847 focuses on the communication requirements between EVs and charging interfaces, while SAE J2931/7 enhances security in plug-in electric vehicle (PEV) communication. Additionally, SAE

J3068 standardises three-phase AC power transfer systems, and SAE J2953 ensures interoperability between EVs and chargers.

The International Electrotechnical Commission (IEC) has contributed significantly to EV charging safety and performance. IEC 61851 governs conductive charging systems, EMC requirements and DC fast charging. IEC 61980 defines wireless power transfer for 1000 V AC and 1500 V DC, while IEC 62196 sets standards for connectors, plugs and sockets used in conductive charging. Furthermore, IEC 61000 addresses EMC compatibility, harmonic emissions and low-frequency conductance. IEC 60038 establishes voltage standards, IEC 60664-1 provides insulation coordination for low-voltage supply, and IEC 62752 ensures cable control and protection.

National Electric Code (NEC) and Underwriters' Laboratories (UL) also play crucial roles in EV safety. NEC 625 and NEC 626 outline safety measures for off-board charging, including conductors, connectors, and inductive charging devices. UL 2231, UL 2251 and UL 2202 provide safety requirements for protection devices in charging circuits, while UL 2594 and UL 1741 establish guidelines for EVSE such as inverters, chargers and output controllers.

Additionally, international standards like ISO 15118 and ISO 17409 focus on V2G communication protocols and external energy source integration. These standards ensure that EVs can interact efficiently with power grids, contributing to energy management and grid stability.

Collectively, these standards facilitate the seamless integration of EVs into the power grid, ensuring compatibility, safety and performance across different charging infrastructures. Compliance with these guidelines is critical for advancing EV adoption, enhancing charging efficiency and maintaining grid reliability.

4.3 Topological fundamentals

A key requirement in V2G and V2X applications, more broadly, is that the charging system must provide an interface between two or more energy sources with fundamentally different characteristics, i.e. the AC grid and the high-voltage DC battery of the vehicle. Power electronics is the critical enabling technology that allows this interface to be realised. Accordingly, to understand development trends in EV charging systems that support V2G and V2X applications, it is important to be familiar with the fundamental concepts and properties of power converter circuit topologies, their behavioural modes and primary operating constraints.

4.3.1 The elementary power electronic switching cell

The essential concept of a power electronic conversion system is to use an array of semiconductor devices to transform electrical energy from one form to another, with the devices switched at high frequency such that on average the output voltage and/or current match a lower frequency target reference. Inductive and capacitive circuit elements are integrated into the array of semiconductor devices to provide filtering action so as to attenuate the harmonic content introduced by the switching

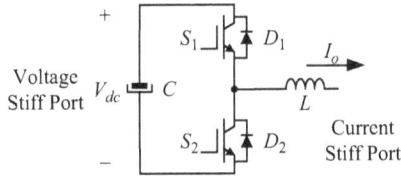

Figure 4.1   Elementary two-level switching cell

action of the devices. An inevitable consequence of this approach is that converter ports can be identified as either voltage-fed (also called voltage-stiff) or current-fed (also called current-stiff).

Figure 4.1 shows the circuit topology of a basic voltage-fed power electronic switching cell, in this case, implemented using insulated-gate bipolar transistors (IGBTs), denoted by S_1 and S_2. These devices form a logical complementary pair, and this ensures that the voltage stiff port of the cell, i.e. V_{dc} formed by the capacitor C, can never be short-circuited. Note that in practice it is necessary to account for the finite rise and fall (i.e. commutation) times of the IGBT devices, which requires the insertion of a 'dead-time' between the turn-off event for one device and the turn-on event for the second device. Depending on the device type being used, this dead-time is typically of the order of a microsecond. The phase leg formed by the series connection of the IGBTs then enables the current stiff port, i.e. I_o formed by the inductor L, to be connected to either the positive or negative rail of V_{dc}. To ensure continuity of current flow in the current stiff port, it is necessary to connect anti-parallel diodes, denoted by D_1 and D_2, in parallel to each IGBT, which support both current polarity reversal and also conduction through dead-time events. Semiconductor manufacturers routinely integrate the diode and IGBT into a single package, while for devices based on metal-oxide-semiconductor-field-effect-transistors the body diode is an intrinsic element to the device structure.

4.3.2   Elementary DC–DC converter topologies

This basic power electronic cell is the foundation topology from which different converter families can be derived, including elementary buck (i.e. step-down) and boost (i.e. step-up) DC–DC converters [15]. This can be illustrated for the buck family, as shown in Figure 4.2. Here, a voltage source V_{dc} is placed across the input

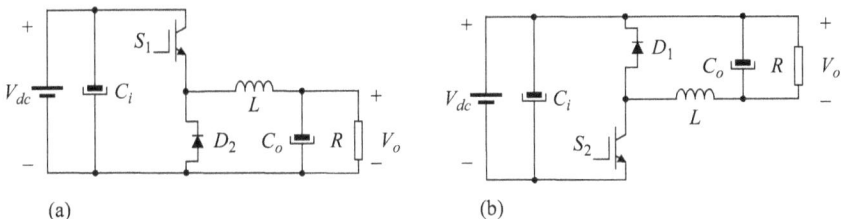

(a)   (b)

Figure 4.2   Buck converter variants: (a) high-side switch with a negative rail-referenced load and (b) low-side switch and positive rail-referenced load

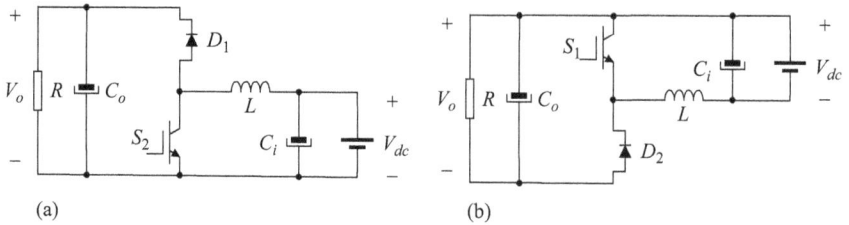

Figure 4.3   *Boost converter variants: (a) low-side switch with a negative*
rail-referenced source and (b) high-side switch and positive
rail-referenced source

capacitor C_i and a load resistance R with parallel output capacitor C_o is connected in series with the switching cell inductance L. This load structure can then be connected to the negative input rail (i.e. a low-side connection) with just the high-side switch S_1 and low-side diode D_2 retained to yield the conventional structure of a buck converter shown in Figure 4.2(a). Alternatively, by connecting the load structure to the positive rail (i.e. high-side connection) and retaining just the low-side switch S_2 and the high-side diode D_1 a low-side buck variant is developed, as shown in Figure 4.2(b).

To develop boost structures, the direction of power flow through the switching cell is reversed from right to left, and the load network with resistance R and parallel output capacitor C_o is placed across the left-hand side voltage stiff port of the switching cell. The source V_{dc} which may include a parallel connected input capacitance C_i is then connected in series with the inductance L on the right-hand side of the switching cell. If this source is then connected to the negative rail of the cell, and only the low-side switch S_2 and high-side diode D_1 are retained, a conventional boost converter results, as shown in Figure 4.3(a). Again, if the alternative connection is employed with the source now connected to the positive rail and with the high-side switch S_1 and low-side diode D_2 retained a high-side boost variant results as shown in Figure 4.3(b).

4.3.3 Bidirectional converter topologies

The buck and boost topologies shown in Figures 4.2 and 4.3 support unidirectional power flow only from the source voltage through to the passive load resistance. To enable V2G/V2X functionality bidirectional power flow is required. Once again, the elementary switching cell is the basis for developing suitable converter topologies [15]. This is illustrated in Figure 4.4, which shows how two DC voltage sources, V_{dc} and V_o, can be connected via the elementary switching cell. This structure now supports a buck or step-down mode from V_{dc} to V_o through the switching action of S_1 and D_2, while a boost or step-up mode from V_o to V_{dc} is realised via the switching action of S_2 and D_1. Operation of the converter is constrained such that the combined interval of the buck and boost modes must not exceed the total switching period of the converter.

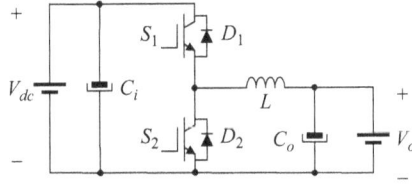

Figure 4.4   Bidirectional buck and boost DC–DC converter

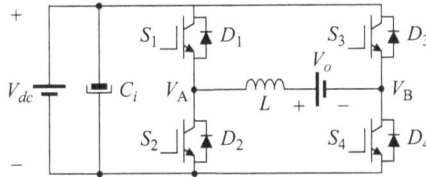

Figure 4.5   Four quadrant buck and boost DC–DC converter

EV charging systems will also need to consider the issue of voltage polarity in addition to the power flow direction. This is in part because of the necessary AC grid interface, but also because AC can be utilised in different stages of a converter to achieve secondary objectives such as providing galvanic isolation. The elementary switching cell is again the basis for such developments, as shown in Figure 4.5. This structure now exploits two switching cells formed by S_1/D_1 and S_2/D_2, and then S_3/D_3 and S_4/D_4, configured as an H-bridge topology. This enables the voltage sources V_{dc} and V_o to be coupled together through the inductance L in a positive or negative polarity depending on the combination of switched phase leg states (e.g. a positive superposition results with S_1 and S_4 gated on, while a negative superposition results when S_2 and S_3 are gated on). This structure therefore supports four quadrants of operation (i.e. both directions for current flow and both voltage polarities), and hence can be used to interface a DC source to an AC source.

4.3.4   Current-fed power converter topologies

The power converter topologies discussed so far are fed from voltage sources. However, in EV applications, there are circumstances whereby the available energy supply will exhibit properties equivalent to that of a current source (e.g. a voltage source with a significant series inductance). When a power converter is fed from a current source, a critical differentiator from voltage-fed systems is that bidirectional power flow manifests through a reversal of voltage polarity rather than the direction of current flow. This means that the elementary switching cell of Figure 4.1 must be modified since the semiconductor devices within the cell support unidirectional voltage blocking but bidirectional current flow. A semiconductor switch capable of reverse voltage blocking and unidirectional current flow is therefore required to enable a current-fed converter [16].

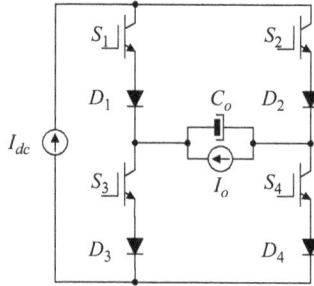

Figure 4.6   Four quadrant current fed converter

Figure 4.6 illustrates such a topology for an H-bridge configuration. The series combination of a diode and IGBT (without the anti-parallel diode) fulfils the reverse blocking need, but suffers from excessive forward voltage drop in the conducting state. Semiconductor manufacturers have developed reverse blocking IGBTs based on a non-punch-through structure, which have better conduction properties, but at the expense of commutation speed. The topology of Figure 4.6 is in fact the topological dual of Figure 4.5, and so the functionally equivalent switching device labels have been relocated (i.e. S_1/S_2 and S_3/S_4 are the logically complementary pairs despite being in separate phase legs) to reflect this. This also means that there will be periods in which the current source is short-circuited by the combination of S_1 and S_3 or S_2 and S_4. This is in fact essential for the operation of the current source inverter, since the current source must never be subjected to an open circuit condition. Similarly, instead of a dead-time event being used to separate the turn-off and turn-on events of logically complementary switch pairs, it is necessary to insert a short interval of overlap between these commutation events.

An additional aspect of a current-fed converter is that the buck and boost modes are reversed when compared to the equivalent voltage-fed converter. That is, the voltage across the parallel combination of the output capacitor C_o and load current I_o must be greater than the switched voltage that will appear across the DC source I_{dc}. Accordingly, if a current-fed converter is configured as a rectifier (i.e. the AC port is where C_o and I_o are located), then the rectified DC voltage must be lower than the input AC voltage creating a buck rectifier. In contrast, rectifiers based on voltage-fed converters such as Figure 4.5 (i.e. the AC port is the series combination of L and V_o) require that the DC voltage must be higher than the incoming AC voltage, creating a boost rectifier. The design of converter platforms for EV chargers must therefore account for these fundamental topological constraints.

4.3.5   Modulation concepts

The process of switching a power electronic converter is called modulation. In general, all modulation strategies have a primary objective, which is to synthesise a target waveform with the same volt-second average as the target reference over the switching interval of the converter. Once this primary objective has been realised, secondary factors such as optimisation of the spectral properties of the waveform or mitigation of common-mode voltage can be considered and integrated into the design of the modulation strategy. For EV charging systems, there are numerous modulation strategies that are intricately linked to the converter platform used to realise the charger. Often these chargers will utilise two distinct variations of modulation pattern, viz. pulse-width modulation (PWM) or fundamental frequency (i.e. quasi-square-wave) modulation.

The simplest PWM strategy is commonly referred to as carrier PWM or, for AC systems sine-triangle PWM [15]. This process is illustrated for an elementary two-level switching cell (i.e. phase leg), as shown in Figure 4.7 for the sine-triangle case. Here, a low-frequency waveform called the reference is compared with a high-frequency triangular carrier waveform. The amplitude of the reference is called the modulation depth M and typically defined as a percentage of the carrier waveform amplitude (i.e. 100% denotes full utilisation of the available DC-link voltage). The frequency of the reference waveform, f_o, is called the fundamental and can be DC*, while the carrier frequency f_c is normally at least an order of magnitude larger than the fundamental frequency. The modulation rules are simply that the phase leg switches high (i.e. $+\frac{V_{dc}}{2}$ relative to the mid-point of the input DC source) when the reference is greater than the carrier, and switches low (i.e. $-\frac{V_{dc}}{2}$ relative to the

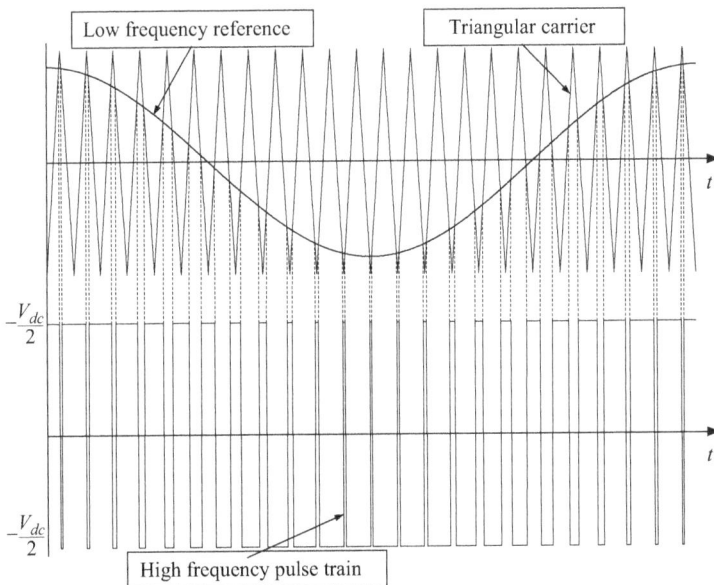

Figure 4.7   PWM of a two-level phase leg

*Often for DC references the carrier waveform is level shifted to occupy the range between zero and one.

mid-point of the input DC source) otherwise. This creates a high-frequency pulse train that spectrally contains the same low-frequency information as the reference waveform. The filtering action of the load attenuates the high-frequency modulation harmonics resulting in the extraction of the low frequency target waveform.

This modulation concept can be readily extended to more advanced topologies, such as the H-bridge of Figure 4.5. However, generally, these topologies have additional operational switched states that enable secondary performance objectives to be realised. In the case of the H-bridge, there are four distinct switched states that allow three possible voltage levels to be realised. That is

- $S_1 = 1$ and $S_3 = 0$: $V_{AB} = V_{dc}$
- $S_1 = 1$ and $S_3 = 1$, or $S_1 = 0$ and $S_3 = 0$: $V_{AB} = 0$
- $S_1 = 0$ and $S_3 = 1$: $V_{AB} = -V_{dc}$

where V_{AB} denotes that switched output voltage across the A-phase and B-phase cells of the H-bridge. To exploit these states and realise harmonic benefits, it is necessary to switch both phase legs at the same frequency, but to interleave the high-frequency pulse trains to allow effective pulse-doubling to occur. This is illustrated in the modulation strategy shown in Figure 4.8. Here, a single carrier waveform is used for the complete bridge, while the phase leg reference waveforms

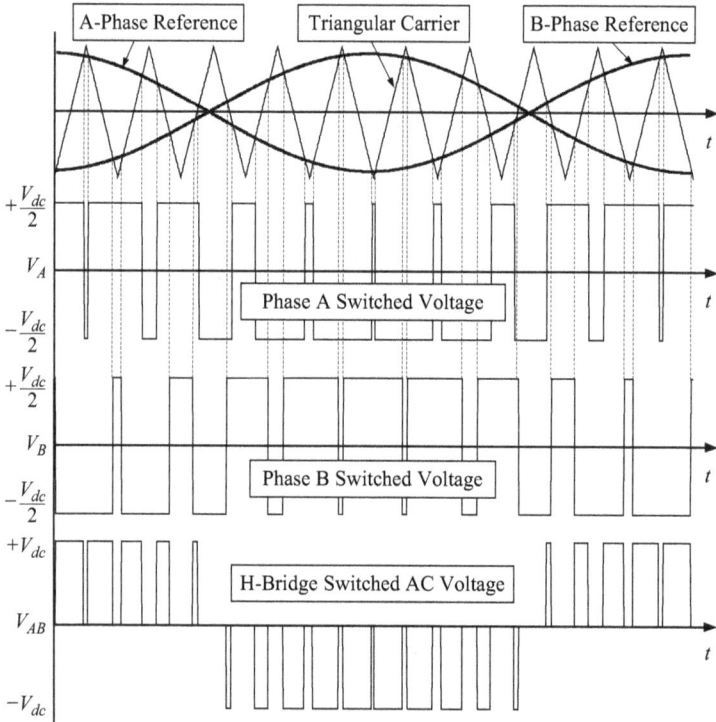

Figure 4.8   Three-level PWM applied to an H-bridge converter

are configured to be equal but opposite. The resulting phase leg switched voltages interleave discrete pulses, such that their difference, which corresponds to the H-bridge output waveform, encodes all three possible voltage levels via the application of all possible switched converter states. This provides significant performance benefits by shifting modulation harmonic distortion to twice the carrier frequency. High-performance EV charging platforms seek to exploit these secondary performance benefits offered by a particular power converter topology.

Fundamental frequency modulation strategies, in contrast to carrier PWM, involve encoding the target fundamental frequency component at the switching frequency of the converter [15]. Such patterns necessarily produce high levels of modulation distortion in the converter waveforms, but they find particular application in high-frequency AC-coupled converter platforms, such as a dual active bridge (DAB) converter where two H-bridges are coupled through a high-frequency transformer. Figure 4.9 shows an example of such a modulation pattern for a single H-bridge. Each phase leg is switched to produce a simple square-wave with a 50% mark-to-space ratio. By phase-shifting the phase-B square wave by the angle $\pi - \alpha$ a three-level quasi-square-wave pulse pattern develops across the H-bridge output. The angle α encodes an effective duty cycle for the quasi-square-wave pulse train (i.e. maximum amplitude occurs when $\alpha = 0$, while the minimum amplitude occurs when $\alpha = \pi$). Certain topologies will also incorporate frequency variation to exploit gain variations due to resonant modes and to provide an additional degree of freedom for an over-arching controller.

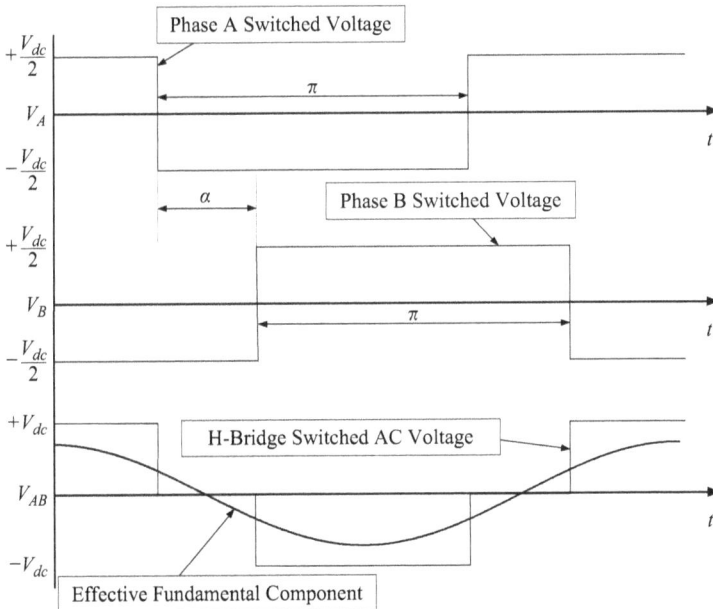

Figure 4.9   Fundamental frequency modulation of an H-bridge using phase-shifted square waves

4.4   Wired EV charger converter topologies

Wired EV chargers can be broadly categorised into two types: AC input OBCs, which convert AC power (single-phase or three-phase) from the grid into DC power to charge the EV battery; and DC input off-board chargers, which provide DC power directly to the vehicle's battery. These chargers can either be unidirectional to charge the EV battery from the grid or bidirectional to charge the EV battery from the grid and also discharge power from the battery to the grid or other loads (V2X), thereby enabling V2G capabilities.

4.4.1   Single-phase on-board EV charger topologies

Single-phase on-board EV chargers typically use a two-stage conversion approach, as shown in Figure 4.10(a), comprising an AC–DC stage operating at unity power factor, followed by a galvanically isolated DC–DC converter that regulates voltage and current to meet the specific needs of the EV battery. Alternatively, a single-stage conversion approach, as can be seen in Figure 4.10(b), can also be achieved by merging the AC–DC and DC–DC stages into a single galvanically isolated AC–DC converter. Embedding galvanic isolation into the on-board EV charger facilitates compliance with safety and leakage current requirements (as specified by the relevant standards, e.g. JEVS G105, IEC 61851-23, SAE J1772 and UL 2231), being a practice currently adopted by most car manufacturers [11,12,17–19].

4.4.1.1   Unidirectional two-stage charging topologies

When implemented as a two-stage charging arrangement, a single-phase on-board EV charger incorporates an AC–DC converter with power factor correction (PFC) capability and a DC–DC converter that is most invariably galvanically isolated.

Examples of the AC–DC stage of unidirectional single-phase two-stage on-board EV chargers are depicted in Figure 4.11. The simplest approach, shown in Figure 4.11(a), is often referred to as a boost PFC converter and employs a diode-bridge rectifier followed by a boost DC–DC converter. The topology is typically operated in continuous conduction mode to synthesise a sinusoidal input current with low total harmonic distortion (THD) and unity power factor operation. However, compared to alternative arrangements, it has the drawback of having three device voltage drops in the current pathway that are detrimental to the

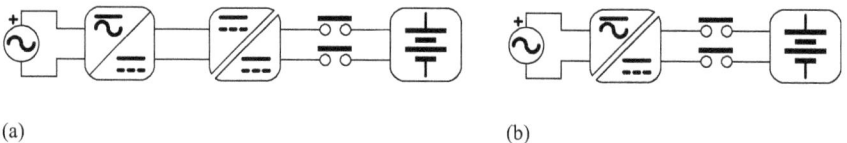

(a)   (b)

Figure 4.10   Power electronic topologies for single-phase on-board EV charging: (a) two-stage, comprising an AC–DC converter followed by a galvanically isolated DC–DC converter; and (b) single-stage, comprising a galvanically isolated AC–DC converter

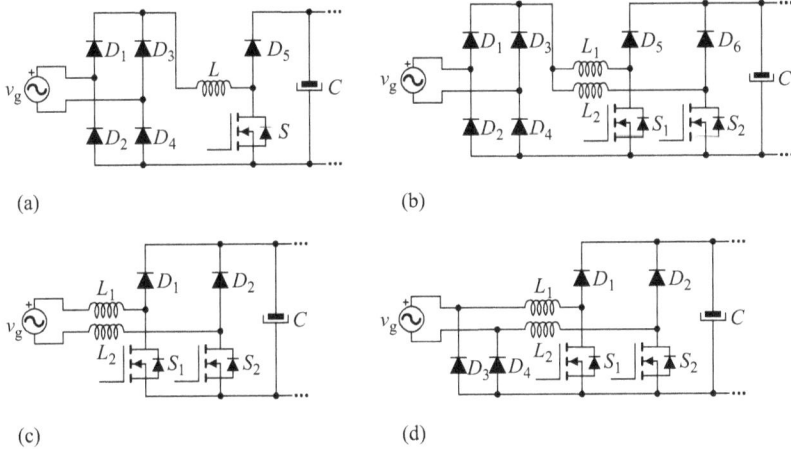

Figure 4.11 Alternative unidirectional topologies for the AC–DC stage of a single-phase two-stage on-board EV charger: (a) boost PFC, (b) interleaved two-phase boost PFC, (c) semi-bridge rectifier (also called bridge-less rectifier) and (d) semi-bridge rectifier with additional low-frequency diodes to minimise common-mode voltage

converter's efficiency. An extension to the boost PFC converter is the interleaved two-phase boost PFC topology shown in Figure 4.11(b), where the two boost DC–DC converters operate 180° out of phase, effectively cancelling the first carrier group of harmonics and doubling the current ripple frequency and thus reducing the size of the input filter components.

The semi-bridge rectifier, also referred to as a bridge-less rectifier and shown in Figure 4.11(c), is an alternative to the boost PFC and interleaved boost PFC topologies. It reduces the number of device voltage drops from three to two and potentially enables higher converter efficiencies. However, it has various drawbacks compared to the boost PFC and its interleaved counterpart. First, the grid voltage terminals are floating relative to the boost PFC ground, requiring isolated voltage measurement circuitry, e.g. a mains-frequency transformer or an optocoupler, instead of a simple voltage divider circuit, as is possible with boost PFC arrangements. Second, as the semi-bridge rectifier has more than one return current path, it requires a current transformer or Hall effect current sensor at the input terminal instead of a single shunt resistor in the return current path as typically used in boost PFC arrangements. Lastly, the semi-bridge rectifier presents an inherently large common-mode (CM) noise component due to having the output rails floating during the negative mains half-cycle, which can be difficult to filter. A topological variation of the semi-bridge rectifier, containing two additional mains-frequency diodes, D_3 and D_4, as shown in Figure 4.11(d), aims to minimise this CM noise, at the expense of additional converter losses and increased circuit complexity. When deciding between these topological variations, General Motors opted for the interleaved two-phase boost PFC for their Volt OBC in 2016, achieving a nominal efficiency of 97.8% [20].

(a)

(b)

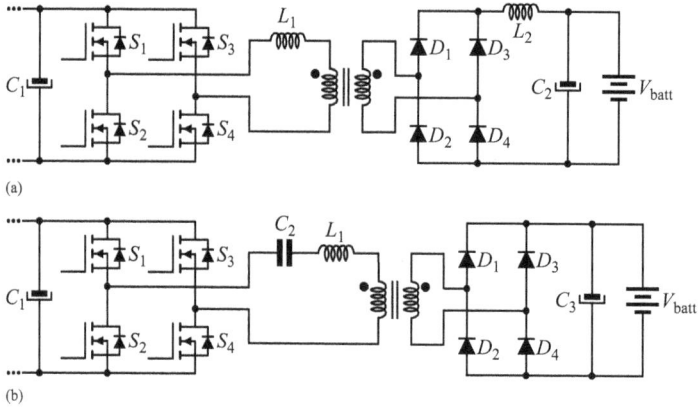

Figure 4.12   Alternative unidirectional topologies for DC–DC stage of single-phase two-stage on-board EV charger: (a) full-bridge phase-shifted converter and (b) full-bridge resonant converter

Figure 4.12 depicts examples of the galvanically isolated DC–DC stage of unidirectional single-phase two-stage on-board EV chargers. The full-bridge phase-shifted converter, shown in Figure 4.12(a), operates with a fixed switching frequency and allows varying the active power transfer by the dynamic phase-shifting of the switching commands of the two primary-side phase legs. The topology allows for zero voltage switching (ZVS) on the primary-side without requiring additional auxiliary circuits [21]. Conversely, the full-bridge resonant converter, illustrated in Figure 4.12(b), operates with a variable switching frequency and uses this mechanism to dynamically vary the active power transfer. While a variety of alternative resonant tanks can be used, the series LC arrangement shown in Figure 4.12(b) is a popular choice, allowing for the use of the transformer leakage inductance as the inductor element and offering an inherent DC current blocking capability. The full-bridge resonant converter was the selected topology for the DC–DC stage of the GM Volt OBC in 2016, achieving a nominal efficiency of 97.7% when operating at full load [20].

4.4.1.2   Bidirectional two-stage charging topologies

Figure 4.13 depicts examples of the AC–DC stages of bidirectional single-phase two-stage on-board EV chargers. The totem-pole bridge-less PFC shown in Figure 4.13(a) incorporates a fast-switching phase leg operating as a boost converter and a slow-switching phase leg operating at the grid frequency and is responsible for the AC–DC conversion. This arrangement presents low conduction losses due to the synchronous rectification, low CM noise, and allows for ZVS over the complete mains period when modulated using triangular current mode (TCM) [22]. However, the topology presents a large input current ripple when operated with TCM, which requires a large differential-mode (DM) input filter. The two-channel and three-channel totem-pole PFCs shown in Figure 4.13(b) and (c) are

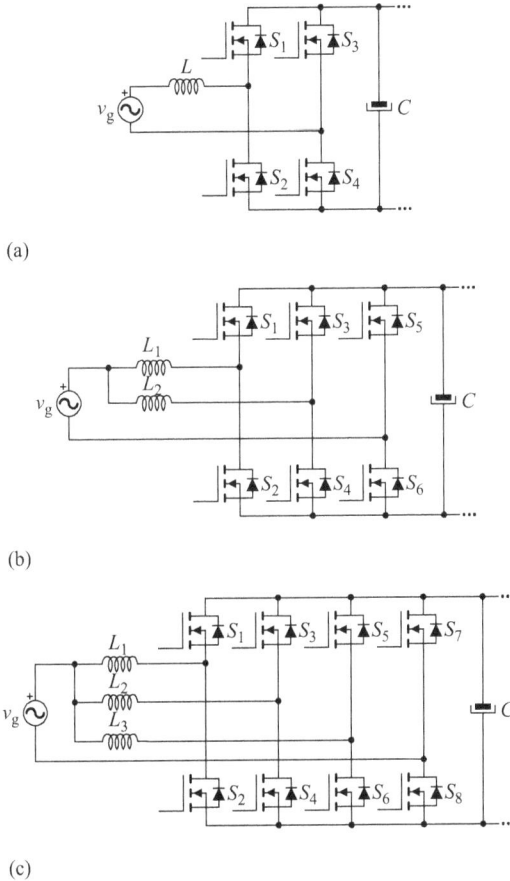

(a)

(b)

(c)

Figure 4.13   Alternative bidirectional topologies for AC–DC stage of single-phase two-stage on-board EV charger: (a) totem-pole bridge-less PFC, (b) two-channel totem-pole bridge-less PFC and (c) three-channel totem-pole bridge-less PFC

topological variants that, despite incorporating additional circuitry complexity, allow for inductor current carrier harmonic cancellation and a reduced DM input filter size.

Figure 4.14 depicts examples of the galvanically isolated DC–DC stages of bidirectional single-phase two-stage on-board EV chargers. The DAB converter, introduced in [23] and shown in Figure 4.14(a), is one of the most popular topologies, fulfilling key functional requirements, e.g. galvanic isolation, high power density and high efficiency, wide input-to-output voltage gain and bidirectional power transfer capability. In this topology, the primary- and secondary-side bridges operate with a fixed switching frequency and a relative phase-shift δ, which regulates the active power transfer direction from primary-to-secondary, when $\delta > 0$,

(a)

(b)

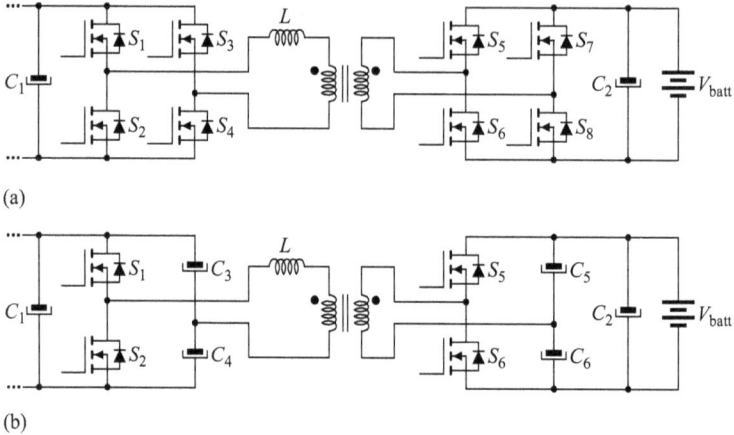

Figure 4.14   Alternative bidirectional topologies for DC–DC stage of single-phase two-stage on-board EV charger: (a) dual active bridge converter and (b) dual active half-bridge converter

or secondary-to-primary, when $\delta < 0$. The primary- and secondary-side switched voltages synthesise a resulting switched voltage across the series inductor L, storing energy in this element, which is then dynamically exchanged with the parasitic capacitance of the power switches to achieve ZVS. However, since ZVS operation largely depends on the current flowing through the series inductor L, its natural range is typically limited [23], particularly at lower power transfer conditions and for non-unity primary-to-secondary voltage transfer ratios, increasing the converter switching losses. Several solutions have been proposed to extend the ZVS range, such as adding external capacitors parallel to the switches, or reducing the coupling factor of the interconnecting transformer [24], or utilising optimised modulation approaches [25,26]. While these methods increase the ZVS range, achieving ZVS operation for the entire operating range remains challenging. Despite these limitations, the DAB topology is a popular choice for EV chargers because of its simple design and control, and its overall performance.

The dual active half-bridge converter, shown in Figure 4.14(b), is formed by replacing one of the phase legs on both of the primary and secondary sides of the converter with capacitors C_3, C_4, C_5 and C_6, as shown in Figure 4.14(b). Although the switch count is reduced in this half-bridge arrangement compared to its full-bridge counterpart, the capacitors required in this topology are significantly large. A full AC line current flows through these capacitors, and therefore, they are rated to handle very large steady-state and ripple currents during operation. Contrarily, in the full-bridge topology, the current flows through the active switches and free-wheeling diodes. This is one of the major drawbacks of the half-bridge topology, particularly because capacitors rated to operate at high power and voltage levels are bulky and costly, overcoming the advantage of the reduced component count of the half-bridge topology [27].

Incorporating a series resonant tank between the two bridges of a DAB converter is an alternative to extend its soft switching capability, with the series CLLC-resonant converter being a popular choice. The transformer's primary- and secondary-side leakage inductances can be used as the series inductors, while the capacitors provide DC blocking capability and mitigate potential transformer core saturation. The full-bridge and half-bridge series resonant DAB converter variants are illustrated in Figure 4.15(a) and (b), respectively. Contrary to the DAB converter, these series-resonant DAB converters operate with a variable switching frequency, which regulates the voltage gain of the resonant tanks. In these topologies, soft switching is attained by exchanging energy stored in the transformer's magnetising inductance to charge and discharge the parasitic capacitance of the power switches. These converters can achieve ZVS over the entire load range. However, operation at a wide voltage range demands switching frequencies well away from the resonant frequency, increasing converter losses [28].

Resonant tanks can also be used between the two bridges of a DAB to reduce the reactive power circulating between the bridges, reducing the required VA rating of each of the converters. Figure 4.16(a) and (b) shows the LCL [29] and CLC [30] variants of this topological arrangement, respectively. The CLC-resonant variant offers the additional advantage that its series capacitors aid in blocking DC bias that may cause magnetising current runaway and transformer core saturation. However, the CLC-resonant DAB may be less desirable than the LCL-resonant counterpart in high-coupling applications due to its greater susceptibility to the frequency response over the coupling factor range [31]. Furthermore, contrary to the series CLC-resonant converters shown in Figure 4.15, the LCL- and CLC-resonant converters depicted in Figure 4.16

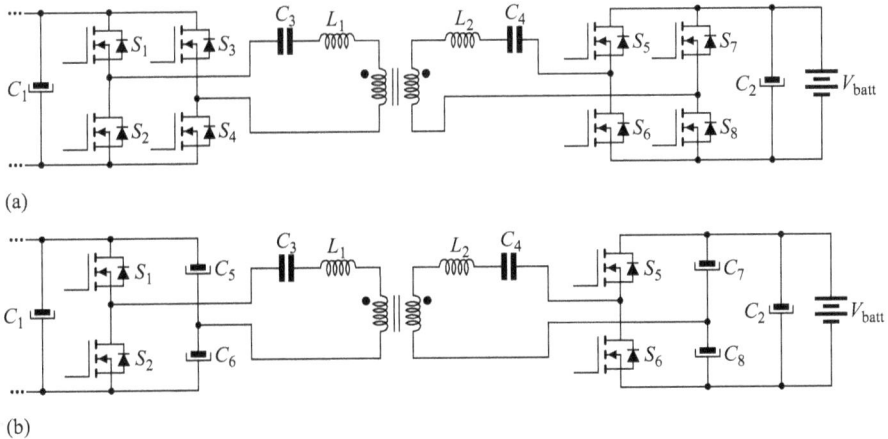

(a)

(b)

Figure 4.15   *Alternative bidirectional resonant topologies for DC–DC stage of single-phase two-stage on-board EV charger: (a) dual active bridge CLLC-resonant converter and (b) dual active half-bridge CLLC-resonant converter*

(a)

(b)

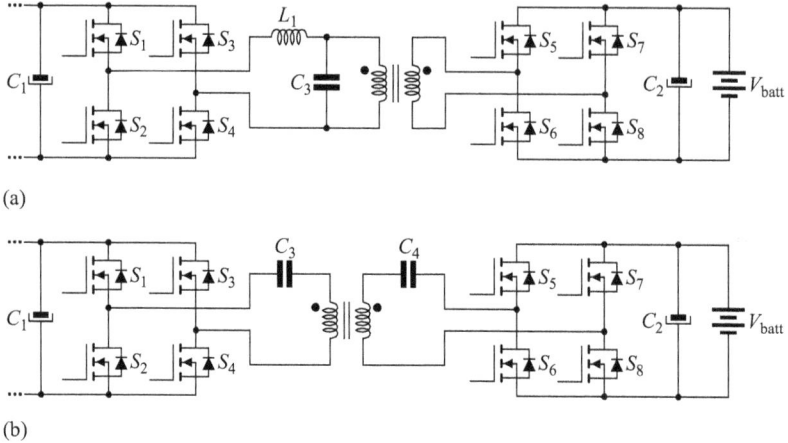

*Figure 4.16   Further bidirectional resonant topologies for DC–DC stage of
single-phase two-stage on-board EV charger: (a) dual active bridge
LCL-resonant converter and (b) dual active bridge CLC-resonant
converter. Half-bridge variants are not shown for briefness.*

operate at a fixed switching frequency in a similar manner to a non-resonant
DAB. Also, achieving matched primary- and secondary-side resonances is
essential for these resonant converters, as discussed in [31].

4.4.1.3   Bidirectional single-stage charging topologies

Single-phase single-stage on-board EV chargers merge the functionalities of the
PFC AC–DC converter stage and the DC–DC converter stage. This offers the
benefit of a potentially more compact and cost-efficient solution that may not
incorporate bulky and lower life expectancy DC-link electrolytic capacitors.

Examples of single-phase single-stage on-board EV chargers are depicted in
Figure 4.17. The interleaved two-channel totem-pole PFC-DAB converter, shown
in Figure 4.17(a), was introduced in [32] and developed further in [33]. It
employs film capacitors for C_1 and C_2 and uses simple modulation and control
approaches, where the primary-side switches S_1 to S_4 operate with a fixed 50%
duty-cycle, the primary-side switches S_5 and S_6 perform the typical unfolding
function of a totem-pole bridge-less rectifier, while the secondary-side switches
S_7 to S_{10} are sinusoidally modulated. This arrangement achieves unity power
factor operation and ripple-free input current. The active power transfer is controlled by phase-shifting the modulation commands of the primary- and
secondary-side power switches. A 3.7 kW prototype reported in [33] achieved a
peak efficiency of 97.2% at the maximum operating power. Alternatively, the
PFC-DAB converter shown in Figure 4.17(b), introduced in [34] and developed
further in [35], comprises a primary-side half-bridge made up of bidirectional
power switches and a typical secondary-side full-bridge. The topology is

(a)

(b)

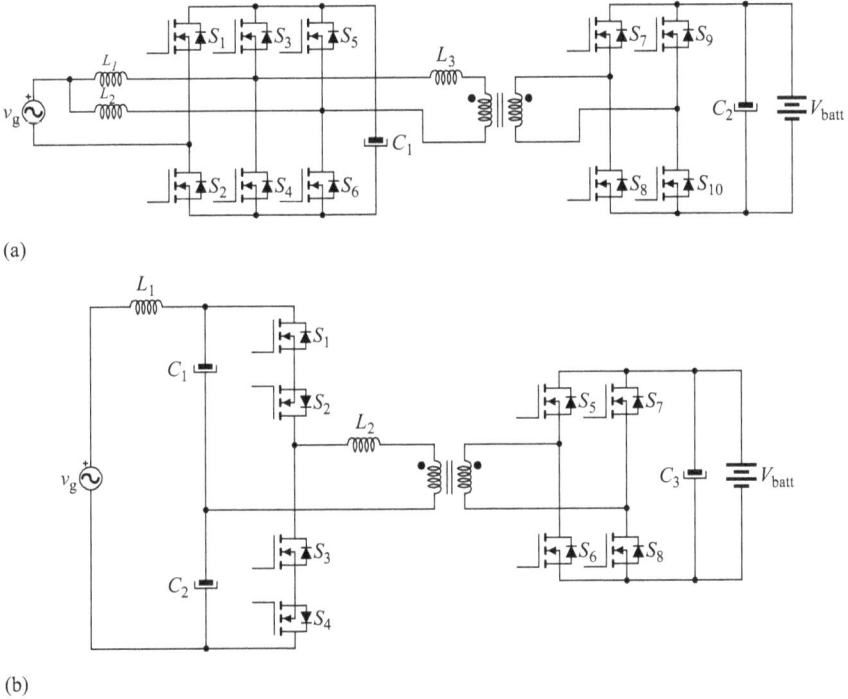

Figure 4.17   Alternative bidirectional topologies for single-phase single-stage on-board EV charger: (a) interleaved two-channel totem-pole PFC-DAB converter and (b) PFC-DAB comprising a primary-side half-bridge made up of bidirectional switches and a secondary-side full-bridge

modulated using a combined phase-shift and frequency approach that achieves ZVS of all switches over the full range of the AC mains voltage. This arrangement achieves unity power factor operation and a mostly ripple-free input current but presents an inherent input current zero-crossing distortion that is caused by the brief turn-off of all switches. Capacitors C_1, C_2 and C_3 are made up of a parallel arrangement of surface-mount X7R ceramic capacitors. A 3.3 kW prototype reported in [35] achieved an efficiency of 95% at a nominal AC mains voltage of 230 V_{rms}, DC-link voltage of 350 V_{dc} and half the operating power, with a calculated efficiency of around 96%.

4.4.2   Three-phase EV charger topologies

There are broadly three types of three-phase fed EV chargers, three-phase two-stage chargers, integrated OBCs and off-board fast chargers [17,18].

Typical three-phase two-stage EV chargers use a two-stage conversion approach, as shown in Figure 4.18(a), consisting of a three-phase AC–DC

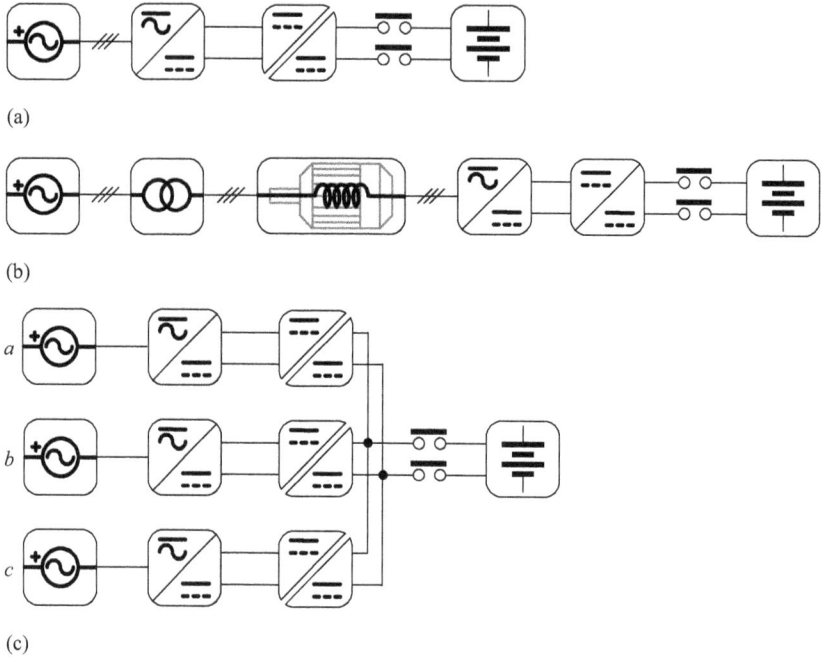

(a)

(b)

(c)

Figure 4.18 *Power electronic topologies for three-phase on-board EV charging:*
(a) two-stage, comprising an AC–DC converter followed by a
galvanically isolated DC–DC converter, (b) EV integrated charger
and (c) two-stage, modular three-phase charger

rectification stage and an isolated DC–DC conversion stage that charges the EV
battery by regulating the voltage and current.

Integrated EV chargers, as shown in Figure 4.18(b), tend to use the on-board
EV motor and drive in a custom configuration to facilitate three-phase AC charging. An external three-phase AC supply is fed through a rectification stage through
the EV motor windings in the reverse direction through the traction inverter, which
regulates the current to charge the EV battery [10,17].

Modular three-phase chargers, as shown in Figure 4.18(c), use modularised
AC–DC stages that incorporate high-frequency galvanic isolation, separated by AC
phase. This allows for the upgrading of separate converter stage modules as new
converter topologies emerge, but also to upgrade the power delivery capability as
new switching device technologies emerge and charger standards and power levels
change [36].

4.4.2.1 Two-stage charging topologies

Two-stage three-phase chargers, shown in Figure 4.18(a), typically consist of an
AC–DC rectification stage followed by a high-frequency galvanically isolated
DC–DC conversion stage to regulate the EV battery charging voltage and current.

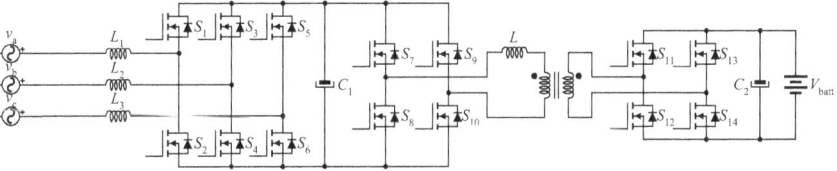

Figure 4.19   Bidirectional three-phase EV charger consisting of an active voltage source AC–DC rectifier and isolated DAB converter for the DC–DC stage

Two-stage charging topologies can be further subdivided into unidirectional and bidirectional topologies.

In the unidirectional case, the AC–DC rectification stage may consist of a three-phase diode- or thyristor rectifier or Vienna rectifier [37]. The Vienna rectifier is a three-phase, three-level rectifier utilising three semiconductor switches that results in low switching losses, low conducted common-mode EMI emissions and low THD. A popular choice for the isolated DC–DC stage is an LLC resonant converter. The full-bridge resonant converter was discussed earlier in Section 4.4.1.1. The LLC resonant converter is typically operated in ZVS mode and is popular due to the fact that the transformer leakage inductance may be used as one of the inductive elements that also inherently blocks DC current [20].

For the bidirectional case, the DC–DC rectification stage typically uses an active voltage source rectifier, with active semiconductor switches replacing the diodes used in a conventional unidirectional three-phase rectifier, which results in a highly controllable rectified voltage with medium THD distortion when using two-level switching. A popular choice for the galvanic isolated DC–DC conversion stage is the DAB converter topology [23], which allows for a high power density, a wide input-to-output voltage gain and ZVS operation. The DAB converter topology was discussed in more detail in Section 4.4.1.2.

An example of a three-phase bidirectional two-stage EV charger is shown in Figure 4.19, where the three-phase AC is fed into the three-phase voltage source rectifier, after which the rectified voltage is fed to a DAB converter to provide galvanic isolation as well as power control to charge the EV battery.

4.4.2.2   Integrated charging topologies

EV manufacturers have been refining and optimising their on-board integrated charger designs in terms of cost, volume and weight to increase the OBC power rating and charging capability [1]. One of the main reasons is that the EV's traction drive system remains idle while the EV battery is charged. Therefore, integrated charger topologies, as shown in Figure 4.18(b), aim at utilising the existing on-board traction system components in the charger architecture. As most current EV traction system ratings exceed 50 kW, these systems can be incorporated in high-power on-board integrated charging architectures to improve charging speed [10].

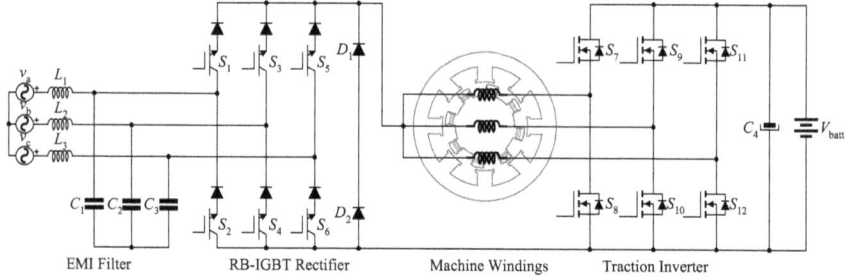

Figure 4.20   Integrated three-phase EV charger consisting of a three-phase current-fed rectifier with interleaved boost stage [17]

An example of an integrated charger, the Renault Chameleon [38,39], is shown in Figure 4.20. This topology uses a reverse blocking IGBT current source rectifier stage that results in a drop in rectified voltage. Consequently, the resulting rectified voltage is then boosted through a DC–DC boost stage, which is achieved by using the EV inverter drive in the reverse direction as an interleaved boost converter that charges the EV battery while regulating the charging current [38].

4.4.2.3   Modular charging topologies

Modular three-phase EV chargers, as shown in Figure 4.18(c), typically use a combination of power electronic building blocks (PEBBs) or converter modules, each consisting of an AC–DC conversion stage that converts a single AC phase to DC. An example of such an approach is shown in Figure 4.21, where each AC phase feeds into an interleaved totem-pole module that converts the AC to DC. The interleaved totem-pole module includes high-frequency galvanic isolation. The DC

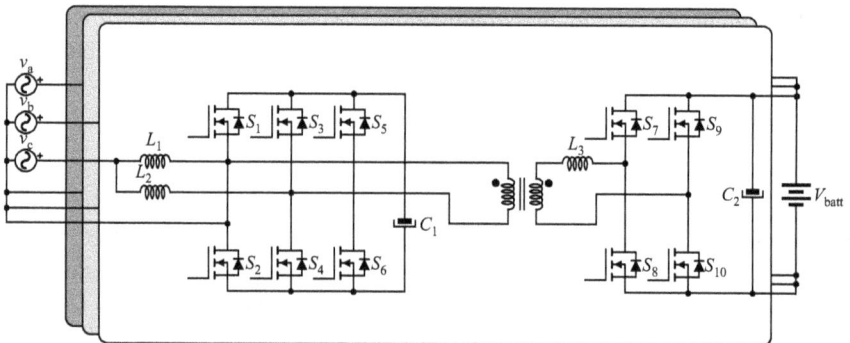

Figure 4.21   Modular three-phase EV charger composed of three single-stage interleaved totem-pole AC–DC converters with high-frequency isolation [36]

outputs of the interleaved modules are connected in parallel to increase the power delivered to the EV battery during charging [36].

4.4.3 Off-board EV charger topologies

High-power off-board EV chargers, also known as ultra and extremely fast chargers, are advanced EV charging systems designed to deliver 150 kW to over 350 kW, significantly reducing charging time [3]. Their modular design allows for flexible power distribution, optimising efficiency at partial loads while benefiting from economies-of-scale effects to lower costs. These chargers support both 400 and 800 V EV architectures by dynamically adjusting voltage and current, ensuring efficient, high-power charging with minimal losses [9]. By reconfiguring power modules, they enhance compatibility, improve thermal design and maximise energy efficiency, making them ideal for next-generation EVs.

In Figure 4.22, the ABB Terra 53/54 Series is a 50 kW high-power DC fast charger designed for efficient and flexible charging of EVs, supporting 400 and 800 V battery packs. It utilises a modular power conversion architecture with 5 × 3 PEBBs to achieve 50 kW output power, ensuring scalability and redundancy. The power conversion system consists of a single-phase rectifier, an LLC resonant half-bridge converter and a full-bridge converter, providing high efficiency and stable operation at different voltage levels [40]. The charger attains 95% efficiency, reducing energy losses and improving thermal performance. Designed for bidirectional capability, it supports V2G applications while offering high-frequency galvanic isolation for safety and EMI reduction. The LLC resonant topology ensures soft switching, minimising losses and improving power density. With its compact and modular design, the Terra 53/54 series provides adaptive power sharing, dynamic voltage regulation and enhanced thermal management, making it a reliable solution for high-power EV charging applications.

The ABB Terra 150 kW high-power fast charger is designed for rapid and efficient charging of EVs, utilising a modular power electronics architecture. Figure 4.23 features a three-phase active bidirectional rectifier for PFC and a three-channel interleaved buck converter as the DC–DC conversion stage [17]. This design improves power quality, reduces harmonics and ensures efficient power transfer to the vehicle battery. The charger incorporates low-frequency galvanic isolation, improving safety and system reliability. With an efficiency of 94% at full load, it minimises energy losses and optimises thermal performance. The modular approach allows for scalability, and in configurations up to 600 kW, it becomes a suitable solution for charging buses and large EVs, ensuring high-power delivery.

The Tesla V2 Supercharger, shown in Figure 4.24, is designed for the Tesla Model S and Model X and utilises a two-stage power conversion circuit to efficiently deliver high power for fast charging. The first stage consists of a boost PFC per channel, followed by a phase-shifted full-bridge DC–DC converter that connects to the vehicle battery [41]. The system is composed of 13 units working together to provide up to 150 kW for a single vehicle with 92% efficiency at full load.

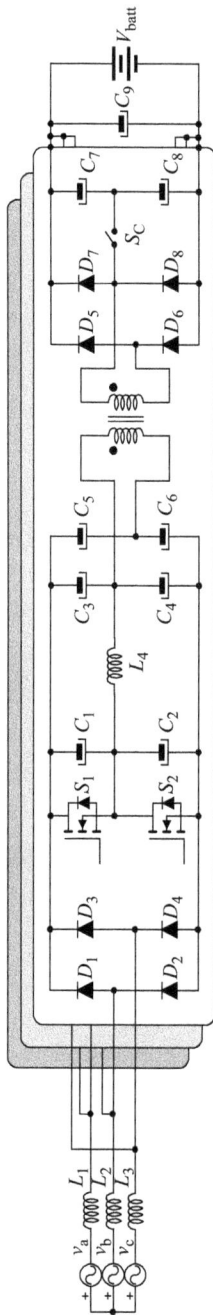

Figure 4.22   Off-board 50kW ABB Terra 53/54 fast EV charger [17]

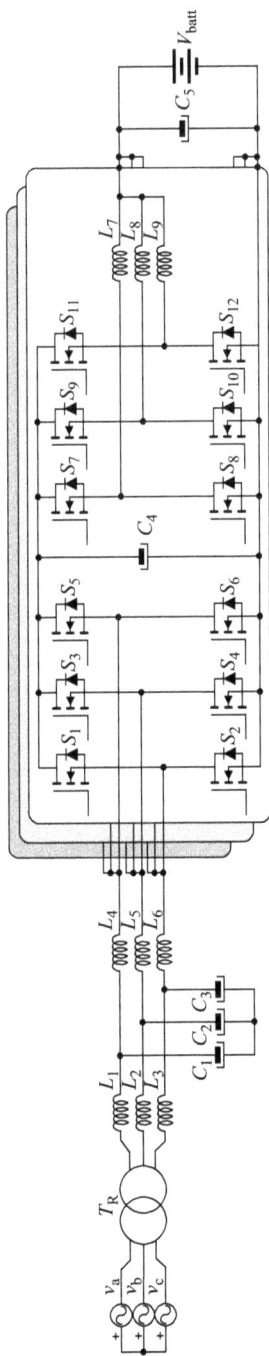

Figure 4.23 Off-board 150 kW ABB Terra high-power fast EV charger [17]

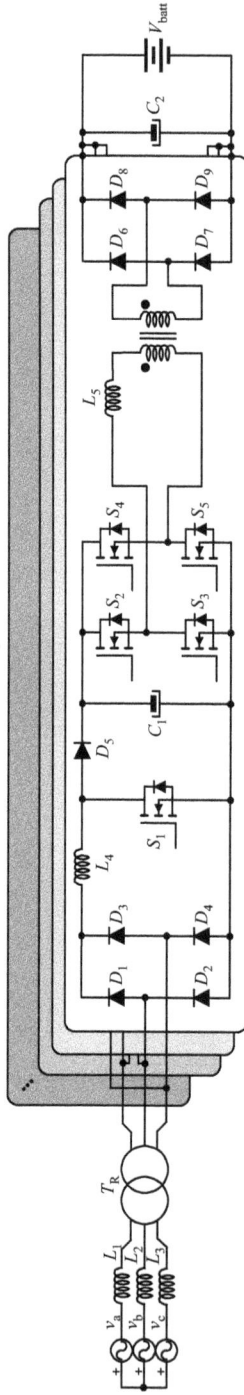

Figure 4.24 Off-board Tesla V2 supercharger [17]

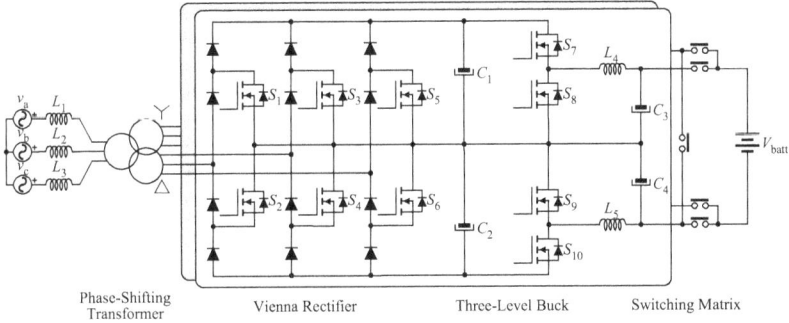

Figure 4.25   Off-board Porsche modular fast charger [17]

Porsche developed their modular fast chargers to provide fast charging services for their EVs equipped with an 800 V battery architecture. The topology of one of their fast chargers is shown in Figure 4.25. The charger also uses a modular architecture using PEBBs. The rectifier stage uses a Vienna rectifier topology that feeds an interleaved three-level DC–DC buck stage to improve the output current ripple charging the EV battery [42].

4.5   Wireless EV charger converter topologies

Current EV chargers connect to EVs by plugging in heavy gauge charging cables. Different charging standards and cable connectors were discussed in Section 4.2.5. These charging cables can cause practical issues, introduce tripping hazards and even be prone to vandalism [1,43]. This has motivated the investigation of wireless- or inductive power transfer, as a means to improve charging aesthetics, convenience, safety and the possibility of process automation [44,45].

Wireless or more generally inductive EV charging approaches tend to focus on using the isolation stage of the high-frequency DC–DC conversion stage within an EV charger for the wireless charging interface [18]. A typical power electronic approach for wireless or inductive EV charging is illustrated in Figure 4.26, where a three-phase AC supply feeds into a rectifier and then into an isolated DC–DC conversion stage. The isolation barrier containing the magnetic coupling of the high-frequency transformer linking the primary and secondary sides of the switching DC–DC conversion stage provides the coupling interface for wireless or

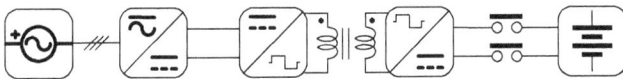

Figure 4.26   High-level power electronic topology for wireless or inductive EV charging

inductive charging. In such an instance, the AC supply, DC rectifier and primary-side high-frequency inversion stage of the DC–DC converter will form part of the off-board EV charger infrastructure, while the rectifying secondary-side of the DC–DC converter, battery management system and EV battery form part of the on-board EV charging infrastructure.

As the wireless coupling stage is contactless, this form of power transfer is considered inherently safer than plugged-in wired chargers due to the lower risk of potential electric shock. Furthermore, a contactless charging interface also allows for the potential of dynamic charging while driving. Research into wireless power transfer has made progress, with the resonant DAB [46], triple active bridge [47] and CLLC converter [48] topologies all shown capable of being used in EV chargers with a wireless charging interface. Currently, wireless charging still needs further development to improve, due to challenges with regard to wireless charging efficiency as well as charging power levels, compared to existing wired charging alternatives [45]. Wireless power transfer for EV applications is covered in more depth in [10,45,49–51].

4.6   Control and management of EV chargers

The typical control and power management architecture of an EV charger within the broader power ecosystem is shown in Figure 4.27. Most chargers are positioned at the prosumer level and are physically connected to the distribution network within the power network.

Modern EV chargers are designed to operate in various modes. These modes include G2V, where the EV battery is charged from the grid; V2G, where the EV battery is discharged to supply power to the grid; and V2X, where the EV battery is discharged to supply power to local loads or other vehicles. Moreover, EV chargers can be operated in different control architectures, such as decentralised, semi-centralised or fully centralised [52]. In a decentralised control architecture, the charging and discharging decisions are made locally in all modes, G2V, V2G and V2X, typically without external communication. This arrangement is commonly referred to as 'uncontrolled charging', where charge or discharge occurs according to the EV owner's preferred schedule or predetermined off-peak hours without coordination or communication with the external systems. However, since EVs are increasingly viewed as valuable energy resources, similar to battery energy storage systems, their charge and discharge processes can be optimised through their aggregation, typically managed by third parties, such as local energy retailers or aggregators, in order to fully exploit their potential. This coordinated charging and discharging of EVs enable them to participate in the energy market, support the grid, e.g. by reducing network congestion and regulating the voltage and reduce charging costs for owners. The aggregators typically use demand forecasts and run optimisation algorithms to achieve a range of techno-economic goals and, accordingly, send set-point signals to the grid-side converter of the EV chargers to charge or discharge at specific rates and times, as shown in Figure 4.27. This arrangement

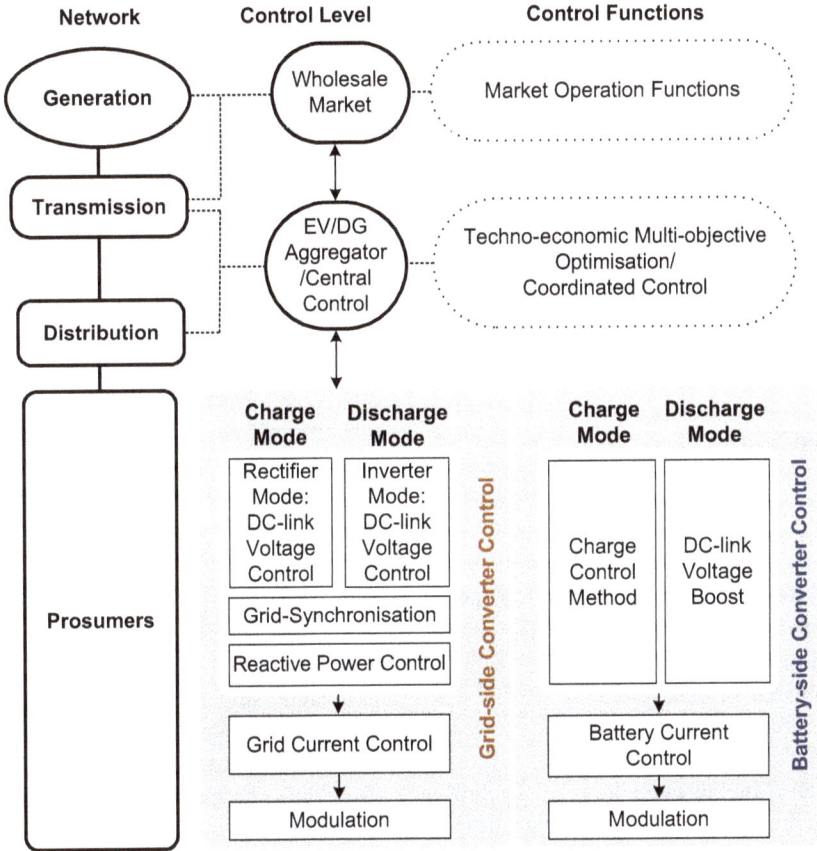

Figure 4.27   Typical control and power management architecture of EV chargers

is referred to as 'semi-centralised', where the EV owners decide based on the signal from the aggregator to charge and discharge their EVs, and referred to as 'fully controlled' where the aggregator directly manages the charge and discharge process of participating EVs.

The internal control of EV chargers can be presented in a generalised form, regardless of the converter topology. A unified control structure, commonly used for dual-stage converters, is illustrated in Figure 4.28. The internal control of the chargers usually comprises multiple loops, with different control objectives achieved at each specific loop. The innermost loop is the modulation control that generates the switching signals based on the input from the outer current control loop, and it varies between converters, as discussed in Section 4.3.5. The current control loop regulates and limits the converter current within the rated value, and it is common in the control of the grid-side and battery-side converters as well as in the single-stage converter.

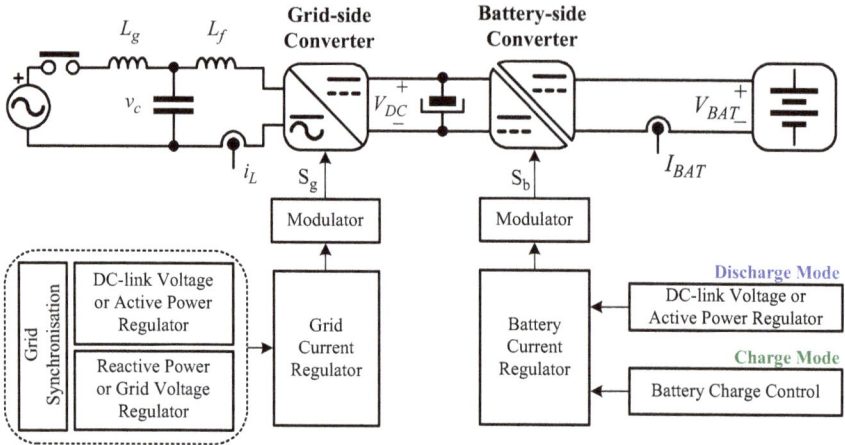

Figure 4.28   Dual-stage charger converter control

The key functional objectives are implemented via an outer loop to the current control. In the outer loop of the battery-side converter, specific charging methods and operating limits are implemented. The charging control strategies, which are discussed in detail later in Section 4.6.3, significantly impact the charging time, operating efficiency and life cycle of the battery. Therefore, deploying proper charging strategies is a crucial task in the overall battery power management ecosystem. In the outer loop of the grid-side converter, charge/discharge rate, grid synchronisation and support functions are implemented.

The specific control and its implementation framework vary depending on the operating mode, converter type and topology, such as whether it is a single-stage or dual-stage converter. The most common control architectures and specific control used in dual-stage and single-stage chargers are discussed in the following sections.

4.6.1   Dual-stage charger control

Dual-stage chargers employ two converter stages: a grid-side AC–DC converter and a DC–DC battery-side converter.

During the G2V mode:

• The grid-side converter operates as a rectifier: it converts AC into DC and draws current from the grid at a unity power factor. Unity power factor operation is achieved by synchronising the grid-side current with the grid voltage.
• The battery-side converter typically functions as a buck converter to charge the battery, except for some integrated converter topologies where it is operated as a boost converter. This converter regulates its output voltage and current according to the specific charge control method discussed later in Section 4.6.3.

During V2G or V2X mode:

- The grid-side converter operates as an inverter: it controls both active and reactive power to the grid or loads.
- The battery-side converter generally functions as a boost converter to regulate the DC-link voltage. This converter regulates and limits its output current to control the discharge rate and the power supplied to the grid or loads.

These control objectives can be implemented through different control architectures, controller types and reference frames [53–56]. The control for the grid-side converter in both the synchronous and stationary reference frame consists of a dual-loop configuration incorporating inner current control and outer DC-link voltage control, as shown in Figures 4.29–4.31.

The control for totem-pole converters, shown in Figure 4.29, differs from buck-boost converter topologies due to its fundamental distinction in employing

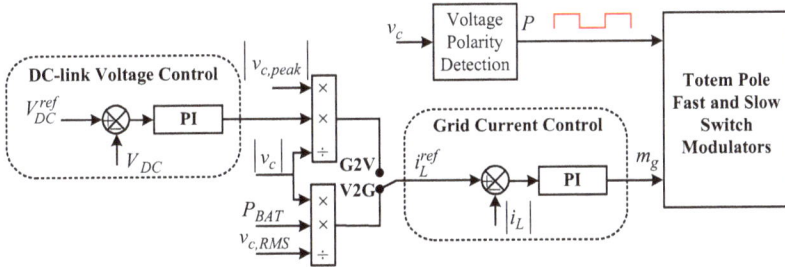

Figure 4.29   Grid-side converter control implemented in the stationary reference frame, used in totem-pole converter rectifiers

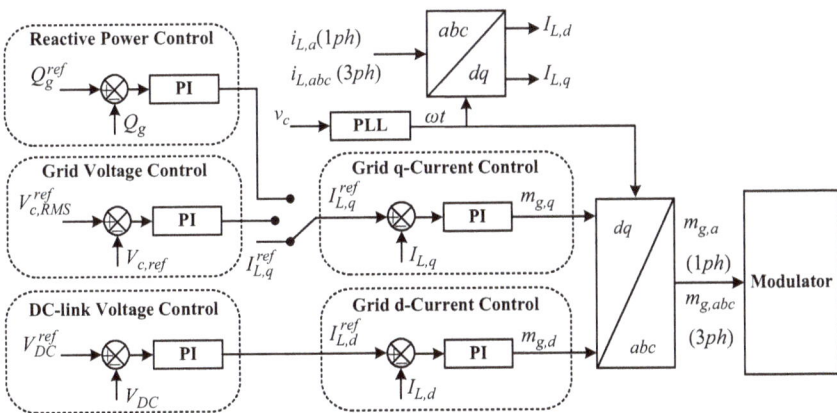

Figure 4.30   Grid-side converter control implemented in the synchronous reference frame, commonly used in single-phase and three-phase boost and three-level neutral-point-clamped rectifiers

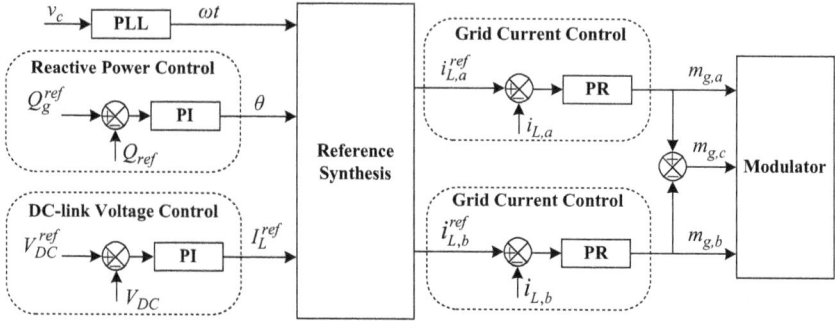

Figure 4.31   *Grid-side converter control implemented in the stationary reference frame, used in various topologies including single-phase and three-phase boost and three-level neutral-point-clamped rectifiers*

both fast- and slow-operating switches [57]. A dual-loop control comprising an outer DC-link control loop and an inner grid-side side current control is used to generate modulation signals for the fast switches of the totem-pole converters. Instead of using the actual AC current, the current fed to the controller is rectified (half-wave pulses), which represents the absolute value of the AC current and the current reference is synchronised with the grid voltage. The use of rectified current in the current control loop simplifies the control process by enabling accurate tracking with a proportional-integral (PI) controller and achieving synchronised operation. An alternative approach in a stationary reference frame involves using a phase-locked loop (PLL) and tracking the AC current without rectification by employing a proportional-resonant (PR) controller [58]. The PR controller is particularly effective for tracking AC signals because it exhibits infinite gain at the fundamental frequency and ensures zero steady-state error. Alternatively, the same control can be implemented in the synchronous reference frame where the AC current is transformed into *dq* components, and then a PI controller can be used to effectively regulate the grid current [32]. The slow switches, on the other hand, are operated at grid-frequency using a polarity signal obtained from the polarity detection block, as shown in Figure 4.29.

The typical control used in other commonly used AC–DC converters, e.g. single-phase H-bridge/boost-, three-phase boost- and three-level neutral-point-clamped converter topologies are shown in Figures 4.30 and 4.31, respectively. In a synchronous reference frame, the outer DC-link voltage controller regulates the converter's DC-link voltage and generates a *d*-axis current reference, which is fed to the *d*-axis current controller. The *q*-axis current reference is typically set to zero, and the converter output current is synchronised with the voltage using a PLL to achieve unity power factor operation. Notably, although the grid-side converter operates as a rectifier during G2V charging mode and as an inverter during V2G discharging mode, the underlying control architecture remains the same. Fundamentally, the same control is used in both single- and three-phase converters, except for the Park transformation and modulator. For the single-phase case, the Park transformation requires an additional second-order generalised integrator to generate the $\alpha - \beta$ components, which are then converted into equivalent dq values. Moreover, in the V2G mode, the grid-side converter can be configured to generate a specified amount of reactive power or regulate the terminal AC voltage. This can be achieved by simply setting the q-axis current reference or by adding an outer control loop for voltage regulation, as shown in Figure 4.30. In a synchronous reference frame, the PI-based current controller is the preferred choice because of its simplicity and precise reference tracking with zero steady-state error.

The same control implemented in a stationary reference frame is shown in Figure 4.31. In the stationary reference frame, a resonant-based current controller, such as the PR controller, is preferred due to its infinite gain at the fundamental AC frequency. This characteristic enables precise tracking of the current reference generated by the outer loop. The outer loop includes a DC-link voltage controller that generates the current magnitude reference, which is then combined with the grid frequency and phase to synthesise the AC current reference(s). In a single-phase converter, only one AC current reference is generated and tracked via a single PR controller. For a three-phase converter, current references for two phases are generated and tracked using two PR controllers, and the modulation reference for the third phase is synthesised from phase 'a' and phase 'b' modulation signals, as shown in Figure 4.31. Optionally, a reactive power controller can be added to regulate reactive power or power factor on the grid-side.

The control of the battery-side converter varies between the G2V and V2G modes, as shown in Figure 4.32. During G2V charging mode, the modulation signal is generated by a specific charge controller, which is discussed in detail in the subsequent sections. In V2G discharging mode, the converter employs dual-loop control, with the outer loop regulating the DC-link voltage to a preset reference and the inner loop regulating the battery current to control the discharge

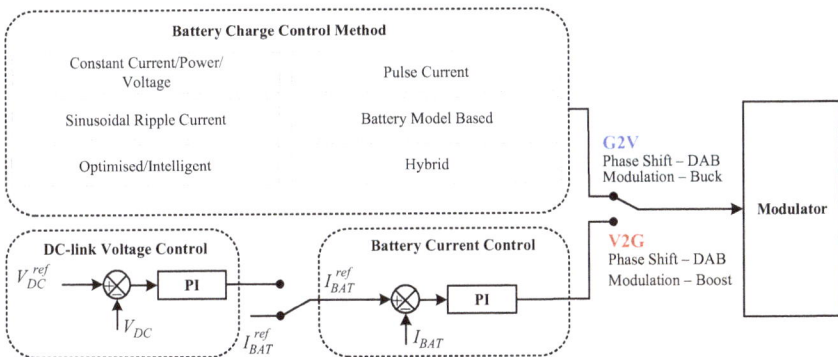

Figure 4.32   Battery-side converter control

rate and, consequently, the active power. The type of reference signals produced by the battery charge controller during G2V and the battery current controller during V2G vary depending on the type of DC–DC converter used. In the DAB family of converters, the controller generates a phase angle reference to regulate the power flow between the grid and the EV battery. This is achieved by adjusting the phase shift between the gate control signals of the two bridges. For the buck-boost family of converters, the controller generates modulation signals of varying amplitudes.

4.6.2   Single-stage charger control

The specific control capabilities and associated architecture for single-stage converters vary depending on the topology. Generally, compared to dual-stage converters, control for single-stage converters is relatively simple and seamless but offers limited flexibility, particularly in achieving synchronisation, controlling both active and reactive power, managing grid-side voltage and ensuring power quality.

For example, the totem-pole PFC-DAB, a key single-stage topology discussed earlier in Section 4.4.1.3, employs a simplified control strategy, as shown in Figure 4.33 [32]. In this topology, the fast switches in the interleaving legs on the primary side are operated using a fixed magnitude modulation signal with a reference carrier. The slow switches on the primary side are turned on and off at the fundamental frequency, synchronised with the grid voltage polarity. On the secondary side, the switches are modulated using a rectified sinusoidal signal of a fixed magnitude and a carrier phase-shifted from the reference carrier used for the interleaving legs. With this control, synchronisation with the grid is achieved by synchronising the secondary-side rectified modulation reference with the grid phase and frequency using a PLL. Active power control is implemented through the phase shift between the primary- and secondary-side carriers, which is generated by a dual-loop control comprising a battery voltage controller and a grid current controller. This simplified control approach enables basic charging and discharging functionalities while maintaining a near-unity power factor.

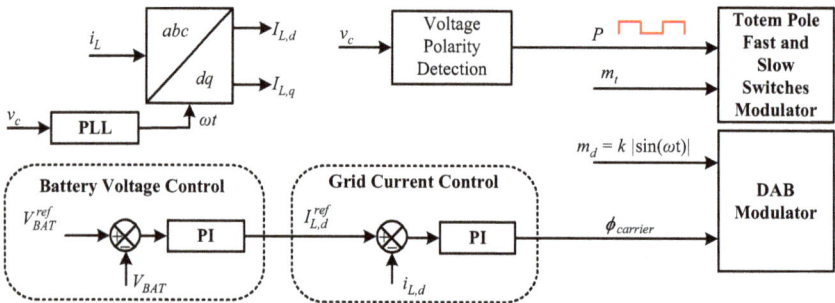

Figure 4.33   Single-stage converter control implemented in the synchronous reference frame, used in totem-pole DAB

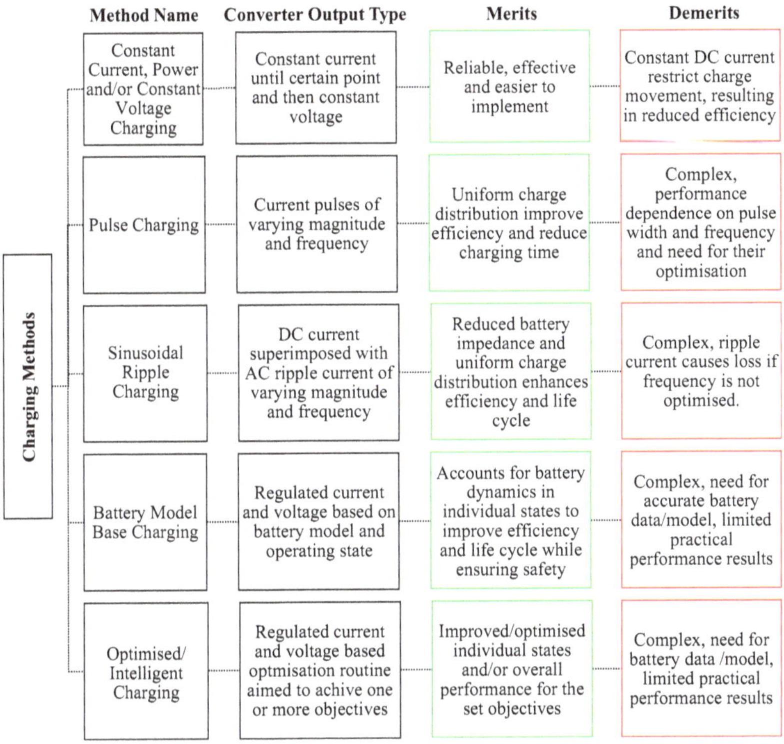

Method Name	Converter Output Type	Merits	Demerits
Constant Current, Power and/or Constant Voltage Charging	Constant current until certain point and then constant voltage	Reliable, effective and easier to implement	Constant DC current restrict charge movement, resulting in reduced efficiency
Pulse Charging	Current pulses of varying magnitude and frequency	Uniform charge distribution improve efficiency and reduce charging time	Complex, performance dependence on pulse width and frequency and need for their optimisation
Sinusoidal Ripple Charging	DC current superimposed with AC ripple current of varying magnitude and frequency	Reduced battery impedance and uniform charge distribution enhances efficiency and life cycle	Complex, ripple current causes loss if frequency is not optimised.
Battery Model Base Charging	Regulated current and voltage based on battery model and operating state	Accounts for battery dynamics in individual states to improve efficiency and life cycle while ensuring safety	Complex, need for accurate battery data/model, limited practical performance results
Optimised/ Intelligent Charging	Regulated current and voltage based optmisation routine aimed to achive one or more objectives	Improved/optimised individual states and/or overall performance for the set objectives	Complex, need for battery data /model, limited practical performance results

(Charging Methods)

Figure 4.34   Comparison of charging methods

4.6.3   Charging methods and controller

In the battery-side converter, the outer loop of the modulation process is the charge control loop. This loop implements key control functions, including specific charge and discharge methods as well as the operating limits. The charging control strategies impact the charging time, operating efficiency and life cycle of the battery. Therefore, proper charging strategies are a crucial task in the overall battery power management system.

Various charge control methods have been developed and are used in practice, as summarised and compared in Figure 4.34.

4.6.3.1   Constant current (CC), constant power (CP) and/or constant voltage (CV) charging

In this set of strategies, the converter produces a constant current until the battery voltage reaches the predetermined threshold. From this point onwards, a constant voltage is produced to charge the battery at a reduced current until it reaches the minimum charge current threshold, such as 0.1 C [59]. This charging method is widely used due to its effectiveness, robustness and ease of design and

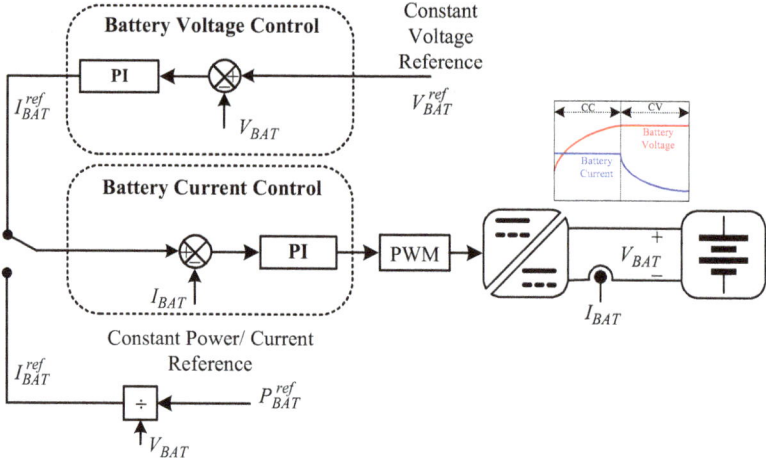

Figure 4.35   Constant power/current–constant voltage charge controller

implementation into various systems. However, the constant voltage process increases the charging time, while the constant current method leads to charge concentration within the battery, affecting the flow and ultimately reducing both charging efficiency and battery lifespan. Therefore, the performance of this method is not comparable to advanced techniques. The control block diagram of the constant current, power and voltage charging strategy is shown in Figure 4.35. To address the charging time limitations, the multi-step constant current (MCC) method is proposed as an alternative where a predefined charging current of different magnitudes is applied in discrete intervals [60]. The level of current is adjusted so that it does not exceed the value used in the preceding stage, and this process continues until the battery is fully charged. This method is more efficient and faster compared to CC–CV.

4.6.3.2   Pulse charging

The pulse charging method is proposed to address ion concentration issues associated with the CC–CV charging method. In this method, the converter produces current pulses of defined frequency and intervals, e.g. on for a second and off for a few milliseconds. The pulse-off period allows for charges to be distributed more evenly, leading to improved efficiency and faster charging. However, the performance improvement with this method depends on optimising the pulse interval and frequency. The authors in [61] analysed the performance of pulsed current charging and concluded that pulsed current, with an average equal to a constant current, does not provide significant benefits. The research further concludes that the form factor is the most influential parameter; a high form factor may cause overpotential in the battery cell and may instead decrease charging efficiency. On the other hand, an optimised pulse interval and frequency can offer significant benefits, such as

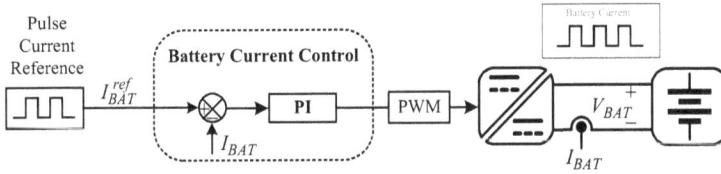

Figure 4.36   Pulse current charge controller

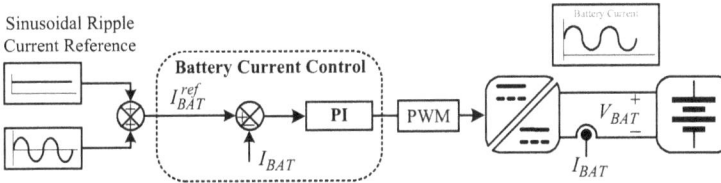

Figure 4.37   Sinusoidal ripple current charge controller

reducing charging time by up to 24% [62]. The control block diagram of the pulse current charging strategy is shown in Figure 4.36.

An advanced variant of pulse charging involves a three-stage charging process in which positive pulse, negative pulse and zero current are applied in a repeated sequence [63]. In this process, during the positive pulse, the battery is actively charged, while during the negative pulse, the battery is actively discharged primarily to depolarise the electrodes. The third zero current stage helps achieve a uniform distribution of ions, thereby providing stability. The magnitude of the positive current pulse is typically the same as that of the negative pulse but of a longer duration, which enhances battery charging speed and efficiency. Since this method involves both active charging and discharging, it requires converter circuitry that supports both forward buck and reverse boost operations.

4.6.3.3   Sinusoidal ripple charging

Sinusoidal ripple charging is another method to enhance charging time and efficiency. In this technique, the converter produces a sinusoidal ripple current, which comprises a superimposed AC on the primary DC charging current. The amplitude of the ripple current determines the charge rate, while the frequency is optimised to maximise performance. This method aims to minimise battery impedance by selecting the appropriate frequency for the output ripple current. This effectively results in a more uniform distribution of the charge and reduces the heat generated during charging. Consequently, this method helps reduce the charge time while improving the efficiency and life cycle of the battery [64]. Despite the advantages, this method is less commonly used in practice due to the complexity associated with its implementation and lack of performance experience. The control block diagram of the sinusoidal ripple current charging strategy is shown in Figure 4.37.

4.6.3.4   Hybrid: two-step or multi-step charging

A hybrid method comprising a constant current and pulse charging technique has been proposed to improve the charging efficiency and lifespan of batteries [65]. In this method, the battery is charged in the first stage using a constant current until the voltage reaches a preset threshold. In the second stage, the battery charge is maintained at a constant level by providing current pulses sufficient to compensate for self-discharge. This process prevents battery self-charging and enhances its lifespan.

The methods discussed above are effective but relatively passive, each having its own merits and demerits associated with charging rate, efficiency, lifespan and practicality. However, due to the complexity of battery technology, which includes coupled non-linear electrical, thermal and ageing dynamics, the methods discussed above cannot guarantee the best performance in all individual states and fully address concerns like overcharging, overheating and safety [66]. These concerns can be addressed by incorporating a detailed battery system model and operating states into the control strategy. The methods to address these concerns are discussed next.

4.6.3.5   Battery model-based charging

This method involves using various types of battery models to monitor, predict and control the converter output in order to optimise the overall charging process. Based on this method, the charge rate is adjusted according to the battery's electrical, thermal and ageing models, as well as the real-time monitoring of the battery operating conditions. This method helps prevent degradation caused by unexpected overheating or overcharging, thereby enhancing the charging efficiency, prolonging battery life and ensuring safety. Among several methods, physics-based models, developed from fundamental principles, are gaining increased popularity due to their ability to provide accurate and reliable predictions of battery behaviour [67]. An electrochemical model has been used in the formulation of a non-linear optimisation problem [68] to minimise the charging duration. A similar electrochemical model, along with the constraints related to battery health, has been used in linear and non-linear model predictive control algorithms [69].

4.6.3.6   Optimised/intelligent charging

Optimised charging strategies mainly leverage real-time battery data and deep learning/artificial intelligence algorithms to minimise one or more objectives [67–70]. These strategies dynamically adjust the charging pattern based on factors like battery state, temperature and usage. For instance, the strategy proposed in [67] uses a physics-based battery model and a genetic algorithm-based optimisation to generate a charging profile composed of voltage intervals with adaptive resolutions. This approach demonstrates improved charging efficiency and reduced battery degradation. Additionally, the authors in [70] utilised an electrothermal-ageing model of a battery and a multi-objective optimisation algorithm to generate a charging pattern based on two passive charging patterns: CC–CV and MCC–CV. The proposed strategy demonstrates a trade-off between charging speed and

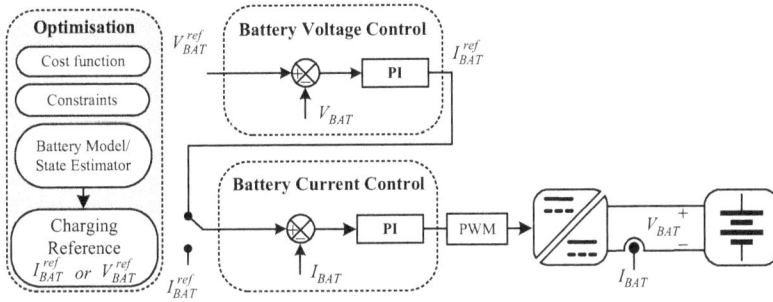

Figure 4.38   Battery model-based charge controller

conversion efficiency. With MCC–CV, the step current decreases when the battery terminal voltage reaches specific thresholds, which reduces battery capacity loss by 16% over 1000 cycles. However, this approach also results in a 2% reduction in charging speed per cycle. A similar approach and findings are reported in [69] for fast-charging applications. This strategy also employs a battery physical model and multi-objective optimisation, taking into account factors such as temperature rise, charging time and energy losses. Compared to the traditional CC–CV method, this strategy reduces charging time and energy losses while maintaining the same temperature rise. However, it is slower than conventional fast-charging methods. The generic control block diagram of the battery model-based and optimised charging strategy is shown in Figure 4.38.

Overall, battery model-based and optimised charging strategies offer significant benefits by effectively balancing charging speed, efficiency and battery lifespan, resulting in superior performance compared to conventional techniques. However, their practical implementation faces several challenges, including reliance on accurate battery models and associated challenges in accurately capturing non-linearities, the need for real-time operational data and the computational complexity of running optimisation algorithms.

4.7   Summary

This chapter has comprehensively reviewed the key aspects of conversion technologies used in modern EV chargers, providing an overview of EV charger types, specifications and standards. A critical evaluation of key converter topologies and their associated performance, used in both on-board and off-board EV chargers, has been presented, together with an exploration of the fundamental switching and modulation concepts employed in these topologies. Several of the key wired EV charger converter topologies, classified according to single-phase AC, three-phase AC and off-board chargers, have been reviewed and discussed in detail. A brief overview of wireless EV charging has also been presented to provide a general understanding of the current field. EV charger control is central to the operation of these charging systems. To facilitate a fundamental understanding, a unified and generalised control approach has been presented to implement these control functions in key converter topologies, followed by a summary of the required control features in modern EV chargers. The research progress has indicated that significant efforts have been devoted to charger control, with a strong focus on converter modulation, efficiency and power quality. On the grid-side, most studies have focused on aspects such as power quality and power factor, while there is a lack of consolidated or comparative performance analysis of different control architectures for emerging grid support features, such as grid-supporting and grid-forming capabilities. The last section of this chapter has thoroughly reviewed and compared various charging methods that play a crucial role in battery life and charging economy. The findings of this review suggest that modern charging techniques, including those incorporating battery models and optimisation algorithms, outperform conventional charging methods. However, their practical adoption is constrained by their complexity, data dependencies and the limited availability of practical performance results.

References

[1] Yilmaz M and Krein PT. Review of battery charger topologies, charging power levels, and infrastructure for plug-in electric and hybrid vehicles. *IEEE Transactions on Power Electronics*. 2013;28(5):2151–2169.
[2] Williamson SS, Rathore AK, and Musavi F. Industrial electronics for electric transportation: current state-of-the-art and future challenges. *IEEE Transactions on Industrial Electronics*. 2015;62(5):3021–3032.
[3] Tu H, Feng H, Srdic S, *et al.* Extreme fast charging of electric vehicles: a technology overview. *IEEE Transactions on Transportation Electrification*. 2019;5(4):861–878.
[4] Liu C, Chau KT, Wu D, *et al.* Opportunities and challenges of vehicle-to-home, vehicle-to-vehicle, and vehicle-to-grid technologies. *Proceedings of the IEEE*. 2013;101(11):2409–2427.
[5] Upputuri RP and Subudhi B. A comprehensive review and performance evaluation of bidirectional charger topologies for V2G/G2V operations in EV applications. *IEEE Transactions on Transportation Electrification*. 2024;10 (1):583–595.
[6] Patel N, Lopes LAC, Rathore A, *et al.* A soft-switched single-stage single-phase PFC converter for bidirectional plug-in EV charger. *IEEE Transactions on Industry Applications*. 2023;59(4):5123–5135.
[7] Li S, Lu S, and Mi CC. Revolution of electric vehicle charging technologies accelerated by wide bandgap devices. *Proceedings of the IEEE*. 2021;109 (6):985–1003.
[8] Patel N, Lopes LAC, Rathore AK, *et al.* High-efficiency single-stage single-phase bidirectional PFC converter for plug-in EV charger. *IEEE Transactions on Transportation Electrification*. 2024;10(3):5636–5649.
[9] Arena G, Chub A, Lukianov M, *et al.* A comprehensive review on DC fast charging stations for electric vehicles: standards, power conversion

technologies, architectures, energy management, and cybersecurity. *IEEE Open Journal of Power Electronics*. 2024;5:1573–1611.

[10]   Khaligh A and D'Antonio M. Global trends in high-power on-board chargers for electric vehicles. *IEEE Transactions on Vehicular Technology*. 2019;68 (4):3306–3324.

[11]   Pradhan R, Keshmiri N, and Emadi A. On-board chargers for high-voltage electric vehicle powertrains: future trends and challenges. *IEEE Open Journal of Power Electronics*. 2023;4:189–207.

[12]   Wouters H and Martinez W. Bidirectional onboard chargers for electric vehicles: state-of-the-art and future trends. *IEEE Transactions on Power Electronics*. 2024;39(1):693–716.

[13]   Shafiqurrahman A, Khadkikar V, and Rathore AK. Electric vehicle-to-vehicle (V2V) power transfer: electrical and communication developments. *IEEE Transactions on Transportation Electrification*. 2024;10(3):6258–6284.

[14]   Acharige SSG, Haque ME, Arif MT, *et al.* Review of electric vehicle charging technologies, standards, architectures, and converter configurations. *IEEE Access*. 2023;11:41218–41255.

[15]   Holmes DG and Lipo TA. *Pulse Width Modulation for Power Converters: Principles and Practice*. Hoboken, NJ: Wiley-IEEE Press; 2003.

[16]   Erickson RW and Maksimović D. *Fundamentals of Power Electronics*. 3rd ed. Cham: Springer International Publishing; 2020.

[17]   Rivera S, Kouro S, Vazquez S, *et al.* Electric vehicle charging infrastructure: from grid to battery. *IEEE Industrial Electronics Magazine*. 2021;15(2):37–51.

[18]   Rivera S, Goetz SM, Kouro S, *et al.* Charging infrastructure and grid integration for electromobility. *Proceedings of the IEEE*. 2023;111(4):371–396.

[19]   Safayatullah M, Elrais MT, Ghosh S, *et al.* A comprehensive review of power converter topologies and control methods for electric vehicle fast charging applications. *IEEE Access*. 2022;10:40753–40793.

[20]   Cesiel D and Zhu C. A closer look at the on-board charger: the development of the second-generation module for the Chevrolet Volt. *IEEE Electrification Magazine*. 2017;5(1):36–42.

[21]   Zhang Z, Xie S, Wu Z, *et al.* Soft-switching and low conduction loss current-fed isolated bidirectional DC–DC converter with PWM plus dual phase-shift control. *Journal of Power Electronics*. 2020;20(3):664–674.

[22]   Marxgut C, Krismer F, Bortis D, *et al.* Ultraflat interleaved triangular current mode (TCM) single-phase PFC rectifier. *IEEE Transactions on Power Electronics*. 2014;29(2):873–882.

[23]   De Doncker RWAA, Divan DM, and Kheraluwala MH. A three-phase soft-switched high-power-density DC/DC converter for high-power applications. *IEEE Transactions on Industry Applications*. 1991;27(1):63–73.

[24]   Riedel J, Holmes DG, McGrath BP, *et al.* Maintaining continuous ZVS operation of a dual active bridge by reduced coupling transformers. *IEEE Transactions on Industrial Electronics*. 2018;65(12):9438–9448.

[25] Oggier G, García GO, and Oliva AR. Modulation strategy to operate the dual active bridge DC–DC converter under soft switching in the whole operating range. *IEEE Transactions on Power Electronics*. 2011;26(4): 1228–1236.

[26] Riedel J, Holmes DG, McGrath BP, *et al.* ZVS soft switching boundaries for dual active bridge DC–DC converters using frequency domain analysis. *IEEE Transactions on Power Electronics*. 2017;32(4):3166–3179.

[27] Mohan N, Undeland TM, and Robbins WP. *Power Electronics: Converters, Applications, and Design*. 3rd ed. New York: Wiley; 2003.

[28] Malan WL, Vilathgamuwa DM, and Walker GR. Modeling and control of a resonant dual active bridge with a tuned CLLC network. *IEEE Transactions on Power Electronics*. 2016;31(10):7297–7310.

[29] Twiname RP, Thrimawithana DJ, Madawala UK, *et al.* A resonant bi-directional dc–dc converter. In: *2014 IEEE International Conference on Industrial Technology (ICIT)*; 2014. pp. 307–311.

[30] Twiname R, Malan W, Minogue J, *et al.* A novel dual active bridge topology with a tuned CLC network. In: *2014 IEEE International Conference on Industrial Technology (ICIT)*; 2014. pp. 895–900.

[31] James LD, Teixeira CA, Wilkinson RH, *et al.* Adaptive modulation of resonant DAB converters for wide range ZVS operation with minimum reactive circulating power. *IEEE Transactions on Industry Applications*. 2022;58(6):7396–7407.

[32] Belkamel H, Kim H, and Choi S. Interleaved Totem-Pole ZVS converter operating in CCM for single-stage bidirectional AC–DC conversion with high-frequency isolation. *IEEE Transactions on Power Electronics*. 2021;36 (3):3486–3495.

[33] Muhammad R, Kieu HP, Park J, *et al.* Integrated grid inductor-transformer structure with reduced core loss and volume for E-capless single-stage EV charger. In: *2022 IEEE Applied Power Electronics Conference and Exposition (APEC)*; 2022. pp. 873–878.

[34] Jauch F and Biela J. Single-phase single-stage bidirectional isolated ZVS AC–DC converter with PFC. In: *2012 15th International Power Electronics and Motion Control Conference (EPE/PEMC)*; 2012. pp. LS5d.1-1–LS5d.1-8.

[35] Jauch F and Biela J. Combined phase-shift and frequency modulation of a dual-active-bridge AC–DC converter with PFC. *IEEE Transactions on Power Electronics*. 2016;31(12):8387–8397.

[36] Kim H, Belkamel H, Park J, *et al.* Modular three-phase single-stage isolated AC–DC converter for electrolytic capacitor-less EV DC charging. In: *2020 IEEE 9th International Power Electronics and Motion Control Conference (IPEMC2020-ECCE Asia)*; 2020. pp. 1573–1578.

[37] Kolar JW and Zach FC. A novel three-phase utility interface minimizing line current harmonics of high-power telecommunications rectifier modules. In: *Proceedings of Intelec'94*; 1994. pp. 367–374.

[38] Loudot S, Briane B, Ploix O, *et al.* Fast charging device for an electric vehicle; U.S. Patent 8,847,555B2, 2014. Available from: https://patents.google.com/patent/US8847555B2.

[39] Briane B and Loudot S. Rapid reversible charging device for an electric vehicle; U.S. Patent 8,917,046B2, 2014. Available from: https://patents.google.com/patent/US8917046B2.

[40] Kardolus M, Schijffelen J, Gröninger M, *et al.* Battery charger for electric vehicles; U.S. Patent 10,166,873B2, 2019. Available from: https://patents.google.com/patent/US10166873B2.

[41] Krauer JP. Charging efficiency using variable isolation; U.S. Patent 9,225,197B2, 2015. Available from: https://patents.google.com/patent/US9225197B2.

[42] Hähre K, Heyne R, Jankovic M, *et al.* Power electronic module for a charging station and corresponding charging station and electricity charging station; U.S. Patent 2019/0190390A1, 2019. Available from: https://patents.google.com/patent/US20190190390A1.

[43] Chinthavali M and Onar OC. Tutorial on wireless power transfer systems. In: *2016 IEEE Transportation Electrification Conference and Expo (ITEC)*; 2016. pp. 1–142.

[44] International Energy Agency (IEA). Global EV Outlook 2013: Understanding the Electric Vehicle Landscape to 2020; 2013. Available from: https://www.iea.org/reports/global-ev-outlook-2013.

[45] Patil D, McDonough MK, Miller JM, *et al.* Wireless power transfer for vehicular applications: overview and challenges. *IEEE Transactions on Transportation Electrification*. 2018;4(1):3–37.

[46] Wu S-T and Chiu YW. Implementation of a bidirectional 400–800 V wireless EV charging system. *IEEE Access*. 2024;12:26667–26682.

[47] Liang Z, Zhu L, Sun Y, *et al.* Full integration of on-board charger, auxiliary power module, and wireless charger for electric vehicles using multipurpose magnetic couplers. *IEEE Transactions on Industrial Electronics*. 2024;71 (8):9962–9967.

[48] Elshaer M, Bell C, Hamid A, *et al.* DC–DC topology for interfacing a wireless power transfer system to an on-board conductive charger for plug-in electric vehicles. *IEEE Transactions on Industry Applications*. 2021;57 (6):5552–5561.

[49] Bai HK, Costinett D, Tolbert LM, *et al.* Charging electric vehicle batteries: wired and wireless power transfer: exploring EV charging technologies. *IEEE Power Electronics Magazine*. 2022;9(2):14–29.

[50] Khalid M, Ahmad F, Panigrahi BK, *et al.* A comprehensive review on advanced charging topologies and methodologies for electric vehicle battery. *Journal of Energy Storage*. 2022;53:105084.

[51] Ali A, Mousa HHH, Shaaban MF, *et al.* A comprehensive review on charging topologies and power electronic converter solutions for electric vehicles. *Journal of Modern Power Systems and Clean Energy*. 2024;12(3):675–694.

[52] Nutkani I, Toole H, Wilkinson RH, *et al.* Assessment of classical demand management strategies and batteries for electric vehicle impact mitigation in distribution networks. *Smart Grids and Sustainable Energy.* 2024;10(1):1–17.

[53] Jarraya F, Khan A, Gastli A, *et al.* Design considerations, modelling, and control of dual-active full bridge for electric vehicles charging applications. *Journal of Engineering.* 2019;2019(12):8439–8447.

[54] Saleeb H, Sayed K, Kassem A, *et al.* Control and analysis of bidirectional interleaved hybrid converter with coupled inductors for electric vehicle applications. *Electrical Engineering.* 2020;102(1):195–222.

[55] Kwon M, Jung S, and Choi S. A high efficiency bi-directional EV charger with seamless mode transfer for V2G and V2H application. In: *2015 IEEE Energy Conversion Congress and Exposition (ECCE)*; 2015. pp. 5394–5399.

[56] Taghizadeh S, Hossain MJ, Poursafar N, *et al.* A multifunctional single-phase EV on-board charger with a new V2V charging assistance capability. *IEEE Access.* 2020;8:116812–116823.

[57] Li H, Zhang Z, Wang S, *et al.* A 300-kHz 6.6-kW SiC bidirectional LLC onboard charger. *IEEE Transactions on Industrial Electronics.* 2020;67 (2):1435–1445.

[58] Fu H, Duan S, Li Y, *et al.* Improved control strategy for zero-crossing distortion elimination in Totem-Pole PFC converter with coupled inductor. *Energies.* 2022;15(15):1–15.

[59] Cope RC and Podrazhansky Y. The art of battery charging. In: *Fourteenth Annual Battery Conference on Applications and Advances. Proceedings of the Conference (Cat. No.99TH8371)*; 1999. pp. 233–235.

[60] Khan AB and Choi W. Optimal charge pattern for the high-performance multistage constant current charge method for the Li-ion batteries. *IEEE Transactions on Energy Conversion.* 2018;33(3):1132–1140.

[61] Savoye F, Venet P, Millet M, *et al.* Impact of periodic current pulses on Li-ion battery performance. *IEEE Transactions on Industrial Electronics.* 2012;59(9):3481–3488.

[62] Chen LR. A design of an optimal battery pulse charge system by frequency-varied technique. *IEEE Transactions on Industrial Electronics.* 2007;54 (1):398–405.

[63] Hua CC and Lin MY. A study of charging control of lead-acid battery for electric vehicles. In: *ISIE'2000. Proceedings of the 2000 IEEE International Symposium on Industrial Electronics (Cat. No.00TH8543).* vol. 1; 2000. pp. 135–140.

[64] Chen LR, Wu SL, Shieh DT, *et al.* Sinusoidal-ripple-current charging strategy and optimal charging frequency study for Li-ion batteries. *IEEE Transactions on Industrial Electronics.* 2013;60(1):88–97.

[65] Yifeng G and Chengning Z. Study on the fast charging method of lead-acid battery with negative pulse discharge. In: *2011 4th International Conference on Power Electronics Systems and Applications*; 2011. pp. 1–4.

[66]   Liu K, Li K, Ma H, *et al.* Multi-objective optimization of charging patterns for lithium-ion battery management. *Energy Conversion and Management*. 2018;159:151–162.

[67]   Liu C, Gao Y, and Liu L. Toward safe and rapid battery charging: design optimal fast charging strategies thorough a physics-based model considering lithium plating. *International Journal of Energy Research*. 2021;45(2):2303–2320.

[68]   Zou C, Hu X, Dey S, *et al.* Nonlinear fractional-order estimator with guaranteed robustness and stability for lithium-ion batteries. *IEEE Transactions on Industrial Electronics*. 2018;65(7):5951–5961.

[69]   Lin X, Wang S, and Kim Y. A framework for charging strategy optimization using a physics-based battery model. *Journal of Applied Electrochemistry*. 2019;49(8):779–793.

[70]   Liu K, Zou C, Li K, *et al.* Charging pattern optimization for lithium-ion batteries with an electrothermal-aging model. *IEEE Transactions on Industrial Informatics*. 2018;14(12):5463–5474.

Chapter 5

V2G developments, market landscape and customer sensitivities: a European case study

Marco Landi[1]

After being the subject of academic research for more than a decade, vehicle-to-everything (V2X) is becoming increasingly popular among energy providers, electric vehicle (EV) manufacturers and consumers. However, the developments and deployments across the world have taken slightly different approaches and directions: while in the US, the focus has been mainly on large fleets, supporting utilities DSR programs, and on energy resilience for home customers, in Europe, commercial applications have seen a slower deployment – but V2X at homes and in public has been at the centre of regulatory discussions and auto/energy industry alignments.

5.1 What is V2X?

Not to be confused with C-V2X 'communications vehicle-to-everything', generally referring to the ability of a vehicle to exchange data with the wider infrastructure and other vehicles, V2X in the energy space is intended as the ability of a battery-equipped vehicle to provide energy back to other devices or systems. The definition is quite loose, as it includes a variety of distinctive features, technical approaches and business models – so it is helpful to set some initial definitions. V2X generally encompasses different applications, all sharing the ability to utilise an EV battery stored energy to power external devices OR networks:

- **Vehicle to vehicle (V2V)** refers to the ability of an EV to charge another battery-powered vehicle, effectively sharing energy among the two traction batteries.
- **Vehicle to load (V2L)** focuses on using the vehicle to power appliances and/or other electrical devices. Together with V2V, its relevance for the integration of EVs into power systems is limited. In fact, while in this application, the vehicles need to be able to convert power from the DC battery to AC (suitable for normal domestic or industrial loads), the output waveform is not synchronised to an external grid setting: in other words, the EV acts as a pure isolated generator.

[1]Jaguar Land Rover Limited, Coventry, UK

- **Vehicle to home (V2H)** aims at integrating an EV into a confined environment like a home. In particular, V2H focuses on the optimisation of electricity generation and consumption *behind the meter* – in other words, without direct access to wholesale or flexible electricity markets. While the whole home might provide energy services, from a vehicle perspective, the goal is to support the home energy 'system'. As one of the largest loads in a home and a sizeable battery at that, EVs can have a considerable impact in favouring energy tariff savings, energy self-sufficiency (e.g. if paired with local generation) and blackout support.

- **Vehicle to building (V2B)** is similar to V2H in terms of the role of EV and support to a local system, but in the context of more complex buildings (e.g. commercial or industrial buildings or multi-dwelling residential buildings with centralised energy management). Generally focusing on fleets of EVs rather than individual vehicles, the focus here is the optimisation of consumptions, costs and fleet operation as well.

- **Vehicle to grid (V2G)** refers to the complete integration with the Power grid – providing services in *front of the meter*. Under V2G operation, an EV can access electricity wholesale and flexibility markets, generating revenue directly from service provision and/or energy trading. Of all the V2X applications, this is the most advanced and complex to implement. While V2H and V2B focus on local system optimisation, V2G depends on the specific market configuration, regulations and value chain applicable where it is implemented. In other words, V2G implementations may differ significantly in different regions due to changing market and regulatory arrangements.

Generally, while discussing vehicle–grid integration (VGI) only grid-interactive services – such as V2H, V2B and V2G – are considered (Figure 5.1).

5.1.1   *Plenty of potential*

V2X has been the centre of increasing interest from researchers, regulators and private enterprises, due to its potential impacts on power system sizing, design and operation. From a system design point of view, the opportunity to rely on a 'network' of distributed storage systems allows for a deferral of required investments into grid upgrades – particularly in the distribution network: by accommodating power spikes from local battery buffers, reinforcements of upstream distribution network can be better planned and allocated in time instead of being required at the same time. Note that this holds true both if the EVs are used to optimise individual home or building consumptions and if they are used for trading energy locally. In the UK, estimates are that the required investments in electricity network to support higher power peaks due to unmanaged EV charging could be as high as £17Bn by 2050; smart charging EVs could already reduce this figure to £8Bn, with further reductions if the vehicles do actively support the grid operations in a V2X setting [1].

From a customer perspective, V2X promises to monetise an asset that sits underutilised for most of its life (in the UK, cars are parked 96% of the time [2]). In addition, V2X applications can increase energy security (e.g. resilience to blackouts) and self-sufficiency (if paired with local energy generation).

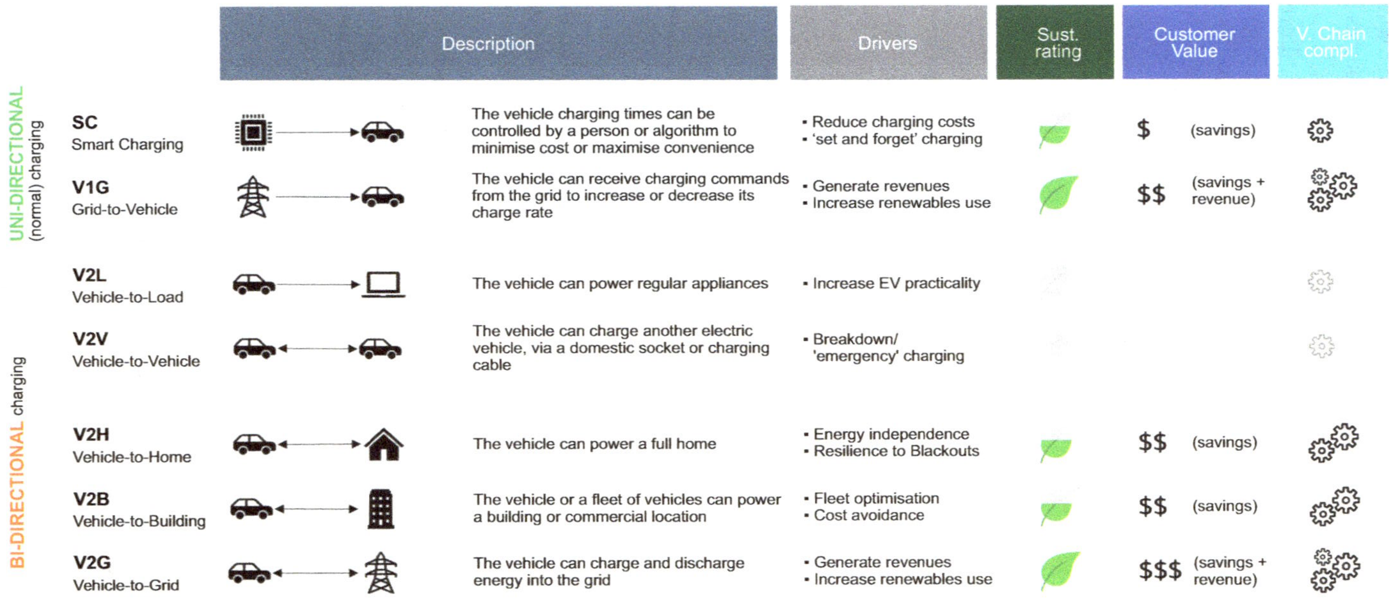

		Description	Drivers	Sust. rating	Customer Value	V. Chain compl.
UNI-DIRECTIONAL (normal) charging	**SC** Smart Charging	The vehicle charging times can be controlled by a person or algorithm to minimise cost or maximise convenience	• Reduce charging costs • 'set and forget' charging	🍃	$ (savings)	⚙
	V1G Grid-to-Vehicle	The vehicle can receive charging commands from the grid to increase or decrease its charge rate	• Generate revenues • Increase renewables use	🍃	$$ (savings + revenue)	⚙⚙
BI-DIRECTIONAL charging	**V2L** Vehicle-to-Load	The vehicle can power regular appliances	• Increase EV practicality			⚙
	V2V Vehicle-to-Vehicle	The vehicle can charge another electric vehicle, via a domestic socket or charging cable	• Breakdown/ 'emergency' charging			⚙
	V2H Vehicle-to-Home	The vehicle can power a full home	• Energy independence • Resilience to Blackouts	🍃	$$ (savings)	⚙⚙
	V2B Vehicle-to-Building	The vehicle or a fleet of vehicles can power a building or commercial location	• Fleet optimisation • Cost avoidance	🍃	$$ (savings)	⚙⚙
	V2G Vehicle-to-Grid	The vehicle can charge and discharge energy into the grid	• Generate revenues • Increase renewables use	🍃	$$$ (savings + revenue)	⚙⚙

Figure 5.1   Different technologies of uni- and bi-directional EV charging

Finally, both at the local and aggregate levels, the availability of widespread storage resources via V2X allows easier integration of renewable sources into the grid; thus, maximising their utilisation and therefore contributing to the decarbonisation of the system. A study from Imperial College and OVO Energy estimated that V2G could bring benefits equating £3.5Bn/year to the UK system by 2040 [3]. Similar studies in Europe have found that V2G may reduce the curtailment of renewable resources – helping the transition to renewables and supporting grid decarbonisation – as well as reducing the overall energy prices [4].

5.1.2   *Which services? Energy savings vs wholesale vs flexibility services*

When discussing V2X, it is important to recognise that the ways in which these technologies can optimise energy consumptions or generate revenues/rewards for EV users are numerous and different across different energy systems.

First of all, different applications support different system configurations. V2H and V2B mainly operate **behind the meter** and focus on the optimisation of energy generation and usage within a local energy system (e.g. a house or a building). However, V2G tends to operate at the **front of the meter**, requiring direct connection to the grid operators and electricity markets. Also, front-of-the-meter operation may focus on either wholesale energy markets (i.e. where the energy from the EVs is traded in wholesale energy markets, for example, in the case of arbitrage) or flexibility services (i.e. supporting the operation of the energy system without necessarily acting as a replacement for bulk power generation, including frequency and voltage regulation). As for the range of services that V2X applications can deliver to EV users, these fall into three broad categories:

1. **Customer programs**. Mainly covering operations behind the meter, this array of services is intended to maximise customer energy savings and optimise consumption. A niche case within this category is the utilisation of V2X technologies to condition and manage EV batteries.
2. **Distribution services**. These focus on supporting local distribution networks and consist of front-of-the-meter operations mainly involving flexibility services. These can be deployed as targeted programs to specific areas (e.g. local load reduction programs, P2P and community energy solutions within local microgrids) or via market arrangements (i.e. local flexibility market) managed by DSOs (Figure 5.2).
3. **Bulk power services**. These involve a wide array of services that can be provided to the electricity system operator, including both wholesale energy and flexibility services [5] (Figure 5.3).

Collectively, these are also referred to as 'flexibility' services. An exemplification is reported in Figure 5.4.

It is worth noting the availability, configuration and accessibility of the aforementioned services are heavily dependent on specific market arrangements. Different markets and regions within these markets (e.g. countries within Europe or regions within the UK) might have different structures, different flexibility

Operating regimes   **Local capacity management mechanisms**

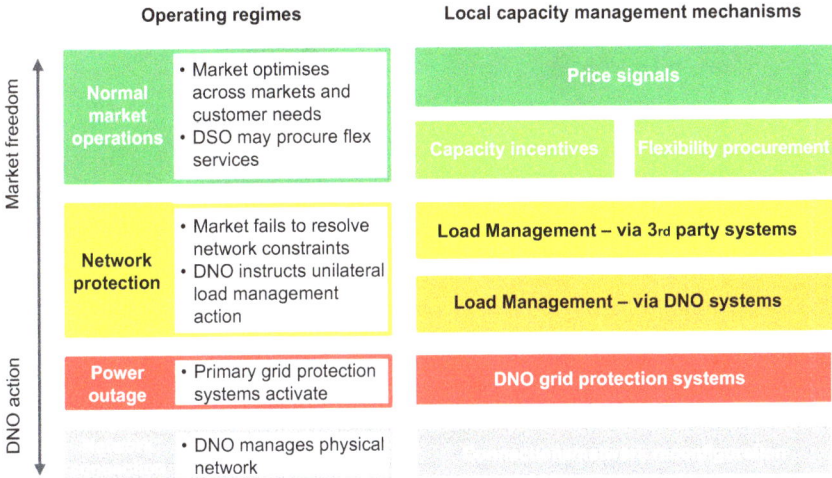

Figure 5.2   Representation of balance between market freedom and direct action in the transition from DNO to DSO in GB (source: UKPN [6])

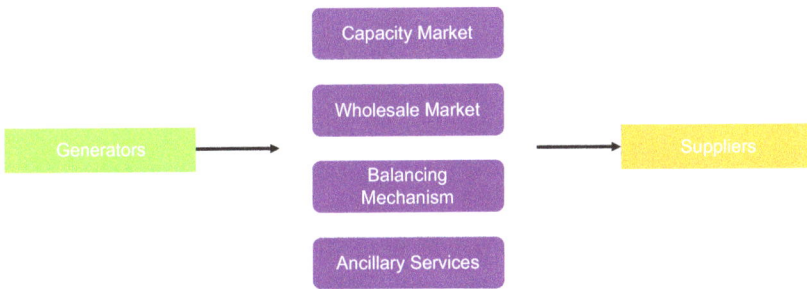

Figure 5.3   Representation of categories for Bulk Power services offered by the System Operator (National Grid ESO) in GB

services, and different processes and access requirements. This by itself is a significant barrier to V2X rollout, as customer and market propositions need to be tuned to the specific market setting.

5.1.3   Linking the V2X value chain

V2X technologies, by their nature, cross a wide array of actors and stakeholders and create extensive value chains across multiple sectors. Therefore, while they hold a lot of potential, their successful implementation depends on effectively creating value chains that not only deliver working technical solutions, but respond to EV drivers' needs and wants and are backed by feasible business models.

In the following sections, we are going to explore further the EV users' perspective, the market and technology angle and finally provide examples of different applicable business models.

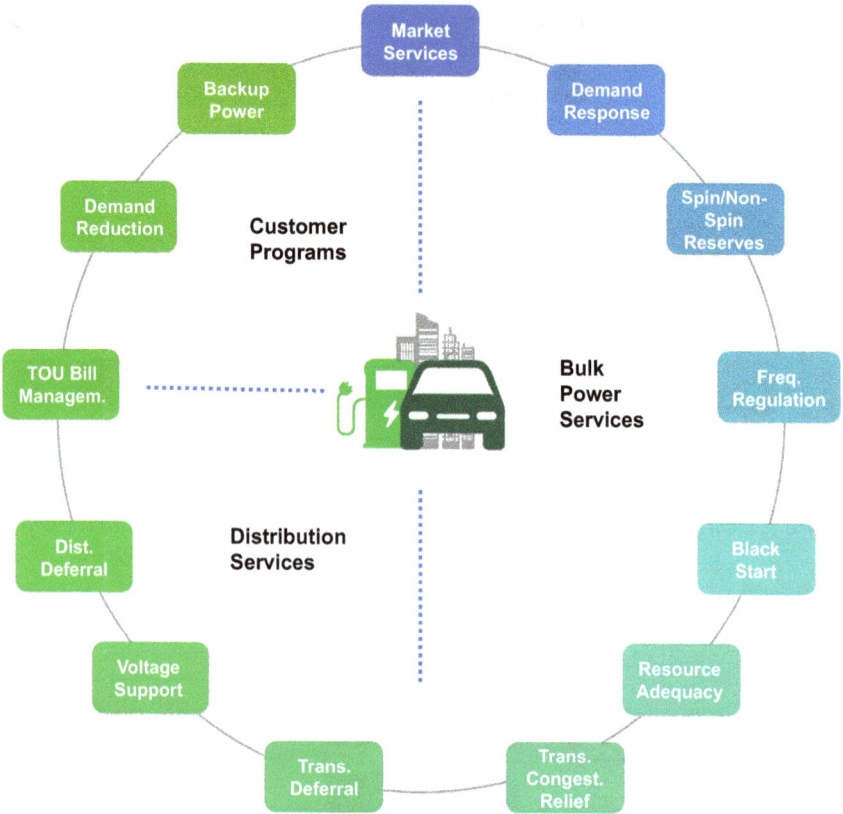

Figure 5.4   Exemplification of V2X-related grid services

5.2   A user-centric deployment of V2X: the EV user angle

The cornerstone of V2X implementation is the availability of vehicles able to exchange energy with the wider power system: ultimately EVs are primarily transportation means, with energy applications being at best secondary uses. Importantly, participation in V2X services should always be a user's voluntary choice and never forced. Therefore, the perspective of EV users – their wants and needs, as well as the benefits that they could gain from allowing their vehicles to participate in V2X services – is paramount.

5.2.1   Customer needs in an evolving electricity system

Today, most EV customers charge their EV at home [7], reaching peaks of 90% of current EV drivers in the UK [8]. For most households, an EV represents the single biggest electrical load; thus, bearing the highest impact on the home's electricity consumption. This is the reason most often transitioning from a combustion engine vehicle to an EV is the first step for consumers to explore more energy products and services – from installation of charging points to smart charging solutions, from changing electricity tariffs to local generation and storage options. Moreover, plugging in EVs connects two different value chains together – energy and automotive – creating opportunities for new relationships, services and products.

However, navigating this complex and fragmented market is today still very confusing for the average EV user. While awareness of energy products and services has increased over the energy crisis, it is still mostly an unfamiliar market for users – with confusion over tariffs and green credentials [9], vast choices of different products (EV and energy-related) with varying features and benefits, and coordination of devices from different brands still challenging.

Additionally, the energy system itself is undergoing a profound evolution: from a centralised and 'top to bottom' vertical system to a decentralised, interconnected and interactive Smart Grid. Consequently, the role of energy users is changing as well, from a passive to an active one. Users are not just consumers anymore, but active participants who can change their consumption, generate energy locally and engage in energy and data exchanges with a wide array of actors and players. So, previously uninvolved – and already at a loss – consumers are now faced with an increasing number of options, choices and opportunities at hand. But what do EV users ultimately expect from smart energy services?

First of all, they seek simplicity of choices and operations. Users want a higher degree of involvement, but with ease of access. Ultimately, users look for personalised services that work seamlessly [10,11].

Affordability is also an important consumer request. Smart energy technologies should make energy systems run smoother, and access to cheaper renewable generation easier.

Additionally, sustainability and environmental benefits for a good range of EV users constitute primary drivers for choosing EVs [12] – and how they use/expect to use them [9]. Therefore, smart energy services that complement their EV should then be equally green.

Given the close link between the Energy and Automotive sectors created by EVs and their need to charge, and the impact that energy products and services have on the EV user customer experience, more and more Automotive OEMs are making significant investments into the energy sector.

5.2.2   Customer benefits of V2X services

Since their inception, V2X solutions have focused on generating financial rewards for participating users. However, as described below, the benefits to customers can be even wider than purely financial returns.

5.2.2.1   Financial incentives

The basic principle behind the V2X concept is straightforward: an EV owner maximises the value of a stationary mobility asset by exchanging energy with the local (e.g. home) or wider energy system, which results in either energy cost savings, or direct revenue generation from the provision of services.

In a V2H/V2B configuration, the energy in the battery can be used to support home/building consumptions and reduce the grid electricity demand at peak times, i.e. when energy is more expensive and less sustainable. Additionally, if paired with local renewable generation, EVs can help decouple electricity generation and utilisation timing – maximising the reliance on green energy. In both cases, a V2X-enabled vehicle operates as an alternative to stationary storage and reduces or eliminates the need for an investment in expensive home batteries. Moreover, EV batteries are normally several times the capacity of commercially available home storage systems. Depending on the specific customer setting and configuration, estimated savings of up to £200/month in the UK can be achieved [13].

While an EV battery may provide significant power and energy compared to typical household demands, if providing power to the whole power system it is individually insignificant (and prevented from accessing directly most electricity markets). Hence, when operating in V2G configuration, the individual EV enters a wider monitoring and control architecture governed by an aggregator and/or VPP, which aggregates a pool of EVs that then is matched with electricity market requirements. It is generally up to the aggregator/VPP to determine the best combination of services to support and generally tends to follow the most remunerative configuration.

Given EV batteries' ability to provide energy quickly, traditionally V2G implementations have targeted ancillary services such as frequency regulation. This choice was encouraged by the high prices the system operator paid for these services. However, albeit the total amount of frequency regulation needed will increase (forecasts for UK's National Grid FFR service place requirement from current 600 MW to over 1 GW in 2030), as more actors able to support the provision of these services enter the market, the value progressively decreases. As an example, in 2015–2016, prices for dynamic FFR stood around £22/MW/h, while reaching values of less than £10/MW/h in the summer of 2018 as the market saturated.

Therefore, more recent trials have targeted mainly power arbitrage, that is effectively operating and trading energy from EV batteries on the wholesale market. In the UK, the Powerloop project from Octopus Energy involved over 100 users and showcased combined savings from V2G of over £800/year compared to unmanaged charging [5] and has resulted in a commercially available tariff for V2G participation [14].

More realistic for widespread commercial implementations is the stacking of different types of services – with the aggregator/VPP optimising the target markets in response to real-time pricing. This is the case of the Scirius project from OVO energy: involving over 300 EV drivers; it estimated combined benefits of up to £720/year [15].

Because the service stacking should be optimised based on current market conditions, not only the same market might see varying prices for services, but different markets (and different regions in the same market) will result in different potential projected rewards from V2G service provision. This in turn means that an aggregator/VPP offering should be calibrated for the specific market and location it is intended to serve – potentially varying the weighting of different services in building the stack. While the inner workings of the service might be transparent to the end user, the practical result is that consumers might see radically different projected benefits based on where they live (or connect the EV to charge).

Additionally, the amount of reward that consumers might expect depends also on the specific business model and composition of the value chain. Different V2X providers and operators might leverage different services, distribute the financial gains differently across the value chain and offer different benefits to EV users.

Finally, it is worth noting that the financial benefits from V2X are mostly aligned with the amount of time the EVs are available to provide services. For example, short-term revenue modelling shows that increasing the plug-in rate from 28% to 75% quadruples the revenues available from grid services [16]. Hence, not all types of EV users are a good match for V2X services. In order to define potential markets for V2G and develop a viable commercial proposition, it is necessary to determine how each customer's behavior influences the potential to generate revenue through V2G. In [16], 34 reference customer archetypes are identified, among both commercial and domestic users. Factors such as type of vehicle, usage patterns of EV and charging point, where users charge are considered to define an index of 'applicability to V2G' for each archetype.

Ultimately, when looking at V2X from a customer perspective and via a purely financial lens, the key question to answer is if the promises of rewards justify the investment in the technology. On one hand, V2X equipment, although still expensive, is quickly reducing in price. A previous study [17] projects a cost for a DC V2X unit of £1000 by 2030 although market developments are accelerating the cost-down trajectory and reaching the same target before the end of the decade. On the other, the rewards offered by V2X are heavily dependent on specific business models and value redistribution – and today suffer from the competition of Smart Charging (which can capture in most markets up to three-quarters of the value of V2X with a much lower initial investment).

In conclusion, while V2X offers a definite potential for financial gains, it is not a panacea for every type of EV user everywhere. However, V2X benefits go beyond the pure financial rewards – and a proposition to consumers should look at the holistic value it can deliver to EV users.

5.2.2.2 Benefits beyond financial rewards

Having a large amount of grid-connected batteries allows to reduce the curtailment from renewable generation [3]: while the benefits are generally quantified financially as cost avoidance compared to alternative solutions that deliver the same results, it is undeniable that V2G leads to lower carbon emissions and widespread environmental benefits across the whole energy system. To some users, the societal benefits that derive from this are a sufficient motivator in itself to engage in V2G operations.

Even applications to the individual household or building can have immediate sustainability benefits when paired with local renewable generation. Moreover, V2X solutions can help EV users become more self-sufficient from an energy perspective and less reliant on grid-transported electricity: this is a particularly attractive proposition in markets that suffer from high cost of energy or have seen incentives for deployment of solar PV among consumers.

Directly related to the self-sufficiency theme is the higher energy resilience that V2X technologies can deliver. V2H has been used for years in Japan as disaster relief in areas hit by natural disasters and with damaged infrastructure. Similarly, in wildfires that hit California and in Texas during extreme winter storms, V2H has been used to provide electricity when the grid infrastructure failed. More in general, even in the absence of natural disasters, V2X technology can be used to compensate for weaknesses in grid supply (e.g. in areas prone to power cuts).

Finally, Cenex [18] describes an added benefit to users: battery/vehicle management. By having control over the charge and discharge process, an aggregator can manage the battery (and the whole EV) in ways that minimise battery aging factors – effectively acting as an additional guarantee for the end user.

All of the benefits above, while quantifiable financially, are generally not included in the estimate of monetary rewards from V2X – but in most cases might have a stronger bearing on the EV users' decision to engage with V2X activities.

5.3 Market context

Electrification of transport is progressing at pace: 2023 recorded a 35% year-on-year increase in sales of EVs, with 18% of cars sold in 2023 being electrified [8] (Figure 5.5). ACEA and McKinsey estimate that only the passenger vehicle parc will see more than 42 M electrified vehicles in Europe, 34 of which will be pure battery electric vehicles [7]. Charging this growing EV parc will pose a significant

Global electric car stock, 2013–2023   **Open** ↗

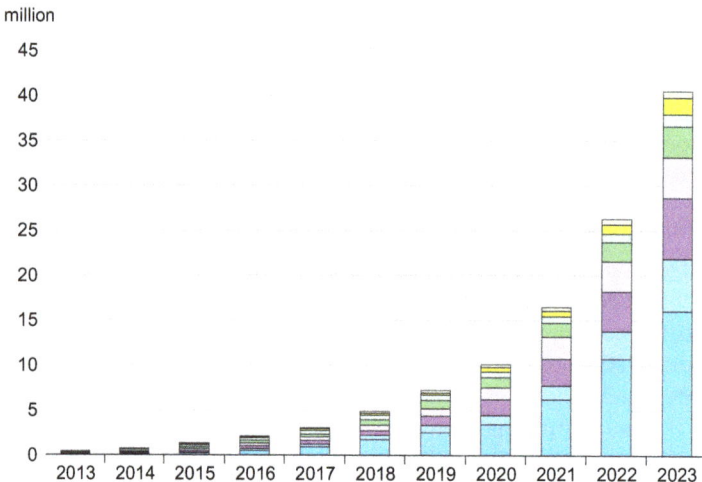

IEA. Licence: CC BY 4.0

○ China BEV   ○ China PHEV   ● Europe BEV   ○ Europe PHEV   ○ United States BEV
○ United States PHEV   ○ Rest of the world BEV   ○ Rest of the world PHEV

Figure 5.5   Electrified vehicles penetration is growing exponentially – with 70% of these being battery electric vehicles (source: IEA [8])

demand on the electricity grid, particularly as taking place at the same time as the transition to electric heating and a wider adoption of renewables. In the EU, estimates are that grid upgrades due to EVs will cost €41Bn by 2030, on top of €69Bn renewable energy investments [7] (Figure 5.6).

If EV charging can be optimised together with the other sources and loads, EVs can turn from a problem into a solution. Already smart charging allows a significant reduction in grid upgrade investments. We have previously seen estimates for the UK case [1]. In Germany, EVs could require up to a 50% cost increase in low voltage grid and transformer by 2035; costs that would be avoided by adopting optimised peak shaving using smart charging [4]. These issues, and

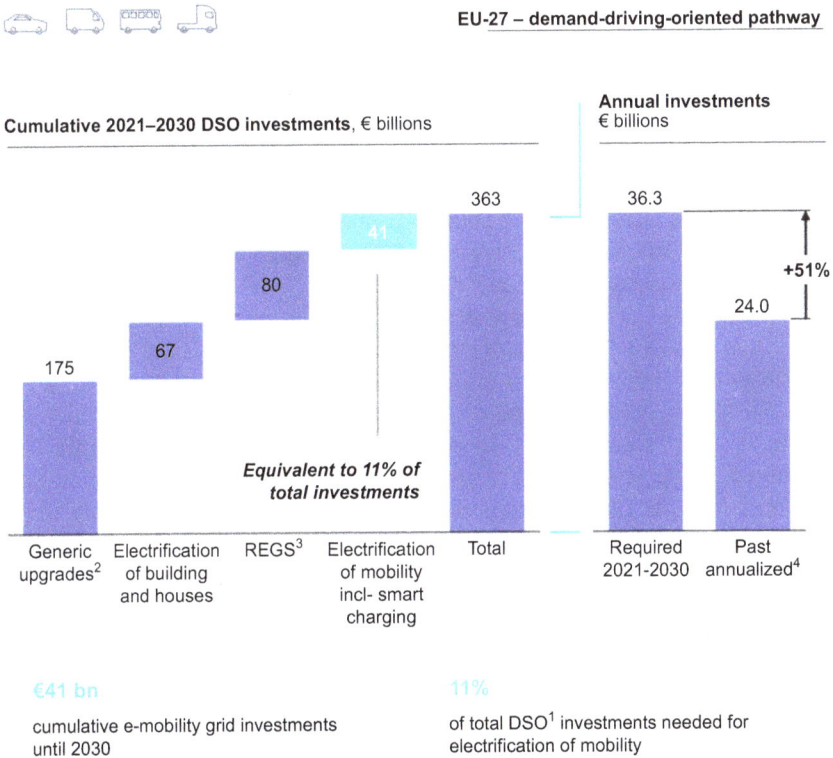

EU-27 – demand-driving-oriented pathway

Cumulative 2021–2030 DSO investments, € billions

Annual investments € billions

175 Generic upgrades[2]

67 Electrification of building and houses

80 REGS[3]

41 Electrification of mobility incl- smart charging

363 Total

Equivalent to 11% of total investments

36.3 Required 2021-2030

24.0 Past annualized[4]

+51%

€41 bn
cumulative e-mobility grid investments until 2030

11%
of total DSO[1] investments needed for electrification of mobility

1. Distribution system operator
2. Such upgrades include smart metering. grid modernisation, digitisation and automation, resilience, and storage
3. Renewable energy generation systems
4. Historicals refer to 2015

Surce: Eurelectric 2020 study, a top-down estimate that uses EVCI model energy demand, and a bottom-up estimate based on EVCI model charging locations, European Commission, expert interviews, EU EV Charging Masterplan

Figure 5.6   E-mobility-driven DSO investments will amount to €41Bn in Europe by 2030 (source: McKinsey [7])

potential solutions, are exacerbated by the growth in the size of EV batteries. As EV customers desire longer ranges from their vehicles, it is now not uncommon to see 100 kWh+ EVs, with this trend set to increase (Figure 5.6).

Naturally, bidirectional energy exchanges constitute an even bigger opportunity for cost avoidance and for smarter management of the network. However, reaching a better integration of EV charging into the grid operation is just one aspect of the evolution of our power systems. The challenge of integrating more renewables and distributed resources requires a transition from a historically vertically integrated system into a decentralised one – with structural changes to how the electricity system operates. An evolution into a Smart Grid is indeed underway: a journey towards a more interconnected, interactive power system at all levels – where all devices are optimised in real time across all actors (represented in Figure 5.7).

Crucial to the success of this transition is digitalisation: effective interactions among transient sources and load can only take place if all devices are monitored and accounted for via real-time data exchanges. An example of this is the ongoing transition to smart meters, which allow more granular monitoring of household loads and sources and open the way to more interactive tariffs and services (e.g. TOU tariffs).

Among the structural reforms underway, one of the most evident changes is the evolution of the DNO into DSO, actively managing loads and sources at local level – most often via market arrangements.

A liberalisation of the power systems is also underway. In a vertically integrated system, geared towards big, centralised generators, there is little scope for the deployment of distributed energy resources. Therefore, in order to move to a more decentralised system, the operations of the power system need to be open to new entrants and new services. In this spirit, energy markets at all levels are being restructured around new types of services and to allow a higher level of participation from small, local (and sometimes specialised) service providers: for example, the UK Electricity System Operator (National Grid) has led a holistic review of its market structure since 2021 [19]; similar initiatives are undergoing in the EU [20].

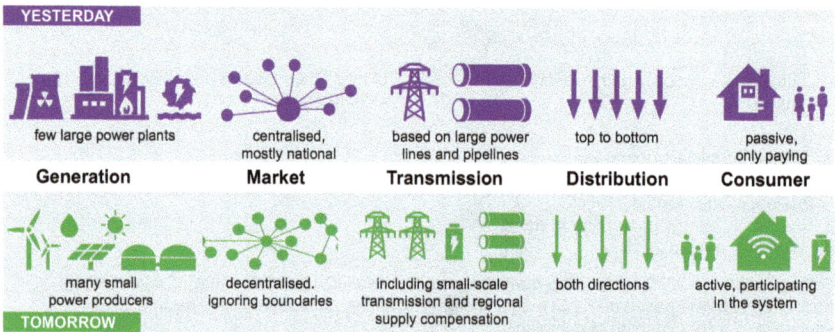

Figure 5.7   Structural changes to the energy system towards a Smart Grid (graphic: Bartz/Stockmar, CC BY 4.0, adapted)

5.3.1   *EU+UK: many different markets*

The market structure may vary significantly across countries in Europe – and most often the differences are not even at the national level but at the regional one. This in turn affects not only the services that could be accessed by V2X-enabled vehicles – and the corresponding benefits – but also the structure of the value chains themselves.

For example, the GB system alone counts eight different DSOs: each has its own roadmap to transition to a local system operator, with different market arrangements (e.g. some implement market-based systems for local ancillary/flexibility services, while others fix a priori the value assigned to flexibility provision). Moreover, the DSOs are also responsible for connection procedures, which also differ by region: the practical result is that even the ability to connect EVs to the grid in a V2X arrangement is not guaranteed and definitely not uniform across regions. More than 2300 DSOs are active in the EU, with efforts to sync approaches across the continent currently underway (e.g. EU DSO Entity and TSO/DSO collaboration with ENTSO-E).

Moreover, specific country legislations may favour or hinder the deployment of V2X technologies: as an example, several countries in the EU still adopt double taxation on energy storage (i.e. both import and export of energy are taxed), reducing the attractiveness of V2X solutions [21].

The degree of liberalisation of each market and access requirements for service providers also affect which type of services is possible to deploy in each region: for example, lower access requirements for grid services may allow a specialised EV aggregator to target these services via V2X alone – while an aggregation with other distributed energy resources may be needed for higher access requirements.

Therefore, the structure of the V2X value chains – and the relationships among their actors – may differ by region and country. For example, the open nature of the UK electricity system means that there are multiple routes to market for V2G solutions: the role of aggregator can be fulfilled by one single service provider or by a charging point operator and a Virtual Power Plant operator, or perhaps directly from the EV manufacturer. In the same way, service provision can be specialised from EVs in a V2G framework, or vehicles can be aggregated together with other loads and sources.

5.3.1.1   Different consumer sensitivities

In the previous section, we discussed the wants and needs of EV drivers when dealing with energy services and value chains. However, geographical and market differences also result in different sensitivities users have to different benefits from V2X technologies.

In markets such as some US states (e.g. California, Texas), which are usually prone to frequent power cuts and supply interruptions, consumers are much more acutely aware of the inconvenience of having to find alternative sources of power during a blackout. Therefore, looking more at EVs with V2X technology, through the lens of energy resilience, is an alternative to local emergency generators.

In the UK, the market structure is already open to the implementation of V2G services. Therefore, a primary driver for V2G with consumers seems to be the potential for financial rewards.

In the EU, several countries have historically had high investment in renewables and local generation – therefore sustainability and self-sufficiency emerge as powerful motivations for V2X technologies.

While the above are generalisations, and each market will inevitably express a variety of drivers among consumers interested in V2X solutions, they highlight the challenge of launching on-the-market technologies that would need to be optimised for different (and sometimes opposing) use cases.

5.4   Technical developments

V2G, as a research topic, has been at the centre of numerous studies in recent years, with the first study on the subject often credited to be [22] dating back to 1997. While the bi-directional energy transfer from the vehicles is also exploited in V2H and V2B contexts, which have been proven to be feasible and deployable on a large scale, as done by Nissan in 2011 when the Leaf was used as relief support to power buildings after the Fukushima earthquake, the focus of V2G has always been about using bi-directional energy (and communication) exchanges to provide grid services.

Over the years, multiple trials and demonstrators have proven the technical feasibility of energy exchanges from EV batteries for grid services. Nissan has successfully operated V2G for frequency regulation services in Denmark. In Utrecht, V2G has been successfully implemented in public charging settings. The UK has held one of the biggest V2X demonstrator programs in the world [23]. These are just some of the multitudes of projects that have taken place on V2X across the world. Innovate UK and UKPN have set up a website to log past and ongoing V2G projects across the world, which can be found at www.v2g-hub.com [24].

While investigations on technology viability have been successful, current research and industrial development studies focus on the scalability of applications and commercial feasibility for large-scale V2G deployments. In this context, large V2G trials have the advantage of both examining the implications of V2G for a wide range of users and settings as well as providing a more realistic platform to evaluate costs and business models. Yet, despite the emergence of technical standards for V2X, not all technical implementations in trials/demonstrators and commercial applications follow the same approach. In the following, we will focus specifically on the topics of AC vs DC implementation and interoperability. Moreover, battery degradation due to V2X usage is also discussed.

5.4.1   AC vs DC V2X

There are two primary configurations for a V2X system (EV and charging point) – depending on whether the power transfer between the two takes place in AC or DC.

Perhaps the simplest of the two, from an architectural perspective, is DC V2X. In this case, the EVSE is responsible for managing the power transfer, as well as accountable for the metering, isolation and grid code requirements. On the other hand, the EV is the controlled party during V2X operation, and virtually 'extends' the onboard DC bus to the EVSE via a DC connection. In other words, the charging point – as the equipment physically connected to the grid – condenses all the requirements that a grid-connected generator (or storage system) needs to abide to. This approach significantly simplifies the certification process – which can include only the EVSE: this is the reason why the standardisation process has started from DC configurations – which in the US can refer to certification according to UL 1771 SA/SB [25] (already recognised for commercial deployments in states such as California). The trade-off is in the high cost of the charging equipment.

An AC V2X system, instead, seeks to minimise the cost of charging equipment. As a power conversion stage is already present in the EV (and potentially already used to export power in a V2L configuration), in this configuration, it is the vehicle that is responsible for adapting power levels and waveforms to be compatible with the power system: in other words, the connection between EV and EVSE is in AC.

Two configurations fall within this category: V2G-AC and V2G-split inverter (Figure 5.8).

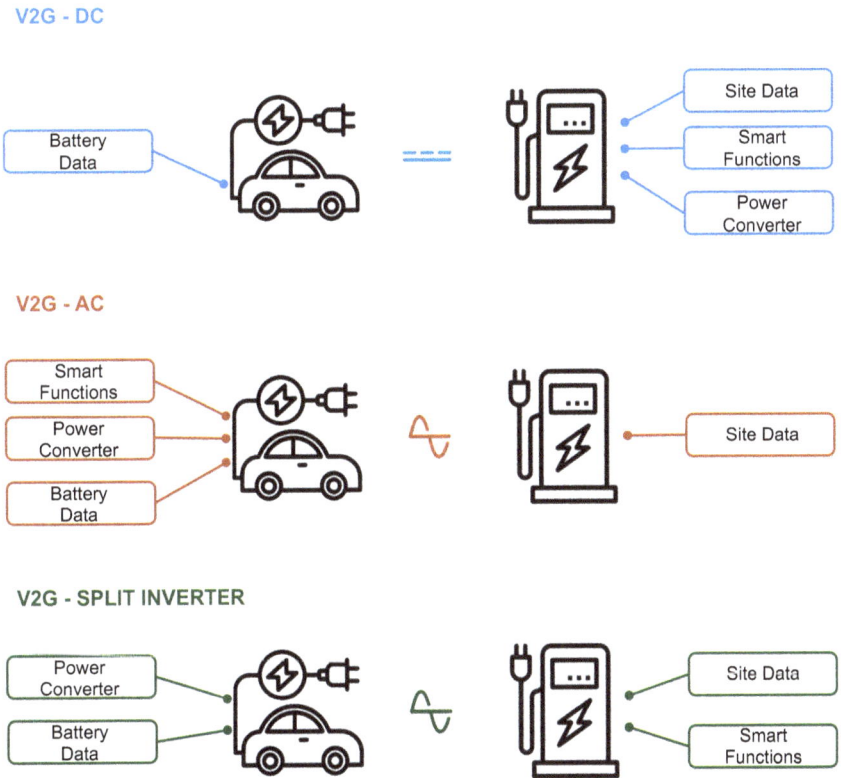

Figure 5.8   V2G configurations (source: IREC [25], adapted)

The former effectively sees the EV responsible for both the power conversion and smart functions (such as metering, isolation, grid code compliance): the EV becomes akin to a mobile smart inverter. The obvious advantage here is that the EVSE becomes a lot simpler – and cheaper – and potentially does not need to be a specific version for V2X operation. The drawback, though, is that the EV design becomes more complicated; moreover, given the local nature of grid systems specifications, it may require the EVs – normally designed to be as 'global' as possible – to have multiple local variants and derivatives compliant with local grid systems. Certifications are in development: in the US for V2G-AC the EVSE can refer to UL1741 SC (once published), while EV could rely on the paired standard SAE J3072 [25].

The second AC V2X configuration is a middle ground between the two approaches above. In this case, the EV is still responsible for the power conversion, but the EVSE retains the smart functions. While in the previous approaches either the EVSE or the EV resembled a smart inverter (in the eye of the power system), in this case, neither can individually be classed as one: EV and EVSE do effectively operate as a single system. From a certification perspective, no single standard has been established for this system configuration as a whole. However, standards intended for the interconnection of DERs can be applied to this use case (example.g. in the US, the IEEE 1547.1 could be used). While making the EVSE cheaper than a DC V2X counterpart, and the EV less exposed to the need for localised versions to comply with local grid codes and specifications, this approach risks sacrificing the interoperability of both EV and EVSE: as they work as part of a single system, it is not guaranteed that an EVSE would work with a different EV than it was designed for (and vice versa).

The two categories also differ in terms of geographical applicability: while in Europe, deployment of both AC and DC V2X solutions is possible under grid operators' rules, in the US at the moment it is still not possible to roll out AC V2X systems.

All three configurations are being trialled and experimented on by EV and EVSE OEMs, researchers and standardisation bodies. In terms of car manufacturers, Nissan has historically worked on DC V2X, as Ford's current solutions in the US; Renault has worked on V2G split inverter solutions and is going to launch an AC V2X offering; Tesla currently supports V2H functionality operating in AC; Volvo has announced support for both on their upcoming flagship vehicle. Moreover, V2L (which also uses EV onboard power conversion) is becoming a widespread feature.

5.4.2   V2X interoperability

Even when an EV and EVSE adopt the same AC or DC configuration, in order to operate together they need to exchange data: for example, this might include the state of charge of the vehicle, the power that needs to be delivered (in either direction), when the vehicle is fully charged or fully depleted, etc. In other words, the two components of the V2X system need an adequate interface and communication protocol.

As the development of V2X solutions started, there was no shortage of ad-hoc solutions to manage the interface and data exchange between EV and EVSE. However, as V2X becomes more prominent in the market, there is the need to guarantee uniformity, and ensure the customers that they are not locked in a specific system. Moreover, as V2X is looked at, by regulators, as a solution for some of the issues on the power system, a series of ad-hoc proprietary implementations reduces the potential available market of the technology.

Probably the most common charging protocol that embeds the necessary communication for V2X implementation is CHAdeMO: widely adopted in Japan, it has been Nissan's protocol of choice and featured in their many V2G trial – and today most of the commercially available V2G solutions are based on CHAdeMO. However, the number of vehicles compatible with this charging protocol is dwindling.

The other communication protocol that is quickly gaining momentum – and will be the solution adopted by virtually all new V2X-enabled vehicles in Europe and the US is ISO 15118-20. The ISO 15118 series manages the high-level communication over the CCS standard – and the −20 revision is an extension specifically intended to enable advanced charging features, including V2X. The ISO 15118 communication protocol is already widespread. The majority of EVSE in the US and Europe, both public and domestic, already supports the ISO 15118 communication, and this communication protocol is already a requirement for IRA-funded public chargers in the US, and debate is ongoing in the EU over a potential mandate. Albeit not all implementations – even to the latest revision – necessarily support bidirectional operation, any widespread V2X implementation in the aforementioned regions would likely need to be based on ISO 15118-20 (or above).

5.4.3 Battery degradation due to V2X usage

Battery degradation, in the context of V2X operation, has been a constant topic of research, with an immediate impact on the feasibility, and public perception, of V2X technologies. It is well established in the literature that the lithium-ion rate of degradation is governed by so-called ageing stress factors, such as resting state of charge, depth of discharge, charging/discharging current, temperature and capacity throughput. Hence, one of the most common criticisms of V2X is that, by driving additional battery cycles, it has to contribute to EV battery wear.

However, as our understanding of the intrinsic mechanisms that determine battery deterioration improves, so does the ability to forecast the impact of specific V2X cycles on battery longevity. For example, simulation-based studies from the National Renewable Energy Laboratory (NREL) in the US determine that V2G high cycle operations only impact 5% of battery health over 10 years. Nissan, one of the most active automotive manufacturers in promoting V2G technologies, extends their EV battery warranty to V2X usage, provided that the cycles the battery is subject to are pre-approved by Nissan themselves.

Multiple real-world trials have been conducted to understand the effect of provision of grid services on battery lifetime, coming in some cases to dramatically different conclusions [26,27]. Key to put these results into perspective is under-standing under which conditions the results were generated – namely, the way the battery is managed during the V2X operation. In fact, as highlighted in [27], there is no doubt that a strategy for V2G operations that only seeks to maximise the grid service provision, with no consideration for the preservation of battery life or EV functionality, is heavily detrimental to battery life. Nevertheless, if V2X control algorithms are set up with consideration of the effects of service provision on the EV battery life, the effect of battery degradation can be mitigated [27,28]. In addition, in case a more in-depth knowledge of the EV battery system structure and operations is available, smart control algorithms can be defined to allow V2X service provision while at the same time operating the EV battery in the best pos-sible condition: in this case, by intelligently controlling if and when to provide energy from the EV to the home or Grid, and vice versa, battery life could even be extended compared to a base case of totally unmanaged EV charging.

5.5 Business models

So, while technical solutions exist and have repeatedly proven the viability of the concept, the bottleneck for commercial deployment has long been the business case associated with V2X propositions.

In earlier sections, we have already covered different services being targeted with V2X solutions. The promising FFR market – the main target market for the majority of first V2G trials – has dropped in value, and recent implementations have targeted arbitrage or DSO services. In reality, given the ongoing transition to a more interactive (and therefore dynamic) grid, the most effective strategy is to stack several types of services and continuously optimise the portfolio of targeted services depending on current market conditions.

Note that the composition of services depends on the specific market structure and scope of the aggregator/VPP. For example, dedicated V2G aggregators may target only services requiring low latency (allowed from Li-Ion batteries), while VPP aggregating different types of resources (e.g. including solar PV, wind gen-eration, DSR, mechanical storage, etc.) may choose more general services – and compensate the shortcoming of one type of resource with other resources within the aggregation.

Moreover, despite V2X bringing significant overall system advantages [3,4], the system-wide savings estimated with a top-down approach are not always reconcilable with the benefits an individual customer may perceive. In fact, it is not yet clear who will benefit from the system-wide savings and how to quantify them. Some of the most tangible advantages derive from maximisation of renewables utilisation: while these sources are cheaper than traditional generators, and the reduced reliance on expensive and polluting reserves does determine savings – which are included in the whole-system analyses – not all these benefits are directly accessible on the energy markets, and therefore might not entirely propagate down the supply chain. Furthermore, some of the biggest advantages from a more flexible system are perceived at a local level in the distribution networks: while it is expected that, with the transition from DNO to DSO and the establishment of local flexibility markets, V2X will be a prime candidate to provide such flexibility, most of the DSO services are still in their initial stages, with pricing not yet defined and potential volume of service required not yet clear.

The composition of services served, the distribution of value across the value chain and the definition of the customer propositions are critical for V2X deployment. EVs are a means of transportation, and that still is their primary use: finding ways to engage and keep engaged the customer is key to the success of the technology. With this outlook, the customer proposition is the biggest uncertainty, but also the area where there are more opportunities and possible choices. In the following, we will explore different approaches adopted from trials and commercial V2X deployment to offer customer propositions.

5.5.1 How to package services to customers?

Perhaps the most immediate customer proposition is to have a dedicated, 'hands on' V2G offering – where the customer has full visibility and control over the energy flows in and out of the EV. This has for example been adopted by companies like Nuvve and Fermata – both in the US and in Europe. In this case, the energy transferred to the EV and the one provided back to the grid are presented to the user as separate flows: the user pays for the energy to the vehicle and receives a reward (minus a fee from the service provider) when participating into V2G services, determining effectively separate transactions. Traceability of both energy flows and revenues is required up to national or local markets, in order to provide valid and updated information to the user. This approach is more indicated for advanced users and might not be applicable to all customers.

A different approach on the opposite side of the spectrum is focused on bundling – applied, for example, by Octopus Energy in their UK trial Powerloop. In this case, the basic principle is to make V2G participation it as easy as possible for the user (specifically, a domestic customer). This 'hands-off' approach presents the customer with an all-inclusive monthly payment figure, which includes lease payments for the EV and its maintenance; repayment for a V2G charger; a renewable energy tariff; cashback when the EV is plugged in during peak hours and available to provide V2G services. The advantage of such proposition is in the appeal to users who do not want to be heavily involved in the technology itself, can easily relate to the monthly payment format and can therefore directly quantify their benefit.

A middle ground is an offering integrating an energy tariff with V2G offering. In this case, the customer would see a single bill containing all energy transactions, and the V2G service provision focuses primarily on arbitrage. This is, for example, what Octopus Energy is proposing with a dedicated V2G tariff, and in line with Renault plan for V2G service launch. In some cases, trials have adopted this approach to offer customers a guarantee of 'free miles': essentially guaranteeing them to drive 'at no additional cost' – as the cost of EV charging is recouped via V2G grid trading.

Other approaches to V2G include a dedicated export tariff (similar to Solar PV dedicated tariffs), or the adoption of DSR-specific programs – for example, the utility-arranged ELR programs in California. Some EV OEMs are also getting directly involved in structuring the platform and service offering for V2X from their products: from Tesla having already a grid services trading platform ('Autobidder', initially developed for their storage solution), to Volvo developing a VPP with DCbel in the US.

It is worth noting that the above models are all focused on V2G. Albeit recent trials (e.g. from Indra in the UK) have focused more on V2H optimisation as a service, normally V2H is still perceived as a 'feature' requiring an initial capital investment (similar to home storage or solar PV), rather than a 'service' – particularly where the focus is on energy independence and resilience (e.g. blackout support or solar PV integration).

5.6 Regulations

Regulations play a crucial role in the development and deployment of V2X by providing a direction of travel in a still nascent and fragmented market.

The European electricity systems are on a journey to modernisation and decarbonisation – increasing the need for system flexibility. While the focus has so far been on large-scale grid-connected storage systems, with growing penetration of EVs these are increasingly seen as part of the energy flexibility landscape [3]. However, from regulators' perspective, this is not purely an opportunity to maximise already deployed assets, but also to evolve to a fairer system – where the end users are direct contributors to and beneficiaries of the much-needed flexibility.

Both the UK and the EU require already that EV charging points need to be connected and smart – allowing the EVs to move the charging away from peak power demand. Regulators in both regions are looking favorably to bi-directional charging as well: while not mandated at the moment, it is seen as a crucial piece of the future energy system.

In the EU, recent regulations are intended to prepare the bloc to achieve a transition to net zero – and explicitly touch upon the role of EVs and EV charging:

- The **Alternative Fuels Infrastructure Regulations** provide targets for the deployment of EV charging points. It crucially also introduces reporting requirements for each Member State – every 3 years starting from 2024 – of assessments of how the EV charging infrastructure allows EVs to contribute to the flexibility of the electricity system: this will need to include also bidirectional charging.
- The **Energy Performance of Buildings Directive** focuses on measures to make buildings more energy efficient. It also includes measure to make deployment of renewable sources and EV infrastructure easier in both residential and non-residential buildings – and requiring that EVSE are smart (and bidirectional if appropriate) and crucially operated on the basis of non-proprietary and non-discriminatory standards and protocols.

- The **Renewable Energy Directive** contains measures to accelerate the large-scale deployment of renewable energy generation. It also makes provisions for facilitating the integration of renewables into the electricity system, aiming at maximising self- and local consumption of locally generated green energy, including via demand response from EVs. Furthermore, introduces requirements for data on renewable generation to be freely available – so that EV chargers, smart meters, and market operators can use them in their operations.

A 2023 policy paper from the UK Government [29] sets out the intention to build an evidence base on how smart and bidirectional could reduce energy costs – and use this to inform future policy. Moreover, it contains the commitment for the energy regulator (OFGEM) to work with the DNOs to streamline – and make consistent across regions – the connection process for V2X.

Despite these advancements, there are still barriers that prevent widespread commercial deployment of V2X solutions. A paper from SmartEN [21] assesses the regulatory framework of several European markets for their ability to support the deployment of V2X solutions with crucial policy and market recommendations. These include:

- Removal of double taxation on energy storage,
- Development of tariffs that encourage the utilisation of renewable energy,
- Open access to energy markets for V2X aggregators,
- Procure flexibility locally in competitive markets open to V2X contributions.

5.6.1 Ongoing regulatory activities

One of the findings of [21] is that there needs to be a harmonisation and a modernisation of grid codes across Europe. Grid codes define the requirements for devices connection to the power system, and, currently in the EU, these are not aligned among Member States nor fully encompass the range of services (from DER with Grid-to-Vehicle (V1G) to storage with V2G) that V2X-enabled vehicles can provide.

Similarly to OFGEM in the UK, ACER (the EU Agency for Cooperation of Energy Regulators) is attempting to establish baseline requirements across the bloc. Furthermore, in 2024, the EU Commission is consulting on updates to the network Code on Demand Connection (DCC) and Requirements for Grid Connection of Generators (RfG) – to introduce requirements for EVs and their ability to contribute to the power system operations [30]. However, the proposed updates are not without issues, and discussions in the industry are still ongoing. Specifically, the new DCC (which look at the demand side, so mainly covers the unidirectional charging process, including V1G) and RfG (looking at generators, so also encompassing V2G) rules look at the EV and EVSE as one system: while this makes sense from electricity system operators' perspective, not defining a functional allocation risks opening interoperability issues, as well as neglecting objective limitations of automotive and communication standards EVs and EVSEs are currently designed to.

Finally, cross-industry collaboration forums are actively moving to define joint vision and principles for the deployment of V2X in Europe. For example, in Germany, automotive, energy and charging players came together to define a roadmap for the introduction and deployment of V2X solutions, including proposals for interoperable and fair data exchange across all players [31]. Similar activities are taking place at the EU level via ACEA.

5.7 V2X in Europe: looking forward

As we have seen, great strides have been made in Europe on V2X technologies, with technologies now ready, regulations being updated to create a more V2X-friendly framework, and commercial propositions reaching the mass market in 2025. However, there are still challenges and open points that will require further consideration going forward.

As a first application in a novel market segment, regulations for V2G technologies are still being written and debated – with an expected rapid iteration and evolution over the next few years. So providers of V2X solutions (and regulators) will need to look at how to ensure compliance, but also consistency – in order not to leave early adopters with stranded equipment and investments.

Moreover, not only the regulations landscape is evolving rapidly, but also the market services and the consumer propositions. This will require EV and EVSE OEMs to provide implementations that not only work today, but are updateable enough to ensure longevity and futureproofing.

Finally, competition in this space is heating up. This on one end would require market players to seek differentiation – while on the other regulations are seeking standardisation in order to ensure interoperability. V2X market players will need to carefully balance these two drivers.

References

[1] Energy Technologies Institute, "Electricity Distribution Network Assessment and Analysis", 2011. URL: https://ukerc8.dl.ac.uk/ETI/PUBLICATIONS/LDV_TR1002_15.pdf. Accessed August 2024.

[2] RAC Foundation, "Standing Still", 2021. URL: https://www.racfoundation.org/wp-content/uploads/standing-still-Nagler-June-2021.pdf. Accessed August 2024.

[3] OVO Energy, Imperial College London, "Blueprint for a Post-Carbon Society", 2018. URL: https://www.ovobyus.com/m/1d7f8a93e4685f9f/original/blueprintforapostcarbonsocietypdf-compressed.pdf. Accessed August 2024.

[4] IRENA, "Innovation Outlook: Smart Charging for Electric Vehicles", 2019. URL: https://www.irena.org/-/media/Files/IRENA/Agency/Publication/2019/May/IRENA_Innovation_Outlook_EV_smart_charging_2019.pdf. Accessed August 2024.

[5]   National Grid ESO, Octopus Energy group, "Powerloop: Trialling Vehicle-to-Grid technology", 2023. URL: https://www.nationalgrideso.com/document/281316/download. Accessed August 2024.

[6]   UKPN, "Smart Charging Architecture Roadmap", 2018. URL: https://d1oyzg0jo3ox9g.cloudfront.net/app/uploads/2023/10/UKPN-Smart-Charging-Architecture-Roadmap-Final-Report.pdf. Accessed August 2024.

[7]   ACEA, "European EV Charging Infrastructure Masterplan", 2022. URL: https://www.acea.auto/files/Research-Whitepaper_A-European-EV-Charging-Infrastructure-Masterplan.pdf. Accessed August 2024.

[8]   IEA, "Global EV Outlook 2024", 2023. URL: https://www.iea.org/reports/global-ev-outlook-2024. Accessed August 2024.

[9]   Centrica, "British Gas Research Shows Brits Confused Over Green Energy", 2021. URL: https://www.centrica.com/media-centre/news/2021/british-gas-research-shows-brits-confused-over-green-energy/. Accessed August 2024.

[10]   Accenture, *The New Energy Consumer: Thriving in the Energy Ecosystem*, 2016.

[11]   EY, "As Consumers Lead the Way, How Can Energy Providers Light the Path?", 2021. URL: https://www.ey.com/en_uk/power-utilities/how-energy-providers-can-light-the-path. Accessed August 2024.

[12]   Plug In America, "The Expanding EV Market: Observations in a Year of Growth", 2022, URL: https://pluginamerica.org/wp-content/uploads/2022/03/2022-PIA-Survey-Report.pdf. Accessed August 2024.

[13]   Indra, "Vehicle-to-Home (V2H) Chargers", 2024. URL: https://www.indra.co.uk/v2h/#:~:text=Our%20vehicle%2Dto%2Dhome%20(,to%20reduce%20your%20carbon%20footprint. Accessed August 2024.

[14]   Octopus Energy, "Octopus Energy Power Pack: The UK's First Vehicle-to-Grid tariff", 2024. URL: https://octopus.energy/power-pack/. Accessed August 2024.

[15]   Cenex, "Project Scirius Trial Insights: Findings from 300 Domestic V2G Units in 2020", 2021. URL: https://www.cenex.co.uk/app/uploads/2021/05/Sciurus-Trial-Insights.pdf. Accessed August 2024.

[16]   Cenex, "Understanding the True Value of V2G", 2019. URL: https://www.cenex.co.uk/app/uploads/2019/10/True-Value-of-V2G-Report.pdf. Accessed August 2024.

[17]   Element Energy, "Vehicle to Grid Britain project report", 2019. URL: https://esc-production-2021.s3.eu-west-2.amazonaws.com/2021/07/V2GB-Public-Report.pdf. Accessed August 2024.

[18]   Cenex, "A Fresh Look at V2G Value Propositions", 2020. URL: https://www.cenex.co.uk/app/uploads/2020/06/Fresh-Look-at-V2G-Value-Propositions.pdf. Accessed August 2024.

[19]   National Grid ESO, "Net Zero Market Reform: Investment Policy", 2023. URL: https://www.nationalgrideso.com/document/299456/download. Accessed August 2024.

[20]   EU Commission, "Green Deal Industrial Plan", 2023. URL: https://commission.europa.eu/strategy-and-policy/priorities-2019-2024/european-green-deal/green-deal-industrial-plan_en. Accessed June 2024.

[21]   SmartEn, "V2X Enablers and Barriers: Assessment of the Regulatory Framework of Bidirectional EV Charging in Europe", 2023. URL: https://smarten.eu/wp-content/uploads/2023/12/V2X-Enables-abd-Barriers-Study_11-2023_DIGITAL.pdf?utm_source=www.readelectricavenue.com&utm_medium=referral&utm_campaign=v2g-in-europe-what-s-the-hold-up. Accessed August 2024.

[22]   W. Kempton and S. Letendre, "Electric Vehicles as New Source of Power for Electric Utilities", *Transportation Research*, vol. 2, pp. 157–175, 1997.

[23]   M. Landi, "Vehicle-to-Grid Developments in the UK", EVS32, 2019.

[24]   Innovate UK, UKPN, "Vehicle to Grid Hub". URL: www.v2g-hub.com. Accessed August 2024.

[25]   IREC, "Vehicle to Grid (V2G) Standards for Electric Vehicles", 2022. URL: https://irecusa.org/wp-content/uploads/2022/01/Paving_the_Way_V2G-Standards_Jan.2022_FINAL.pdf. Accessed August 2024.

[26]   N. S. Pearre and H. Ribberink, "Review of Research on V2X Technologies, Strategies, and Operations", *Renewable and Sustainable Energy Reviews*, vol. 106, pp. 61–70, 2019.

[27]   K. Uddin, M. Dubarry, and M. B. Glick, "The Viability of Vehicle-to-Grid Operations from a Battery Technology and Policy Perspective", *Energy Policy*, vol. 113, pp. 342–347, 2018.

[28]   M. Landi and G. Gross, "Battery Management in V2G-Based Aggregations", *Power Systems Computation Conference*, 2014.

[29]   UK Government, "Electric Vehicle Smart Charging Action Plan", 2023. URL: https://www.gov.uk/government/publications/electric-vehicle-smart-charging-action-plan/electric-vehicle-smart-charging-action-plan. Accessed August 2024.

[30]   ACER, "Connection Codes", 2024. URL: https://www.acer.europa.eu/electricity/connection-codes. Accessed August 2024.

[31]   NOW Gmbh, "Facilitating Bidirectional Charging Nationwide", 2024. URL: https://www.now-gmbh.de/en/news/pressreleases/facilitating-bidirectional-charging-nationwide/. Accessed August 2024.

Chapter 6

Law and regulation of V2G interactions

Zsuzsanna Csereklyei[1] and Anne Kallies[2]

6.1 Introduction

To address increasing environmental and economic challenges in the wake of global warming, governments across the world began ratifying pledges and regulations to limit global warming. The Paris Agreement (COP21) [1] aims to limit global average temperature rise to no more than 1.5 °C of pre-industrial levels. The agreement has introduced the novel concept of nationally determined contributions, through which states pledge their greenhouse gas mitigation targets. Noteworthy, the total mitigation efforts pledged so far do not add up to the reductions needed to limit temperature increases to no more than 1.5 °C. While a 'ratcheting up' mechanism requires a gradual tightening of the pledges over time, states will have to considerably upgrade their mitigation commitments to align with Paris goals [2].

The decarbonisation of the transportation sector, which accounts globally for one-quarter of greenhouse gas emissions [3], is now increasingly in the focus of policymakers as a central part of their decarbonisation efforts. Transport sector electrification provides new opportunities to not only lower emissions due to transportation, but also to interact with and support the decarbonisation of the electric grid. The introduction of vehicle-to-grid (V2G) applications and the bi-directional flows to and from the grid associated with it provide an opportunity to harness the capabilities of decentralised storage in the form of electric vehicle (EV) batteries.

The uptake of EVs, including V2G-enabled EVs, can be supported through multiple instruments by policymakers, including tax relief or other financial support through vehicle purchasing credits, vehicle emission performance standards and other environmental mandates, or EV targets. States may also support investment into public charging infrastructure or require labelling of vehicle fuel economy and emissions. For a summary and overview of possible policy measures, see [4].

Given the increasing interconnectedness of sectors, decarbonisation measures will have to be carried out simultaneously in the electricity generation and

[1]School of Economics, Finance and Marketing, RMIT University, Australia
[2]Graduate School of Business and Law, RMIT University, Australia

transportation sectors [5]. Infrastructure investments in both sectors will have to be planned jointly to enable widespread and affordable transport electrification in a manner that best suits the electricity system, avoiding, among others, worsening peak demand, higher capacity requirements and higher electricity prices.

At the time of writing this book, most countries are in the early phases of EV adoption. While home and some public charging is enabled and usually available (though the presence of public charging stations, which act as key enablers of EV uptake, is still limited), the regulation, registration and administrative requirements of EVs and their interaction with the grid are diverse across different jurisdictions. Discharging to the grid is not widespread or universally enabled, and regulatory barriers are central to the slow uptake of these capabilities for EVs. There are also major differences in whether and how EV owners can participate in electricity wholesale and ancillary markets directly or through an intermediary.

This chapter introduces an overview of regulatory frameworks enabling or providing barriers to the introduction of V2G interactions for EVs with a focus on their interaction with the shared electricity system. The chapter also explores key regulatory questions and dilemmas that will have to be addressed in the future.

6.2   EVs and the electricity system

How EVs (both privately and commercially operated) will interact with the electricity system in the future will have a significant impact on the intraday distribution of electricity demand as well as on generation and transmission requirements. The charging (and potentially discharging) behaviour of individual EVs will influence the electricity market, and the market settings, in turn, will influence these behaviours. Most EV owners will interact with the system through an intermediary, such as an aggregator or a retailer. EVs connect to the distribution grid and therefore also interface with grid operators. The next sections give an overview of the most common market frameworks, institutions and market participants and how these settings provide opportunities or barriers to the use of V2G services by EVs.

6.2.1   Standard market frameworks, institutions and participants

Electricity market regulatory frameworks and their interactions with end users and owners of grid-connected (connectible) assets can vary widely. Electricity market design is often jurisdictionally unique but can be broadly categorised into rate-of-return markets (non-liberalised markets), liberalised markets or a combination of these. The underlying design of the market sets forth the fundamental framework within which market participants such as generators, batteries or other grid-connected assets such as EVs may or may not interact with the market.

Traditionally, rate-of-return markets (also called cost-of-service models) were widespread and often characterised by vertically integrated utilities, generating and supplying electricity to customers for a price that allowed utilities to recover their 'prudently incurred operating costs and a regulated return on capital investment' [6, p. 438].

The 1990s brought sweeping regulatory changes in the electricity and energy sector in most Western countries, including the member states of the European Union, Australia or parts of the United States. This included a push towards the liberalisation of electricity markets with the fundamental goal of increasing efficiency. Liberalisation was characterised by the legal and administrative unbundling of network functions from generation and retail markets (or supply) [7]. This process however was not uniform across countries. Within liberalised (or in the US often called 'restructured') wholesale electricity markets (WEM), several designs emerged, especially with respect to the presence of capacity markets or other forms of capacity mechanisms. Liberalised wholesale markets theoretically allow unhindered market entry into the electricity generation landscape.

Retail market restructuring or liberalisation has been much slower and only partially successful. Importantly, retail competition is often considered a precondition for innovative new products and approaches, including aggregation, demand-side participation and other flexibility measures [8], which would include services facilitating EV V2G participation. For example, EU directive 2019/944 states in its preamble para (1) that 'Healthy competition in retail markets is essential to ensuring the market-driven deployment of innovative new services that address consumers' changing needs and abilities, while increasing system flexibility'. Retail competition allows retailers to differentiate their offerings by providing additional services beyond electricity supply.

Case study: Examples of different market designs across the world

In the **United States**, wholesale market liberalisation or restructuring has been regulated at the federal level, with FERC orders 888 and 899 introducing the concept of independent system operators (ISOs), followed by FERC Order 2000, which encouraged utilities to join Regional Transmission Organisations (RTOs). A number of ISOs/RTOs were established in the 1990s–2000s. Csereklyei and Stern [9] noted the influence of continued strong state control over utilities, stating that 'the transition to wholesale competition was most likely to occur in regions or states where the average cost of generation was above marginal cost, such as the North-East, Illinois, Texas, California, and Montana (Borenstein and Bushnell, 2015). Regions with comparatively low electricity rates at the end of the 1990s, such as the Southeast or the Northwest largely resisted restructuring efforts' [9, p. 161]. The choice of capacity mechanisms or the absence thereof within particular ISO/RTOs also differs by region. Retail market liberalisation on the other hand is regulated at a state level and so far only offered in a limited number of states [10].

In the **European Union**, market liberalisation was introduced in a step-wise fashion through a series of European Directives with the aim of creating a harmonised internal European energy market. These included three so-called energy packages, which contained the following central electricity directives: Directive 96/92/EC, Directive 2003/54/EC and Directive 2019/944. While large differences between member states' markets persist, member states do now have to provide non-discriminatory third-party access to networks, separation of networks from up- and downstream markets, and a free choice of supplier. In practice, retail competition has been slow to emerge in some member states [11]. A number of member states continue to support capacity mechanisms [12].

Australia, a federated country, displays a variety of market designs, including the National Electricity Market (NEM), which serves the vast majority of Australian energy consumers on the Eastern Coast, the Western Australian Wholesale Electricity Market (WEM), and the Interim Northern Territory Energy Market. The NEM is an energy-only wholesale (generation) market without a capacity market, with a regulated network component and retail choice in most areas. The WEM is a liberalised wholesale market with a capacity market, while the Northern Territory operates three regulated power systems, which facilitate bilateral contracts between generators and suppliers, and has recently introduced full retail competition.

6.2.1.1   Wholesale electricity markets

Liberalised wholesale markets, with the objective to ensure and increase the efficiency of the short- and long-run allocation of resources [13], rely fundamentally on the concept of private investments in electricity generation. Generators and other participants in wholesale markets – such as batteries charging and discharging, including registered utility-scale or other batteries (including EVs), or owners of distributed energy resources (DER) that are able to participate in the market through some mechanism – place bids to supply electricity to the market. The electricity price will be determined by the last bid (marginal bid) that is required to meet demand. While the price theoretically is the 'marginal price' of generation, in practice the clearing price may be below, at or above the marginal price. Liberalised wholesale markets thus work through 'price signals' to ensure both the short-run (efficient real-time operation and market clearing) and long-run (long-run profit expectations, efficient investments and retirements) resource allocation functions [13].

Despite its liberalised nature, there are frequent government interventions in liberalised electricity markets, such as price caps, cumulative price caps and out-of-market operations by the system operator. Interventions are usually designed to prevent or are undertaken if the market regulator or the operator perceives a threat to the safe, secure and affordable provision of essential services. The presence of price caps however may result in insufficient investment incentives, especially for investment in peaking generation. The appearance of new technologies and participants, such as the widespread introduction of zero-marginal cost intermittent renewable technologies, or the appearance of prosumers ('people who both consume and produce energy' [14]), and in the future EV owners are significantly disrupting these markets.

Understanding how different wholesale market designs may or may not facilitate the uptake of electricity from grid-connected assets (such as EVs) is crucial. In liberalised markets, which are built on the premise of non-discriminatory access to markets for generation and load, EV market participation can be facilitated for example through an aggregator. In contrast, in rate-of-return markets, vertically integrated utilities responsible for generation, transport and supply to the consumer can act as gatekeepers to new participants. Under particular regulatory settings, EVs would be able to provide demand response in both markets (either through demand response mechanisms or reacting to retailer incentives and signals).

6.2.1.2   Markets for network services

Networks in a liberalised electricity system – as natural monopolies – are usually regulated businesses around the world. Grid operators are tasked with ensuring that demand is met at all times, and that the stability of the grid is maintained. They can do this by investing in the necessary grid infrastructure and, if necessary, may contract network services or rely on ancillary markets 'to manage the power system safely, securely, and reliably' [15].

A lack of monetisation for ancillary services has been identified in the literature as limiting the ability of V2G applications to deliver benefits [16]. This issue is impacting the economic viability not only of V2G interactions but also of other battery storage technologies [17].

V2G interactions provide both a challenge and an opportunity for grid operators. Network investment requirements will be impacted by EV operations, both charging and discharging, at both the transmission and the distribution network level. Depending on the regulatory settings in the relevant jurisdiction, and especially the degree of unbundling between supply and distribution network functions, EV owners may interface directly with the distribution system operator (DSO) or act through an intermediary, such as an aggregator or retailer. These options are expanded on in section 6.2.3.

6.2.1.3   Retail markets

Retail market liberalisation refers to the ability of customers to choose their retailers and, if available, a customised bundle of services these retailers may be able to provide.

Prosumers, and owners of grid-connectible assets, such as V2G-enabled EVs, will most likely interact with the retailer itself, or potentially with another intermediary. Retailer tariffs and contractual arrangements will therefore determine at what rate EVs can charge from the grid and feedback into the grid. Tariffs are considered a central instrument to incentivise a system-friendly timing of charging from and feeding to the shared network [18]. Their design and impact will be discussed further in Section 6.2.3.

6.2.1.4   Market institutions and participants

A modern liberalised electricity market relies on several market institutions to support its operation, management and timely investment. As electricity supply is considered to be an essential service, tight regulatory frameworks ensure safe, secure and reliable supply, in particular of network-bound electricity. Consequently, institutional frameworks will usually include an economic regulator to ensure that network investment is prudent and a market operator to match demand and supply and, potentially, also operate ancillary markets. Competition watchdogs and consumer protection entities further ensure that the powerful position of a retailer or network is not abused. These functions can be provided by a number of different public or private entities.

EVs seeking to use V2G capabilities face similar challenges to other DER seeking to interact with the shared network, such as prosumer resources including home batteries and rooftop solar generation.

Under the concept of V2G, EVs will export electricity to the distribution system. Thus, the role of the DSO or distribution network operator (DNO) will be central to the successful integration of any DER. The integration of new energy sources at the distribution level will shift the responsibilities of DSOs to include balancing and frequency control services. Traditionally, these services are the responsibility of the transmission system operators (TSOs) [19], whereas DSOs, overseeing a one-directional flow of electricity, did not have an active role in system management [20]. The new reality of bi-directional flows in the distribution systems, accelerated by the uptake of rooftop solar panels, is now requiring a rethink of regulatory arrangements. Burger *et al.* [21] discuss two main options of either making the DSO responsible for 'distribution-level balancing' or expand the scope of the TSO to include these responsibilities but conclude that a mixed model with increased communication between the two levels was the most likely [21]. Different jurisdictions seek to incentivise distribution networks to facilitate these future electricity flows in different ways.

Case study: Network innovation mechanisms

The UK has introduced network innovation funding for distribution network operators with their RIIO (Revenue = Incentives + Innovation + Output) approach to network price controls [22].

In **the US state of New York**, active roles for DNOs are enabled through incentivising distribution network utilities to provide better customer engagement, peak load reduction, and, through this, energy efficiency and affordability [23].

In **the Australian NEM**, the regulator seeks to incentivise innovation in the distribution network through a demand management incentive scheme, which could lead to the emergence of competitive markets for network services [24] – markets that could be accessed by EVs providing V2G services.

6.2.2   *Defining storage and its impact on V2G capabilities*

In order for V2G-enabled EVs to interact with electricity markets, whether wholesale or ancillary, or to provide network services, they need to be recognised as a participant in these markets. V2G-enabled EVs are multi-functional assets. A current German expert study sees potential for participation in intraday and day-ahead electricity markets, as well as network service provision for transmission and distribution networks, all facilitated through intermediaries, such as aggregators [25]. Alternatively, entirely new categories may be created to facilitate market participation for EVs [16].

Electricity markets traditionally clearly distinguish between electricity generation and consumption (load). Storage technologies, more generally, with their ability to provide multiple services across a number of traditional asset classes, struggle to fit into these systems. Csereklyei *et al.* [17] and others [26] writing on utility-scale storage point out how the need to register across multiple categories is associated with additional administrative, financial and operative burdens. Similar burdens may be faced by V2G-enabled EVs, if they seek to interact with a number of markets. Therefore, clear definitions for storage technologies in the regulatory system would be conducive to enabling their market participation [16,17,26].

Case study: Defining storage in regulatory frameworks

Different jurisdictions have several categories for electricity storage.

The UK regulator, Ofgem, is treating storage as a defined subset of generation. The licence exemption regime for small generation also includes small batteries operating at a domestic level [27].

In Australia, batteries smaller than 5 MW are exempt from a requirement to register as a generator. Nevertheless, participation in the market through a small generation aggregator does require classification as 'market generating unit' or 'market load'. From June 2024, the aggregator is replaced by a new category of market participant, the integrated resource provider (IRP), which can submit load and generation at the same time, as well as participate in ancillary service provision. An IRP will be able to aggregate multiple EVs (to be classified as non-scheduled bidirectional units) to provide V2G services in any of the service categories [28].

6.2.3   *Interaction of electric vehicles with the electricity grid and market*

This section will provide a more detailed overview of the interactions that V2G-enabled EVs will have with the grid and different available markets.

According to Amani *et al.* [5], behavioural choices of EV owners with respect to the timing of their charging and discharging actions will have a significant impact on the intraday distribution of demand worldwide, which in turn will impact electricity prices, future generation capacity and mix (including storage) requirements as well as network (transmission and distribution) infrastructure requirements. The location of EV demand will have a significant bearing on distribution networks and can create potential congestion if not handled correctly.

It is important to remember that EVs are not just storage assets, but also a means of transport. These two potential purposes of an EV are in competition and will depend on the user profile and requirements of the EV owner. For example, a business fleet vehicle used by a staff member providing in-house nursing services during the day is likely not available for V2G charging during daytime. A private EV, mostly used for weekend trips and school pick-ups, could provide V2G services at peak times.

One of the central tasks of policymakers will be to steer clean energy transitions in transport electrification and incentivise the right charging (discharging) behaviour both locationally and temporally.

6.2.3.1   The role of tariffs in charging and discharging

Widespread EV adoption in the future by both residential or small electricity users (households) and large electricity users (e.g. organisations likely operating a fleet) is expected to result in a significant additional demand in peak electricity use, assuming convenience charging (6–12 p.m.).

Charging behaviour

As most customers will interact with retailers while charging (or purchasing electricity), the key economic and policy tool to incentivise an intraday shift in demand is customer tariff regulation. Customer tariff regulation differs widely around the world and can range from flat tariffs (usually for residential customers) to time-of-use (ToU) tariffs (usually for industrial customers). In some jurisdictions, residential customers have the choice between flat and ToU tariffs or can opt to receive wholesale market prices if the regulation allows retailers to pass this on. Many jurisdictions implemented a ceiling on electricity end-user prices – such as the Default Market Offer in Australia – to protect customers.

Residential EV owners who are not required to adopt ToU tariffs would receive a flat electricity price throughout the day, which could incentivise charging at 'convenient times', usually adding to peak demand after 6 p.m. Therefore, it is important that future tariff design should take into account the generation mix, the marginal emissions intensity, as well as the infrastructure requirements of the electricity system. For example, in a state or country with abundant solar energy generation, charging during the middle of the day is both the cheapest (from a wholesale market point of view) as well as the least emissions intensive. Despite this, tariff design may incentivise night-time energy use, which can be suboptimal in the face of a rapidly changing electricity mix [29].

Case study: Public transport electrification in Victoria, Australia

In their study on the electrification of Victorian public transport, Say *et al.* [29] find that the current large customer tariff system predominantly incentivises night-time charging, which is fundamentally more carbon-intensive than day-time charging in Victoria, Australia. Furthermore, prevailing tariff systems were found to mute wholesale market signals (of supply availability and scarcity). They recommend to co-optimise power and transport sector transitions by ensuring greater pass-through of wholesale market pricing signals.

Therefore, future regulation should pay particular attention to tariff design both for residential and large customers [5], to enable temporally optimal charging patterns. The availability of public charging infrastructure throughout key locations is however a pre-requisite for the shift in intraday charging patterns. Without the option to charge during the day (due to the lack of infrastructure), charging will still happen at home during peak hours. Therefore, infrastructure investments must be implemented hand in hand with tariff design to enable its effects.

Discharging behaviour

EVs discharging into the grid is a concept that is not yet enabled in most countries. Exemptions include countries, where EVs may participate in the market through private markets (see case study on private markets for network services below). The owners of grid-connectible assets, such as EVs, in the future, may be in contact either with

- retailers (who may offer a feed-in-tariff of some sort),
- with an aggregator who will act on their behalf to participate in the wholesale or in ancillary markets, or
- the DSO

Their charging and discharging behaviour is likely to be incentivised by economic rationale if arbitrage between these is enabled. However, how much an EV owner will be able to receive for the electricity discharged into the system will depend on regulatory settings, and whether they are interacting with the retailer/aggregator or with the market directly or with the DSO.

6.2.3.2   Interactions with wholesale and ancillary markets

In their assessment of the Australian regulatory framework for the future enablement of vehicle-to-everything operations, Amani *et al.* [5] state that the first step in understanding how an EV (or a grid-connectible asset) is interacting with the system is to define the fundamental purpose of the technology. They pose a number of additional questions, including whether the 'use of EV as battery is feasible and desirable from a system and stakeholder point of view' and if 'EVs can be effectively incentivized to participate in demand side response (e.g. load shifting)?' [5, p. 16].

Amani *et al.* [5] identified several key regulatory issues regarding the interaction of EVs with the wholesale market, including whether EVs should interface with the market in any capacity other than a consumer, whether and how these interactions need to be facilitated and which entity could facilitate, as well as the differences between charging locations, such as workplaces, freestanding homes or apartment buildings. They also raise the impact of locational charging, including differences between charge and discharge location as a potential future regulatory query. They finally raise questions on the ability of market operators to plan for and (if necessary) constrain EV interaction with the electricity system. From the viewpoint of the wholesale market EVs may appear theoretically either as:

- Behind the meter or
- In front of the meter (if aggregated).

Most EVs will always be physically connected to the system in a behind-the-meter manner. This means that they will either be perceived as load (while charging) in the system or potentially negative load (similar to small-scale generation) whilst discharging. Behind-the-meter assets are normally not controlled by the market operator and cannot participate in wholesale market activities. Their primary contact will be the retailer, who might offer incentives for demand management, such as shifting the timing of charging or discharging activities within the household.

The legal and economic question remains whether EVs could act through an intermediary similar to in front of the meter actors. This would only be possible in jurisdictions that enable both free wholesale market participation, and the participation of EVs through a retailer/aggregator.

6.2.3.3   Interactions with networks

EVs will interact (either through residential or public charging locations) with the distribution network. Distribution networks around the world were built with one-way electricity flow in mind and are already grappling with accommodating electricity uptake from DER (such as solar rooftop PV) [5]. The projected increase in residential electricity demand, as well as the future necessity to accommodate grid-connected assets charging back into the grid, will require an upgrade and expansion of existing distribution and transmission networks [5].

Fundamental questions which will need to be addressed concern the design of a future grid that is able to accommodate EVs and their locational and temporal charging demands and discharging, as well as which instruments can incentivise these developments [5]. In addition, the role of the network providers and DSOs needs to be reconceived [30]. Currently, the role of smart charging (and discharging) as well as the role of private markets enabling economic interactions between EV owners and networks is at different stages of development. One such example is provided in the case study at the end of this section.

Future distribution grid planning around the world will need to anticipate an increase in the number of EVs. Generally, network costs are recouped through all consumer electricity bills. Given that the network cost component of residential electricity bills is substantial, any developments and enhancements to the distribution grid should be planned in a way that minimises costs while maintaining system stability and security. This could include examining the role of locationally well-placed (either utility-scale or community-scale) batteries, which may act as avoided network investment [17]. Information on where DER including EVs are located in the grid can support grid management by operators.

Case study: Visibility of distributed energy resources

In the Australian NEM, a DER register is required by the National Electricity Rules (Rule 3.7E) to allow the system operator visibility of behind-the-meter DER in the system [31]. The stated purpose of the register is to allow for better management of the grid. It requires network service providers and installers to provide DER generation information, which is defined as data in relation to a small generating unit of a nameplate rating less than 30 MW which is owned, controlled or operated by a person, which is exempted from registration as a generator.

This includes storage devices such as home batteries as well as small-scale solar rooftop PVs. The determining factor is to capture potential flows of energy to the grid. Consequently, EV batteries with bi-directional capabilities are captured in the register. In addition, a current rule change proposal by the Australian Energy Market Operator is proposing to also collect standing data by EV supply equipment, in particular charging equipment [32].

EVs can be viewed as small-scale storage. If V2G interactions are enabled, they can provide a number of network services, as well as ensure electricity supply to households during blackouts. Further benefit to the grid can arise when discharging to the grid is enabled and incentivised at times of supply scarcity. In this case, potential exports from EV batteries could be aggregated and fulfil a similar role to community-scale batteries. Coordinating EV users however at this scale may have substantial administrative costs, compared to for example the building of a utility-scale battery.

If EV charging and discharging is optimised through tariffs or private markets, infrastructure investment into distribution networks can also be optimised, to avoid overinvestment, thus saving money for end-users.

Case study: Private markets for network services

Recently, private market solutions surfaced that allow connecting flexible sellers such as the owners of EVs and batteries, to electricity markets. Piclo (https://www.piclo.energy/), which operates a flexibility marketplace, for example, claims to 'enable flex buyers (system operators) to source flexible electricity from flex sellers during times of high demand or low supply'. The private company works with DSOs in several countries around the world, including the UK, US, several EU countries and Australia.

6.3   Other legal considerations

Several additional regulatory barriers and open challenges to the uptake of V2G applications have been discussed in policy and literature. This section shortly introduces the most important ones – in particular, the standardisation of charging infrastructure, legal considerations around battery degradation and privacy, as well as risks of double taxation and tariff raising associated with the bidirectional nature of V2G applications.

6.3.1   Charging infrastructure and legal considerations

The lack of standardisation of the technical requirements for metering, charging and connection to the grid has been identified as a regulatory hurdle by a number of commentators [18,33,34]. In Australia, for example, a rule change [32] to enable flexible trading arrangements was introduced, which includes simplifications of metering requirements, as well as allowing households to establish additional settlement points to separate flexible consumer resources, such as EVs from passive consumer loads.

In addition, the installation of vehicle charging infrastructure in apartment buildings or for renters remains a contentious area. Whether and how a resident of an apartment building or a renter (whether of a house or apartment) will be able to access future V2G capabilities of their EV, will depend on the availability and management of charging points in apartment buildings. This will require a considerable retrofit of existing building stock as well as new requirements for newly built homes and apartment buildings. Shared charging arrangements and how these allow users to reap the benefits from their EVs in a V2G charging scenario will also need to be developed. Solutions to these scenarios, for example, answering what happens to a renter-installed charging station at the end of the rental contract, are likely to be found in contract law.

> **Case study: Privileging charging infrastructure on common property in Germany**
> Germany has legislated in 2020 that renters and owners of apartments have a right to install charging infrastructure on common property, such as car parks, without the agreement of other owners [s20 of the Condominium Act (Wohnungseigentumsgesetz) and s554 of the German Civil Code (Bürgerliches Gesetzbuch)]. The costs of the installation are borne by the renter or owner seeking the installation. New builds will require the installation of charging infrastructure by law from 2025 as prescribed by the Building-Electromobility-Infrastructure Act (Gebäude-Elektromobilitätsinfrastruktur-Gesetz).

6.3.2   Consumer protections: warranties and reliability

Battery degradation from charging and recharging has been identified as a concern for users of V2G services [18,25]. EV battery degradation associated with frequent charging and discharging is tightly bound up with technical capabilities. Legally, a question arises on whether using an EV battery for services other than solely powering (charging) the EV could have an impact on warranties and the insurability of a vehicle. A recent expert report for the German government [25] expressly identifies the need for the standardisation of warranty conditions for EV batteries used for bidirectional capabilities.

6.3.3   *Taxation and double charging*

Finally, double taxation or double application of network tariffs when charging and discharging has been identified as a concern by a number of commentators [18], [25] and is a common concern for any type of electricity storage [16,18,26,35]. Gschwendtner *et al.* [18] consider this a question that may be avoided through a better definition of batteries – rather than defining them as generators and consumers, with associated taxation impacts. In Germany, one suggestion is to privilege mobile storage in a similar way to stationary storage (for which the network tariffs are only applied when discharging to the shared network according to s118 (6) of the German *Energy Industry Act*) [25]. The difficulty here is that charging and discharging V2G-enabled EVs may not happen at the same charging station, and innovative regulatory approaches to consider this challenge will have to be developed.

6.4   Conclusion

Regulatory frameworks to successfully integrate V2G-enabled EVs into shared networks are still developing. The key to the future use of V2G applications will depend on the ability to integrate these in a system-friendly way, while simultaneously acknowledging their primary use as a mode of transport.

The fundamental structure and regulatory design of electricity markets with which EVs are expected to interact determine the avenues through which EVs would be able to feed back into the grid. New products and services, supported by new market participants such as aggregators, and enabled by access to increasingly sophisticated smart equipment, will provide future owners of EVs with opportunities to use their vehicles not only for a more environmentally-friendly transport but also to support the rapid decarbonisation of the electricity sector.

Simultaneously, given the substantial additional electricity demand, the widespread uptake of EVs would be associated with, it is particularly important to manage and incentivise both the charging and discharging behaviour of EV owners through policies to avoid major increases in peak demand, prices and generation and storage infrastructure requirements.

Mechanisms most likely to influence consumer behaviour – conditional on the availability of sufficient locationally well-placed public charging infrastructure – are tariff structures, the incentivisation of flexibility markets, as well as the facilitation of participation in these markets through aggregators or similar models, providing for a two-sided energy market.

Both electricity supply and access to transport opportunities are considered to be essential services. Future research on regulatory models should also consider how regulatory choices can empower consumers without creating or perpetuating inequalities.

References

[1]   Paris Agreement under the United Nations Framework Convention on Climate Change, opened for signature 22 April 2016, [2016] ATS 24 (entered into force 4 November 2016).

[2]   United Nations Climate Change, 'Climate Plans Remain Insufficient' (Press Release, 26 October 2022), https://unfccc.int/news/climate-plans-remain-insufficient-more-ambitious-action-needed-now, accessed April 2024.

[3]   United Nations, 'Fact Sheet: Climate Change' (2021), https://www.un.org/sites/un2.un.org/files/media_gstc/FACT_SHEET_Climate_Change.pdf, accessed April 2024.

[4]   Kreeft, G., and Kuiken, D.: 'Chapter IX.49: Greening the transport sector: promoting 'zero emissions vehicles' in the EU and US', in Roggenkamp, M., de Graaf, K., and Fleming, R. (eds.), *Energy Law, Climate Change and the Environment: Volume IX of the Elgar Encyclopedia of Environmental Law* (Edward Elgar Publishing, Cheltenham, 2021), pp. 584–598.

[5]   Amani, A.M., Csereklyei, Z., Dwyer, S., *et al.*: *Opportunity Assessment: My V2X EV: Informing Strategic Electric Vehicle Integration* (RACE for 2030 CRC, 2023), https://www.racefor2030.com.au/content/uploads/V2X-Stage_1_Final-Report.pdf, accessed May 2025.

[6]   Borenstein, S., and Bushnell, J.B.: 'The U.S. electricity industry after 20 years of restructuring', *The Annual Review of Economics*, 2015, 7, pp. 437–463.

[7]   Joskow, P.L.: 'Lessons learned from electricity market liberalization', *The Energy Journal*, 2008, 29, (SI2), pp. 9–42.

[8]   Directive (EU) 2019/944 of the European Parliament and Council of 5 June 2019 on common rules for the internal market for electricity and amending Directive 2012/27/EU.

[9]   Csereklyei, Z., and Stern, D.: 'Technology choices in the U.S. electricity industry before and after market restructuring', *The Energy Journal*, 2018, 39, (5), pp. 157–182.

[10]   U.S. Energy Information Administration, 'Residential retail electric choice participation rate has leveled off since 2019', (15 March 2023), https://www.eia.gov/todayinenergy/detail.php?id=55820, accessed April 2024.

[11]   Klopčič, A., Hojnik, J., and Bojnec, S.: 'What is the state of development of retail electricity markets in the EU?', *The Electricity Journal*, 2022, 35, (3), p. 107092.

[12]   ACER, European Union, Agency for the Cooperation of Energy Regulators, 'ACER's Final Assessment of the EU Wholesale Market Design', (April 2022),

https://www.acer.europa.eu/sites/default/files/documents/Publications/Final_Assessment_EU_Wholesale_Electricity_Market_Design.pdf (figure 20, page 34).

[13]   Joskow, P.L.: 'Challenges for the wholesale electricity markets with intermittent renewable generation at scale: the US experience', *Oxford Review of Economic Policy*, 2019, 35, pp. 291–331.

[14]   International Energy Agency, 'Energy efficiency and digitalisation', (2019), https://www.iea.org/articles/energy-efficiency-and-digitalisation, accessed April 2024.

[15]   Australian Energy Market Operator, 'Ancillary services', https://aemo.com.au/en/energy-systems/electricity/national-electricity-market-nem/system-operations/ancillary-services, accessed April 2024.

[16]   Noel, L., Zarazua de Rubens, G., Kester, J., and Sovacool, B.K.: 'The regulatory and political challenges to V2G', in Noel, L., Zarazua de Rubens, G., Kester, J., and Sovacool, B.K. (eds.), *Vehicle-to-Grid: A Sociotechnical Transition Beyond Electric Mobility* (Palgrave Macmillan, Cham, 2019), pp. 117–139.

[17]   Csereklyei Z., Kallies, A., and Diaz Valdivia, A.: 'The status of and opportunities for utility-scale battery storage in Australia: A regulatory and market perspective', *Utilities Policy*, 2021, 73, 101313.

[18]   Gschwendtner, C., Sinsel, S., and Stephan, A.: 'Vehicle-to-X (V2X) implementation: An overview of predominate trial configurations and technical, social and regulatory challenges', *Renewable and Sustainable Energy Reviews*, 2021, 145, 110977.

[19]   Burger, S., Jenkins, J., Batlle, C., *et al.*: 'Restructuring revisited part 1: Competition in electricity distribution systems', *The Energy Journal*, 2019, 40, (3), pp. 31–54.

[20]   MIT, *Utility of the Future: An MIT Energy Initiative Response to an Industry in Transition* (MIT Press, Cambridge, MA, 2016).

[21]   Burger, S., Jenkins, J., Batlle, C., *et al.*: 'Restructuring revisited part 2: Coordination in electricity distribution systems', *The Energy Journal*, 2019, 40, (3), pp. 55–76.

[22]   Ofgem, '*Guide to the RIIO-ED1 Electricity Distribution Price Control*' (Ofgem, London, 2017).

[23]   New York State, 'Reforming the energy vision (REV)' https://rev.ny.gov/, accessed April 2024.

[24]   Australian Energy Regulator, 'Demand management incentive scheme and innovation allowance mechanism', https://www.aer.gov.au/industry/registers/resources/schemes/demand-management-incentive-scheme-and-innovation-allowance-mechanism, accessed April 2024.

[25]   Nationale Leitstelle Ladeinfrastruktur, *Positionspapier Bidirektionales Laden Diskriminierungsfrei Ermöglichen* (NOW GmbH, Munich, 2024).

[26]   Anuta, O., Taylor, P., Jones, D., *et al.*: 'An international review of the implications of regulatory and electricity market structures on the emergence of utility-scale electricity storage', *Renewable and Sustainable Energy Reviews*, 2014, 38 (C), pp. 489–508.

[27]   Ofgem, *Ofgem Decision on Clarifying the Regulatory Framework for Electricity Storage: Changes to the Electricity Generation Licence* (Ofgem, London, 2020).

[28]   Australian Energy Market Commission, *Integrating Energy Storage Systems into the NEM, Rule Determination* (AEMC, Melbourne, 2021).

[29]   Say, K., Csereklyei, Z., Brown, F., *et al.*: 'The economics of public transport electrification: A case study from Victoria, Australia', *Energy Economics*, 2023, 120, 106599.

[30]   Cantley-Smith, R., Ben-David, R., Briggs, C., *et al.*: *Opportunity Assessment: DSO and Beyond: Optimising Planning and Regulation for DM and DER* (RACE for 2030 CRC, 2023), https://www.racefor2030.com.au/content/uploads/N4-OA-Final-Summary-Report-2.pdf, accessed May 2025.

[31]   Australian Energy Market Operator, 'Distributed energy resource register', https://aemo.com.au/energy-systems/electricity/der-register/about-the-der-register, accessed April 2024.

[32]   Australian Energy Market Commission, *Unlocking CER Benefits through Flexible Trading, Rule Determination* (AEMC, Melbourne, 2024).

[33]   European Commission, EU Smart City Information System, 'Electric vehicles and the grid: Solutions booklet', https://smart-cities-marketplace.ec.europa.eu/sites/default/files/2021-02/D32.1D3_Solution%20Booklet_EVs%20and%20the%20Grid.pdf, accessed April 2024.

[34]   Knezović, K., Marinelli, M., Zecchino, A., *et al.*: 'Supporting involvement of electric vehicles in distribution grids: Lowering the barriers for a proactive integration', *Energy*, 2017, 134, (Suppl. C), pp. 458–468.

[35]   Castagneto-Gissey, G., Dodds, P., and Radcliffe, J.: 'Market and regulatory barriers to electrical energy storage innovation', *Renewable and Sustainable Energy Reviews*, 2018, 82, (P1), pp. 781–790.

Chapter 7

Microgrid-based V2G-enabled charging stations

Mirsaeed Mousavizade[1], Foad Taghizadeh[2], Feifei Bai[3], Mohammad J. Sanjari[1] and Junwei Lu[1]

Vehicle-to-grid (V2G) technology is expected to revolutionize the way we consume and utilize energy by enabling electric vehicles (EVs) to not only draw power from the grid but also feed excess energy back into it, thus creating a dynamic and bidirectional flow of electricity. Performing thorough research in this area is crucial for establishing effective regulations and standards and formulating forward-thinking policies that facilitate the seamless integration of V2G systems into power grids. As the world moves toward cleaner and more efficient energy solutions, comprehensive study in these areas will pave the way for a more resilient, responsive, and environmentally friendly energy system while simultaneously unlocking new revenue streams and benefits for both consumers and the broader energy industry. In this regard, this chapter discusses the infrastructure, regulations, and standards associated with V2G technology, which is crucial to the development of sustainable transportation and energy systems.

7.1 V2G technology development and policies

V2G technology has emerged as a transformative innovation at the intersection of transportation and energy sectors, offering a solution to both the challenges of EV integration and grid management [1]. The development of V2G technology has gained significant momentum worldwide over the past decade. Initially conceived as a way to enhance the sustainability of EVs by allowing them to feed excess energy back into the grid, V2G technology has evolved into a multifaceted system that benefits both EV owners and grid operators. In the early stages of its development, V2G faced technical hurdles and regulatory complexities that hindered its widespread adoption [2]. However, as advancements in battery technology, communication protocols, and grid infrastructure unfolded, V2G systems became

[1]School of Engineering and Built Environment, Griffith University, Brisbane, Australia
[2]School of Engineering, Macquarie University, Sydney, Australia
[3]School of Electrical Engineering and Computer Science, The University of Queensland, Brisbane, Australia

increasingly feasible and attractive. Countries like Japan, the United States, and several European nations took the lead in research and implementation. The integration of renewable energy sources (RES) into the grid further amplified the appeal of V2G, enabling EVs not only to store and release energy but also to act as dynamic grid assets that support demand response (DR) and grid stabilization [3,4]. As a result, the technology has garnered interest from automakers, utilities, and policymakers alike, propelling V2G into a pivotal role in the ongoing transition toward a more resilient and sustainable energy system. Figure 7.1 demonstrates that the global V2G technology market size was US$ 1.77 billion in 2021, and it is expected to reach US$ 17.43 billion by 2027, with an annual growth rate of 48% [5]. The increasing use of EVs across the globe is one of the primary factors driving the growth of the V2G market, which affects the demand for charging infrastructures such as uni-directional and bidirectional systems. In addition, the North American and European regions account for the largest revenue share to the V2G market. A major factor that propels market growth in these regions is the government's initiatives to encourage the use of battery-powered vehicles to reduce carbon emissions. A significant innovation in the manufacturing processes of vehicle batteries is another component that contributes to growth. Auto manufacturers are promoting V2G as it enhances their value chain by improving battery life. The Asia Pacific region is also expected to experience significant growth for V2G during the forecast period. The green revolution efforts in Asian countries like China, Japan, India, and South Korea are one contributing factor. For instance, China is committed to making the region fully electric by 2025, and it is investing heavily in this area [6].

As discussed, the development of V2G technology varies across regions and countries, driven by their specific energy challenges, policy priorities, and EV adoption rates [7]. Governments and regulatory bodies worldwide are recognizing the potential of V2G to enhance grid resilience, support renewable energy integration, and create new revenue streams for EV owners [8]. As V2G technology continues to evolve, global collaboration and knowledge-sharing will be crucial for its widespread adoption and successful integration into energy systems. Various countries around the world have been actively developing and implementing V2G

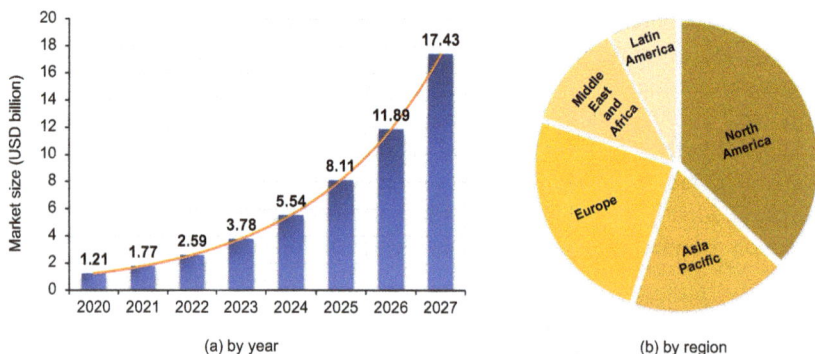

Figure 7.1   *V2G technology market size: (a) by year and (b) by region*

technology, often accompanied by specific policies and regulations to accelerate its adoption and integration into their energy systems [9]. The following provides an overview and a general perspective of V2G technology development and associated policies in different countries.

- **United States:** In the United States, the development of V2G technology has been driven by a combination of environmental concerns, energy security, and the potential for grid stability. Policymakers and regulators have taken steps to incentivize the adoption of V2G technology. The Federal Energy Regulatory Commission (FERC) issued Order 719 in 2008, which enabled electric utilities to compensate EV owners for providing grid services through V2G systems [10]. Moreover, the FERC Order 2222, implemented in 2020, allowed distributed energy resources (DERs), including EVs, to participate in wholesale energy markets [11]. This opened up avenues for EV owners to sell excess energy back to the grid during peak demand periods. Additionally, federal tax credits and grants have supported research and development in this field. State-level initiatives, such as California's Zero Emission Vehicle mandate [12] and the New York VDER tariff, encourage the integration of V2G-enabled EVs into the energy system [13]. These policies aim to not only reduce greenhouse gas emissions but also enhance grid resiliency by utilizing EVs to store and supply electricity during peak demand.

- **Japan:** As a pioneer in EV technology, Japan has made significant strides in advancing V2G systems. The country's commitment to reducing dependence on fossil fuels and its vulnerability to natural disasters have spurred interest in V2G technology to enhance energy resilience. The Japanese government has implemented supportive policies, including subsidies and tax incentives for EV purchases, as well as research funding for V2G development. Regulatory efforts, such as amending the Electricity Business Act to allow EVs to supply electricity to the grid, have facilitated V2G integration [14]. Moreover, collaboration between automakers, utilities, and government agencies has resulted in research, pilot projects, and real-world demonstrations of V2G capabilities to improve the infrastructure and create standardized communication protocols. By aligning technological innovation with supportive policies, Japan has laid the groundwork for a sustainable energy system.

- **European Union:** Across the European Union (EU), V2G technology has gained traction as countries strive to achieve their renewable energy targets and reduce reliance on fossil fuels [15]. The EU has funded research initiatives like the INVADE project, which focuses on integrating EVs and batteries into the grid. National governments have also played a significant role in promoting V2G adoption. For instance, Denmark offers tax incentives for EVs equipped with bidirectional charging capabilities. Furthermore, regulatory frameworks like the Clean Energy Package emphasize the role of energy communities and empower consumers to become prosumers, engaging in both consumption and production of energy [16]. As the EU continues to prioritize clean energy transitions, V2G technology serves as a bridge between sustainable transportation and a resilient, decentralized power grid.

- **China:** China has been a pioneer in the development and adoption of V2G technology as part of its ambitious efforts to promote clean energy and reduce carbon emissions. The Chinese government has implemented a series of policies and regulations to accelerate the penetration of V2G technology in its energy systems. These policies include incentives such as subsidies for EV purchases, tax breaks, and grants for V2G infrastructure deployment [17]. Additionally, China has established a robust charging infrastructure network to support V2G, making it convenient for EV owners to participate in grid services. The government's strong commitment to promoting electric mobility and grid integration has created a conducive environment for the growth of V2G technology, helping to balance electricity demand and supply while reducing the carbon footprint of the transportation and energy sectors.
- **Australia:** Australia has also been actively exploring the potential of V2G technology to enhance grid stability and accommodate the increasing share of RES [18]. The country faces unique energy challenges due to its vast geography and dispersed population centers. To promote V2G technology, Australia has introduced innovative policies and regulations. These include financial incentives such as feed-in tariffs for EV owners who supply excess energy back to the grid, as well as grants for research and development in V2G technology [19]. Also, regulatory frameworks have been established to enable bidirectional power flow and ensure safety and technical standards are met. By leveraging V2G, Australia aims to create a more resilient and adaptable energy system while encouraging EV adoption.
- **Netherlands:** The Netherlands has emerged as a leader in promoting V2G technology, driven by its ambitious renewable energy targets and commitment to reducing carbon emissions [20]. The Dutch government has implemented a range of policies to encourage V2G adoption, including tax incentives, grants for EV charging infrastructure, and subsidies for V2G-enabled vehicles. The Dutch Charging Act mandates that all new parking spaces in non-residential buildings must have EV charging points, further promoting EV adoption and the potential for V2G participation. Moreover, the Netherlands has established collaborative platforms involving stakeholders from the energy, automotive, and technology sectors to foster innovation and address regulatory challenges [21]. These efforts are underpinned by the country's goal to achieve a carbon-neutral energy system by 2050, positioning V2G technology as a key component of its sustainable energy transition.
- **Norway:** Norway, known for its leadership in electric mobility, has also embraced V2G technology to optimize its energy systems. The Norwegian government has implemented policies that prioritize EV adoption through tax incentives, toll exemptions, and reduced registration fees, creating a conducive environment for V2G integration [22]. To increase the penetration of V2G, Norway has invested in smart grid infrastructure and established clear standards for grid interaction and communication protocols. Additionally, the country has introduced dynamic pricing mechanisms that incentivize EV owners to supply energy to the grid during peak demand periods, promoting

grid stability and minimizing reliance on fossil fuels [23]. Norway's comprehensive approach to policy and infrastructure development has positioned it as a frontrunner in utilizing V2G technology to advance its clean energy goals.

• **United Kingdom:** The UK has been a pioneer in the development and deployment of V2G as part of its broader efforts to decarbonize the transportation and energy sectors [24]. The UK government has implemented a range of policies and regulations to foster the growth of V2G technology. One of the key initiatives is the OLEV (office for low emission vehicles) grant scheme, which provides financial incentives to encourage the adoption of EVs equipped with V2G capabilities [25]. Additionally, the UK has introduced regulatory frameworks that enable EV owners to participate in grid services, such as providing energy back to the grid during peak demand periods. This has led to collaborations between automakers, energy companies, and technology providers to develop V2G infrastructure and compatible vehicles.

• **Denmark:** Denmark, known for its strong commitment to renewable energy and sustainability, has also been at the forefront of V2G technology development. The Danish government has implemented policies that encourage the integration of EVs into the energy system. One notable approach is the establishment of partnerships between utilities and EV manufacturers to promote V2G technology [26]. Denmark has also introduced favorable grid regulations that allow EV owners to sell surplus energy back to the grid, thus creating an incentive for wider V2G adoption. Moreover, Denmark's comprehensive approach includes research and development funding to support advancements in V2G technology and its seamless integration into the national grid [27].

• **Germany:** Germany, a leader in both the automotive and renewable energy industries, has been actively working on integrating V2G technology into its energy system. The German government has provided incentives for EV adoption and the development of V2G infrastructure through subsidies and tax breaks. The country has also introduced policies that facilitate the participation of EVs in grid services and dynamic pricing mechanisms [28]. Germany's commitment to innovation is evident through collaborations between automakers, utilities, and research institutions to refine V2G technology. Furthermore, the German energy transition strategy aligns V2G development with its renewable energy goals, aiming to use EVs as flexible energy storage solutions to balance intermittent RES.

7.2   Microgrid-based V2G-enabled charging system

A microgrid-based V2G-enabled charging system combines the concepts of microgrids and V2G technology to create a dynamic energy system where EVs can both charge from and discharge electricity back to the grid [29]. This system is designed to enhance system reliability, efficiency, and sustainability by leveraging the flexibility of EV batteries and integrating them into the local electricity infrastructure. Moreover, the microgrid-based V2G-enabled charging system operates within the framework of a microgrid, which is a localized energy network capable of operating independently or in conjunction with the main grid. Within the microgrid,

V2G-enabled EV charging stations are strategically located to facilitate charging for EVs and bidirectional power exchange with the grid. Energy management systems (EMS) play a crucial role in orchestrating the operation of the system, managing energy flows, optimizing charging schedules, and coordinating V2G interactions with the grid. This system consists of the following critical elements:

- **EV charging stations**: These stations are equipped with bidirectional charging capabilities, allowing EVs to both charge from and discharge electricity back to the grid. They include hardware such as charging plugs, sockets, cables, and communication interfaces for data exchange with EVs and the EMS.
- **V2G-enabled EVs**: EVs participating in the system are equipped with V2G technology, enabling them to communicate with charging stations and the EMS and to provide grid support services by adjusting their charging and discharging patterns.
- **Energy management system**: The EMS monitors and controls energy flows within the microgrid, optimizing the operation of EV charging stations, RES, energy storage systems, and other DERs. It coordinates V2G interactions with the grid to maximize system efficiency and reliability.
- **Grid connection points**: These points serve as interfaces between the microgrid-based V2G-enabled charging system and the main grid or other external energy sources. They facilitate bidirectional power exchange and ensure seamless integration with the broader energy infrastructure.

As shown in Figure 7.2, microgrid-based V2G-enabled charging systems differ from typical V2G-enabled chargers in the following ways:

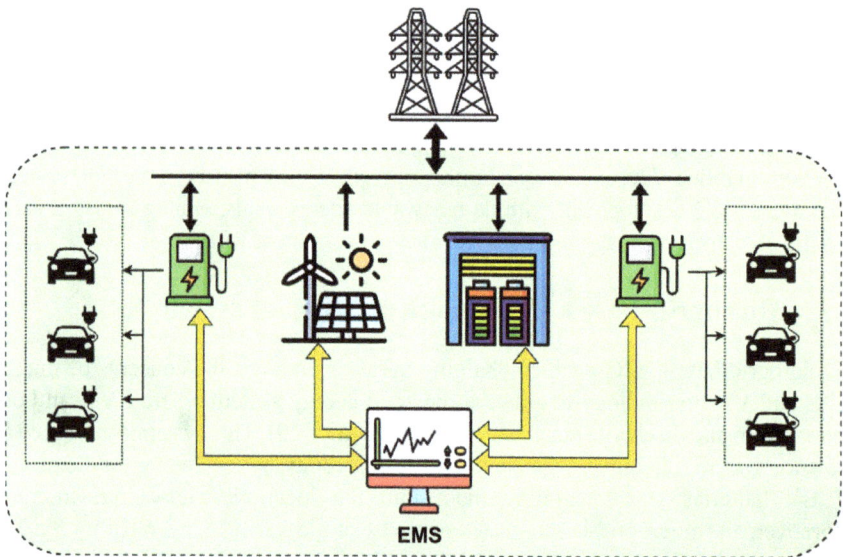

Figure 7.2   The configuration of a microgrid-based V2G-enabled charging system

- **Integration into a microgrid**: Unlike a typical charging system, which operates independently within the broader grid infrastructure, a microgrid-based V2G-enabled charging system is integrated into a localized energy network. This integration enables more efficient energy management, greater system resilience, and enhanced grid stability.
- **Coordination with DERs**: In a microgrid-based V2G-enabled charging system, EV charging and V2G interactions are coordinated with other DERs such as RES (e.g., solar panels, wind turbines) and energy storage systems. This coordination optimizes energy utilization and enhances the overall performance of the microgrid.
- **Localized energy management**: The EMS in a microgrid-based V2G-enabled charging system operates within the context of the microgrid, considering local energy supply and demand dynamics, grid constraints, and system objectives. This localized energy management approach allows for more tailored and responsive control of EV charging and V2G operations.

7.3   Codes and standards for EVs

Establishing standard protocols across all parts of the integrated EV infrastructure is crucial for maximizing its operational efficiency. Different aspects of EV technology may be subject to standardization, including charging components, data and communication frameworks, grid integration processes, and safety protocols. Figure 7.3 shows internationally used standards and the applied areas [30].

7.3.1   EV charging infrastructure

The international landscape of EV charging infrastructure standards is diverse, with different standards prevailing in various regions. In the US, standards from SAE (Society of Automotive Engineers) and IEEE (Institute of Electrical and Electronics Engineers) are commonly utilized, while Europe predominantly adheres to IEC (International Electrotechnical Commission) standards. Japan has its own unique EV charging standard called CHAdeMO. China follows Guobiao (GB/T) standards for both AC and DC charging, with AC standards resembling

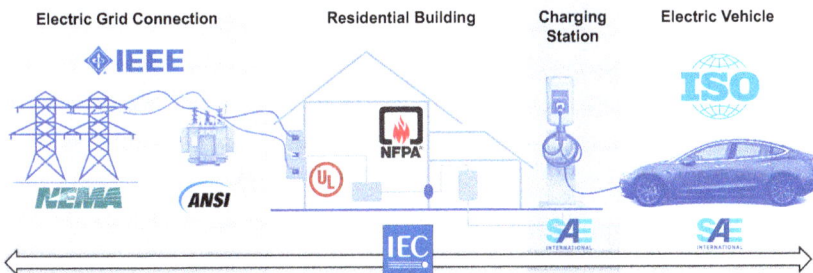

Figure 7.3   EV charging standards and their application areas

those of IEC. These standards exhibit notable distinctions in port and connector design. In the US, the SAE J1772 connector is widely used, offering compatibility with both AC and DC charging. Tesla has developed its proprietary fast-charging connector, which supports AC and DC charging as well. Europe commonly employs the "combo" connector, which integrates separate DC charging pins into existing AC charging connectors. Efforts have been made in recent years to converge these diverse standards toward a universal solution.

Manufacturers of charging equipment are collaborating to harmonize charging standards, aiming for greater compatibility and interoperability across different regions [31]. Establishing standardized EV charging protocols not only facilitates efficient investments for manufacturers and infrastructure providers but also mitigates risks associated with future EV-related product launches. Legislation mandating appropriate standards and regulations for charging systems empowers car manufacturers to import EVs equipped with standard plugs, enables infrastructure providers to deploy uniform charging stations, and facilitates consumers' access to consistent and reliable charging infrastructures. Below are the widely recognized standards governing EV charging configurations and plugs:

- IEC 62196-1: General requirements for conductive charging of EVs, covering plugs, socket outlets, vehicle connectors, and vehicle inlets.
- IEC 62196-2: Dimensional compatibility and interchangeability requirements for AC pin and contact-tube accessories within plugs, socket outlets, vehicle connectors, and vehicle inlets.
- IEC 62196-3: Dimensional compatibility and interchangeability requirements for DC and AC/DC pin and contact-tube vehicle couplers within plugs, socket outlets, vehicle connectors, and vehicle inlets.
- IEC 61851-1: General requirements for different charging modes within electrical vehicle conductive charging systems.
- IEC 61851-21: Requirements for conductive connection to AC/DC supply for both on-board and off-board charging systems in electrical vehicle conductive charging systems.
- IEC 61851-22: Specifications for AC charging stations within electrical vehicle conductive charging systems.
- IEC 61851-23: Specifications for DC charging stations within electrical vehicle conductive charging systems.
- UL 9741: Standards governing bidirectional electric vehicle charging system equipment.
- SAE J1772: Specifications for electrical connectors designed for electric vehicles.
- SAE J3072: Interconnection requirements for onboard, utility-interactive inverter systems within the context of electric vehicles.

7.3.2   Data and communication infrastructures

A communication infrastructure will facilitate efficient data exchange and collaboration among many instruments, components, and systems. Standardization should explore the potential limitations of different charging methods on the compatibility and interoperability between different types of equipment. This area aims to determine communication protocols, data formats, and mechanisms for exchanging information between EVs and electric vehicle supply equipment (EVSE) and involved stakeholders. Also, determining the details of the information shared among the participants and identifying the operational limits and technical requirements for implementing effective control will be considered by these types of standards [32]. There are several international standards that describe the communication channels between the grid and EVs. Recognizing that standardizing communication links to EVs could potentially limit the availability of EV models, it is important to address a significant challenge: the proliferation of diverse standards, each with distinct capabilities and varying degrees of alignment. This wide range of candidate standards presents a major obstacle, as not all standards are well-aligned, and many possess overlapping functionalities [33]. Comparing standards poses challenges due to variations in coverage and intended applications. Therefore, a thorough assessment is essential before adopting any standard. Below are some notable candidates for consideration:

- Standards addressing communication between EVs and EVSE, including ISO 15118, IEC 61851, and SAE J3072.
- Standards focusing on communication between the electricity system and EVSE, such as IEEE 2030.5, OpenADR, and IEC 61850.
- Standards governing communication between a charging station operator and EVSE, such as OCPP (Open Charge Point Protocol) and IEC 63110.
- ISO 15118-20 (2022): Standardized interface for V2G communication.
- IEC 61850: Communication protocols for electrical substation systems and networks.

7.3.3   Integrating EVs with the power grid

This segment delves into the distinctive characteristics and operational aspects of EVs and EVSE under normal grid conditions, as well as their capacity to withstand power system disturbances and mitigate associated impacts. Moreover, it addresses grid-support functionalities, encompassing voltage and frequency response mechanisms, voltage phase balance, voltage level maintenance within specified parameters, provision of fault ride-through capability, and recovery performance delineating the response of EVs/EVSE following an interruption. Consequently, the grid integration standards govern the interaction of EV charging/discharging with the grid, with EVs assuming the role of DERs during these operations [34]. Thus, the grid interconnection standards established for DERs also extend to EV grid integration. The grid-support framework for DERs is typically outlined by dedicated grid codes in numerous countries. Standards such as IEEE 1547, UL 1741, and UL 62109 specify the criteria for inverters, converters, controllers, and interconnection system equipment utilized in integrating DER assets into the power grid [35]. Specifically, SAE J3072 outlines the requirements for on-board V2G inverters installed in EVs.

7.3.4   Safety issues

EVs are mandated to adhere to safety requirements stipulated by state laws or local regulations. Furthermore, EV batteries must undergo testing according to standards that encompass various conditions, including overcharge, temperature extremes, short circuits, fire hazards, collision impacts, vibration susceptibility, humidity exposure, and resistance to water immersion [36]. Safety features must be integrated into the design of EVs, including collision detection, short circuit prevention, and insulation of high voltage lines. Various organizations establish safety standards for EV charging and grid integration. Notably, the National Fire Protection Association and the National Electrical Code prioritize safety measures. A charging system for EVs should adhere to the following safety standards:

- IEC 61140: Standards for protection against electric shock, covering installation and equipment aspects.
- IEC 62040: Specifications for Uninterruptible Power Systems (UPS).
- IEC 60529: Classification and ratings for degrees of protection provided by enclosures.

7.4   Realizing interoperability for V2G functioning in microgrids

Interoperability for V2G technology is a crucial step toward maximizing the potential benefits of EVs and renewable energy integration in microgrids. Interoperability refers to the ability to exchange and use data securely between components, devices, systems, and applications. Achieving seamless communication and interaction between diverse EV models, charging infrastructure, and grid systems requires a multi-faceted approach. The most important initiatives to enable interoperability in the V2G adoption process include designing appropriate standards, incorporating V2Gs into grid codes, enabling industry collaborations, and establishing supportive regulatory platforms [37].

7.4.1   Standardization and harmonization

EVs are not always connected to the grid, and they are not always connected at the same location. As a result, standard plugs, standard infrastructures, and standard communication platforms are required. The adoption of unified standards will allow operators to take full potential of V2G, such as its fast response time, high power load, and flexible support. V2G deployment is significantly hindered by the absence of effective standards between EVSEs and upstream electric networks [38]. The standardization of equipment and communication protocols can be considered a key element in enabling V2G and accomplishing interoperability. By establishing universally accepted protocols, such as the Open Charge Point Interface (OCPI) or the Combined Charging System (CCS), different EVs can communicate effectively with various charging stations and grid operators. This standardization enables plug-and-play functionality, simplifying the user experience and encouraging widespread V2G adoption. Also, consensus on unified data formats is crucial for smooth data sharing between different components of the V2G system. This common language ensures that information is accurately interpreted and used across various stakeholders, such as EV owners, utilities, and grid operators. Unified data formats streamline energy management, billing, and monitoring processes. In addition, as data exchange increases, ensuring data security and privacy becomes paramount. Encryption, secure access controls, and anonymization techniques safeguard sensitive information while enabling seamless interoperability.

Beyond regulatory compliance, developing technologies are essential. In the context of V2G, one of the primary challenges is that industries developed products independently for many years are now being asked to become compatible. By establishing standards collaboratively, EVs may make cross-industry interaction possible. The presence of multiple interface points makes it clear that harmonization is necessary for vehicles to be able to connect to any charging station. Depending on the location and application, there may be a variety of interface standards. Thus, establishing hardware compatibility standards is essential to ensure EVs and charging stations can seamlessly interact with one another and the grid. Standardized connectors, communication modules, and power ratings enable EVs to connect to different charging infrastructure, promoting a user-friendly experience and widespread adoption of V2G services.

Moreover, collaboration among automakers, charging infrastructure providers, utility companies, and technology developers is pivotal. This collaboration involves developing cross-compatible hardware and software solutions that adhere to the established standards. Open-source platforms could play a vital role in fostering innovation and cooperation within the V2G system. By sharing knowledge and resources, stakeholders can collectively address technical challenges and promote a more seamless integration of V2G systems with the existing energy infrastructure.

7.4.2   Including V2G in grid codes

Integrating V2G technology into grid codes is a complex yet crucial endeavor that requires a meticulous approach to ensure seamless and efficient operation. Grid codes should be designed to support V2G applications at the distribution level [39]. To incorporate this technology effectively, grid codes need to be updated to accommodate the bidirectional flow of energy and maintain grid stability. First, the grid codes must outline the technical requirements and standards for V2G-enabled EVs and the associated charging infrastructure. To minimize the potential adverse effects on the local network, these requirements should cover aspects such as communication protocols, power quality, voltage regulation, protection, frequency stability, and synchronization mechanisms. Defining specific technical specifications will ensure that V2G systems interact harmoniously with the grid and adhere to established reliability and safety standards.

Grid connection is a particularly challenging issue for EVs that have V2G capabilities (having an on-board DC/AC bidirectional converter), as the vehicles are able to inject power at various points in the network. Bidirectional EVSEs are currently not widely recognized by DSOs as potential generators, and demonstration projects have repeatedly been hampered by slow, expensive, and, in some countries, unattractive grid connection procedures [40]. Moreover, grid codes should establish guidelines for monitoring and controlling V2G operations, enabling grid operators to manage energy flows in real-time, anticipate potential issues, and respond swiftly to maintain grid stability.

Furthermore, to facilitate the process of connecting V2G technology, equipment certification is a crucial step [41]. Developing testing and certification procedures ensures that V2G components meet interoperability standards. Certified components inspire trust among stakeholders and simplify the integration process. Rigorous testing validates the compatibility and reliability of EVs, charging stations, and grid systems, minimizing potential issues during real-world implementation. The certifying process should be based on a well-defined framework and consistent across DSOs. It is likely that V2G equipment will qualify for connection under existing procedures for other distributed generation (DG) and storage equipment, including rooftop PV systems.

Moreover, smart metering and authentication is another important issue regarding EVs integration into the grid [42]. Smart metering systems with accurate measurement capabilities are essential for tracking energy flow between EVs and the grid. Additionally, robust authentication mechanisms ensure secure access to the grid, guarding against unauthorized usage. Also, the grid codes should emphasize the importance of cybersecurity and data privacy to safeguard both the grid and the participants in the V2G system. As V2G systems involve bidirectional communication between EVs and the grid, stringent cybersecurity measures need to be implemented to prevent unauthorized access, data breaches, and potential disruptions. Grid codes should mandate encryption protocols, authentication mechanisms, and regular security audits to mitigate risks. Additionally, clear guidelines must be established for data sharing, ensuring that EV owners' privacy rights are respected, and their personal information is protected.

7.4.3   Establishing supportive regulatory platforms

Regulatory support and policy frameworks need to be established to encourage interoperability. This administrative platform should address the economic incentives and regulatory frameworks required to encourage widespread adoption. This may involve tariff structures that incentivize EV owners to participate in V2G programs by offering compensation for the energy they supply to the grid. Regulatory measures should promote fair compensation, protect against potential misuse of vehicle batteries, and ensure that grid operators prioritize grid stability over V2G operations. Also, governments and regulatory bodies can incentivize V2G deployment by creating favorable conditions, such as offering tax incentives or subsidies for V2G-enabled vehicles and charging infrastructure [43]. Clear guidelines for grid connection and electricity pricing can also contribute to the widespread adoption of V2G. Also, as V2G systems involve bidirectional energy flow and potential grid services, regulatory frameworks should address issues related to energy pricing, grid stability, and data privacy to ensure a fair and efficient system. Collaborative efforts between regulatory bodies, automakers, charging infrastructure providers, and grid operators are essential to establish a cohesive framework that encourages investment and promotes V2G integration.

7.5   Technical requirements to implement V2G technology

An effective V2G implementation hinges on an intricate convergence of hardware and software components that harmonize EVs with the power grid, paving the way for more sustainable and adaptive energy systems. To achieve the maximum benefits of EV technology and realize the potential opportunities of V2G, it is necessary to alleviate certain technical difficulties and challenges. The following are the most critical functional requirements for V2G integration:

- charging infrastructures capable of V2G
- intelligent control systems
- data collection, security, and privacy for EVs
- availability of EVs with V2G capability
- Electricity markets supporting V2G hosting capacity of the power grid for V2G integration
- adaptive and intelligent protection systems

7.5.1   Charging infrastructures capable of V2G

The EV is recharged by plugging it into an electricity source. As shown in Figure 7.4, there are two types of charging configurations: AC charging and DC

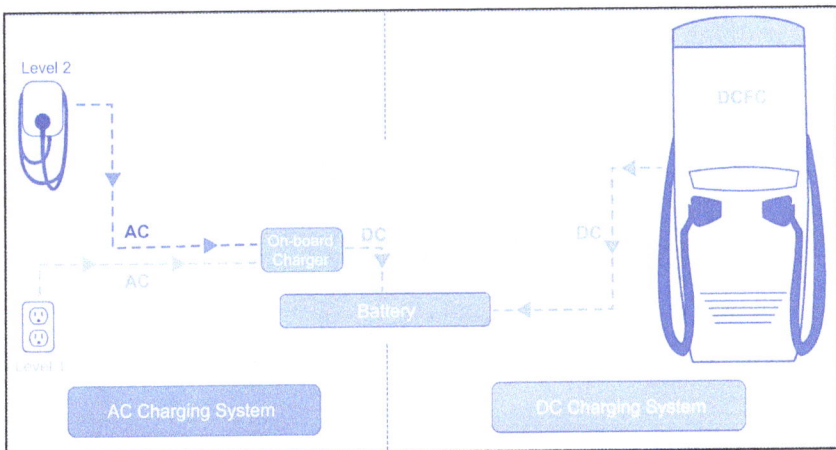

Figure 7.4   AC and DC charging schemes

charging [44]. Applying the first method, the AC to DC power conversion is accomplished by using the onboard chargers of the vehicles. The charging capacity of the onboard charger is limited due to size, heat dissipation, and weight restrictions. This limitation varies between models and ranges from 7 to 43 kW [45]. This is powerful enough to recharge an EV overnight or for daily top-ups. In DC charging, a high-powered public charging station is used to convert AC to DC power outside the vehicle. Using this technology, the onboard charger of the vehicle will be bypassed, allowing DC power to be delivered directly to the battery. The DC charging equipment has been installed in power ranges from 25 to 350 kW, and 50 kW is considered the standard size for fast chargers [46].

Moreover, EVs can be charged in three different ways: through conductive charging, wireless charging, and battery swapping [47]. Conductive charging is the most common method. In conductive charging, the power supply and battery are physically connected (via cable), but in wireless charging, there is no physical contact. Further, charge speed refers to the rate at which power is delivered to an EV. The classification of chargers into levels is commonly used to describe their speed. Moreover, many charger manufacturers have adopted specialized terms (such as Tesla's Supercharger) to describe their high-power chargers [45]. Table 7.1 outlines the definitions of levels and common names used throughout the report. Note that, to enable V2G and bidirectional power flow between the EV and

Table 7.1 Specifications of EV charging levels [45]

Power level	Common name	Charger type	Power	Time to achieve 100 km of range[1]	Application
Level 1	Slow charging	Wall socket	2.3 kW	8 h 42 min	Home charging, emergency charging
Level 2		AC charger	3.5 kW	5 h 43 min	Workplace charging, All day/night parking
	AC fast charging		7.4 kW	2 h 42 min	Public destination charging
			22.1 kW	54 min	
Level 3	DC fast charging, rapid charging	DC wall charger	25 kW	48 min	Public multi-purpose charging
		DC charger	50 kW	24 min	Public journey-enablement
			100 kW	12 min	
	Tesla supercharging		120 kW	10 min	
	Ultra-fast charging		Up to 350 kW	Less than 10 min	

[1]Considering driving energy efficiency of 20 kWh/100 km

the power grid, additional features need to be added to the EV charger [48]. A V2G system consists of three key components: a connection to the electricity grid, a communication system for controlling the charge and discharge process, and a capability to monitor, measure, and track the services provided to the grid [49].

7.5.1.1   EV chargers capable of V2G

To connect to the grid, an EV requires a charger. The key characteristics of a charger include its power capacity, whether it is installed on the vehicle or at the charging station, and its bidirectional power and communication capabilities. Despite the fact that any of the charge levels may be used for V2G, Level 2 chargers are likely to be utilized in most V2G projects in the short term; however, as the cost of high-capacity chargers (both AC and DC) decreases, Level 3 chargers may become more prevalent. After the EV is connected to the power grid, bidirectional power flow and communication capabilities must be provided. They are integral features that must be incorporated at the design stage. To realize the V2G concept, both the EV and the EVSE must be capable of bidirectional operation. An advanced charger that is capable of both charging and discharging of an EV battery is known as a bidirectional charger. In comparison to regular EV chargers, bidirectional chargers are considerably more complicated and expensive since advanced electronics-based power conversions are employed to handle the energy flow between the charger and the vehicle [50]. With the use of bidirectional chargers, EV batteries can serve a variety of purposes, including providing power for a household, allowing the energy to be returned to the grid, and even serving as a backup source during blackouts or emergencies [51]. As depicted in Figure 7.5, a standard bidirectional charger comprises an AC/DC converter and a DC/DC converter. The AC/DC converter is responsible for converting AC power from the grid into DC power during the EV's charging mode. Conversely, it converts DC power into AC power before injecting it back into the grid during the discharging mode. On the other hand, the DC/DC converter regulates bidirectional power flow through control techniques. It functions as a buck converter during the charging mode and as a boost converter during the discharging mode [52].

Although bidirectional converter technologies have developed and become more widely available, most EVs sold today lack this capability, with a few notable exceptions. In the current market, only a small number of EVs are equipped with V2G technology and bidirectional charging; among these are Nissan Leaf later models and Mitsubishi Outlander and Eclipse plug-in hybrids which both use the

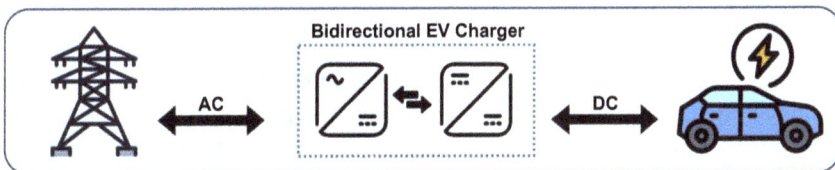

Figure 7.5   Schematic diagram of an EV charging system that is enabled by V2G

CHAdeMO connector. There are several car manufacturers, including Ford, that have developed bidirectional chargers that are exclusively designed to be used with their EVs. There are also some other manufacturers, such as Nissan, that utilize universal bidirectional chargers such as the Wallbox Quasar. The Hyundai Ioniq 5, which is being sold in Australia, offers vehicle-to-load (V2L), which allows owners to run tools, appliances, etc., using 240-V electricity directly from the battery (ideal for camping). The only EV which utilizes a CCS port for bidirectional charging is the Ford F-150 Lightning. By 2025, all EV manufacturers worldwide will be able to provide V2G connectivity [53]. Therefore, to effectively realize the potential of V2G services, there should be a sufficient number of charging stations and EVSE equipped with bidirectional chargers. Active players must be provided with the necessary charging hardware and ample support to participate effectively. In the current market, there are a limited number of EVs capable of providing bidirectional power flow between the vehicle and the charging station. The availability of V2G-ready EVSEs is also limited. The future requires significant investment in charging infrastructure, especially in bidirectional chargers, to alleviate consumer concerns and facilitate the V2G uptake.

7.5.1.2   Communication channel

Aside from the physical flow of power, controlling the charge and discharge process requires a communication pathway. Once the capability of controlling bidirectional power has been established, the next step is to inform the EV of the power flows that are currently required by the grid. Hence, a communication channel is imperative between the EVSE, which connects to the Internet and receives instructions from an aggregator or utility, and the vehicle. This necessitates the adoption of secure and low-latency communication protocols to regulate charging schedules, enhance grid stability, and accommodate user preferences. Depending on the intended services, such as frequency regulation, and the requisite reaction speed for their execution, real-time or online telecommunication implementation may be indispensable. In many instances, enabling the EV to communicate entails integrating an additional module into its system [54]. Various methods exist for enabling communication capabilities in EVs. Past projects have utilized control pilot line communication as specified in IEC 61851 or power line communication via protocols such as OCPP and Smart Energy Profile 2.0 (SEP 2.0). Future standards are currently under debate, with contenders including ISO 15118 and SAE J2847. Wide adoption of these standards is crucial for the widespread availability of V2G services [55].

7.5.1.3   Smart metering

The implementation of smart metering is imperative for the successful integration of V2G systems into the energy infrastructure. Traditional energy meters are unidirectional and designed to measure energy consumption only, lacking the capability to accurately track bidirectional energy flow necessary for V2G operations. Smart metering addresses this limitation by enabling precise measurement of energy both drawn from the grid during EV charging and injected back into the grid during discharging [56]. This accurate monitoring is essential for billing purposes, ensuring that EV owners are fairly compensated for the energy they contribute to the grid and enabling grid operators to manage grid stability and energy distribution effectively. Moreover, smart meters provide real-time data on energy consumption patterns, charging rates, and grid demand, enabling informed decision-making for optimizing V2G services and contributing to a more resilient and sustainable energy system. Additionally, smart meters with DR capabilities enable grid operators to adjust charging schedules during periods of high electricity demand or supply fluctuations, enhancing grid stability. Finally, home EMS integrated with smart meters allow users to manage their EV charging preferences and energy consumption, further promoting efficient V2G integration while empowering consumers to make informed energy decisions [57]. Together, these smart metering technologies lay the foundation for the successful deployment of V2G systems, offering both grid reliability and user convenience.

7.5.1.4   Future trend: DC fast charging hubs

DC fast charging hubs have emerged as a pivotal infrastructure component in the EV ecosystem, addressing the need for rapid and convenient charging to promote widespread EV adoption. These hubs provide a significantly faster charging experience compared to traditional AC charging stations, enabling drivers to recharge their vehicles quickly and continue their journeys with minimal downtime. DC fast charging technology plays a crucial role in reducing range anxiety, one of the primary concerns for potential EV owners, by offering a reliable solution for long-distance travel. One noteworthy advancement in the realm of DC fast charging is the development of Extreme Fast Chargers (XFCs). These chargers are designed to push the boundaries of charging speeds, providing an even faster and more efficient solution for EV users. XFCs boast specifications such as ultra-high power charging rates, cutting-edge battery management systems, and enhanced safety features. The integration of innovative cooling technologies ensures that these chargers can sustain high power levels over extended periods, contributing to a seamless charging experience for users. One significant trend in this trajectory is the emphasis on pushing the limits of charging speeds, and this is where the adoption of phase-shift full-bridge (PSFB) converters comes into play. In this context, the choice of DC/DC converter technology is crucial for optimizing charging efficiency. The transition from LLC resonant converters to PSFB converters in ultra-fast chargers represents a strategic move to enhance charging efficiency and reliability [58]. PSFB converters offer several advantages that make them well-suited for the demanding requirements of XFC scenarios. One primary advantage is the superior control over power flow that PSFB converters provide, enabling precise regulation of the charging process. The improved control contributes to minimized switching losses, thereby optimizing overall efficiency.

Furthermore, the use of PSFB converters aligns with the need for robust thermal performance, a critical consideration when dealing with high-power applications. The design of PSFB converters allows for effective thermal management, ensuring that the charging infrastructure can sustain ultra-fast charging rates over extended periods without compromising reliability [59]. This becomes particularly crucial as the automotive industry strives to deliver not only faster but also more durable and dependable charging solutions. In contrast to LLC resonant converters, which may have limitations in terms of control and thermal performance under extreme conditions, PSFB converters emerge as a technology choice that complements the ambitious goals of achieving ultra-fast charging. As the industry continues to innovate and refine the technologies powering DC fast charging hubs, the strategic adoption of PSFB converters underscores a commitment to providing users with an efficient, reliable, and sustainable charging experience. As research and development efforts progress, the integration of PSFB converters is likely to play a pivotal role in defining the next generation of ultra-fast charging infrastructure, contributing to the acceleration of the global transition toward sustainable and electrified transportation.

Looking toward the future, the trend in DC fast charging hubs is expected to focus on scalability, interoperability, and increased energy efficiency. As EV adoption continues to rise, the demand for a standardized and user-friendly charging infrastructure will become more pronounced. Additionally, advancements in energy storage technologies and renewable energy integration are likely to play a pivotal role in shaping the sustainability of these charging hubs. In the United States, projects such as the XFC and Future Renewable Electric Energy Delivery and Management (FREEDM) are at the forefront of pushing the boundaries of DC fast-charging technology. The XFC project [60] aims to develop and deploy charging stations capable of delivering power at unprecedented rates, addressing the need for ultra-fast charging to support long-distance EV travel. On the other hand, the FREEDM project [61] focuses on developing a smart and resilient electric grid infrastructure, with an emphasis on integrating RES and promoting sustainability. These projects have received significant funding to support their research and development efforts, reflecting the growing importance of advancing charging infrastructure in the transition to a more sustainable transportation ecosystem. The aims of these initiatives align with broader industry goals of reducing carbon emissions, enhancing grid reliability, and promoting the widespread adoption of electric vehicles.

7.5.2   Intelligent control systems

The uncontrolled management of EVs, particularly those equipped with V2G capability, can have detrimental effects on the power grid. To mitigate these undesirable impacts and address associated challenges and issues, it is imperative to introduce suitable and effective control algorithms for managing the charging and discharging of EVs. These approaches must consider various factors such as grid demand, electricity prices, and user preferences [49,62]. These algorithms must be integrated with real-time communication systems that allow EVs and charging infrastructure to communicate with grid management systems effectively.

Implementing coordinated management of EVs can facilitate various benefits including frequency regulation, loss reduction, voltage support, improvement of power quality, energy arbitrage, and load management through V2G services.

According to the literature, a wide range of approaches have been developed regarding technical and financial aspects, attempting to address grid stability by managing EVs and encouraging the owners and aggregators to participate in providing ancillary services [63]. These strategies can be classified into several categories based on control structures, mechanisms, optimization objectives, etc.

7.5.2.1 Control structures

In general, EV management in the power grid can be implemented by applying three control structures: centralized, decentralized, and hierarchical methods.

- **Centralized control structure:** With centralized control, the charging and discharging are typically managed directly by a central controller. As shown in Figure 7.6, the central unit initially collects information and charging requirements from each vehicle. It then calculates the optimal charging/discharging rate for EVs based on network conditions and specific objectives. Finally, the unit transmits a signal to the chargers to optimize the charging process accordingly. As part of the optimization process, the charging and discharging power of the battery must always conform to the constraints of the EV. To avoid overcharging and over-discharging of the battery, some approaches limit the SoC. Because all the information is collected by the unit, it would be able to provide the grid with the optimal solution and offer ancillary services. However, this method has several limitations and challenges [64]. The efficiency of the task is affected by the response of EVs to the signals. Some approaches have been explored for determining the

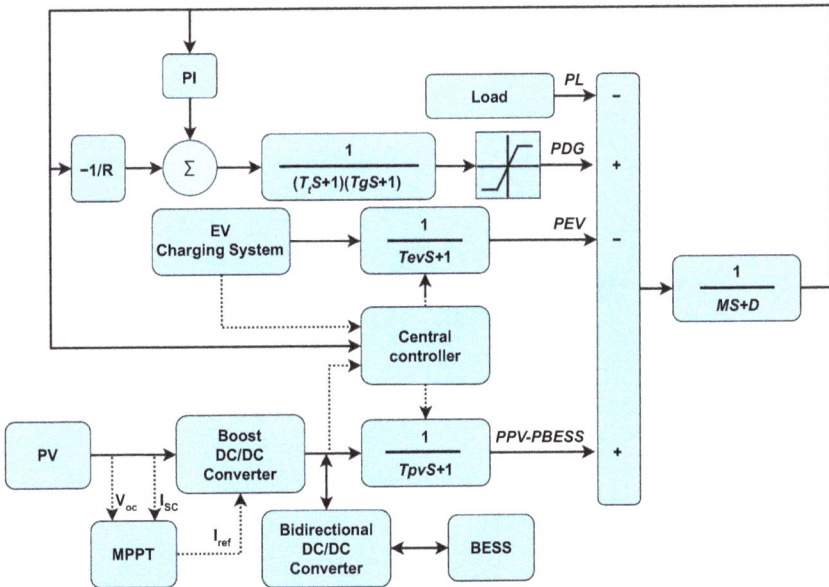

Figure 7.6 Centralized control structure

response of EVs to regulation signals, such as giving priority to each vehicle, considering the uncertainty of signals, etc. Moreover, the whole system will be affected if the unit cannot optimize the EV management problem. As a result, a backup will be required, but this will incur additional costs. Scalability is also a criticism of the centralized approach. Increasing the number of EVs will result in a larger computational burden, making it more difficult to solve the EV control problem. Moreover, when it comes to real-time applications, the speed of problem-solving is crucial. Thus, centralized control may not be practical when dealing with large-scale and real-time problems.

- **Decentralized control structure**: As shown in Figure 7.7, the decentralized control method empowers EV owners to autonomously manage the charging process according to their individual preferences. Strategies like price incentives enable system operators and aggregators to indirectly influence EV behavior. However, decentralized methods lack direct control, making it

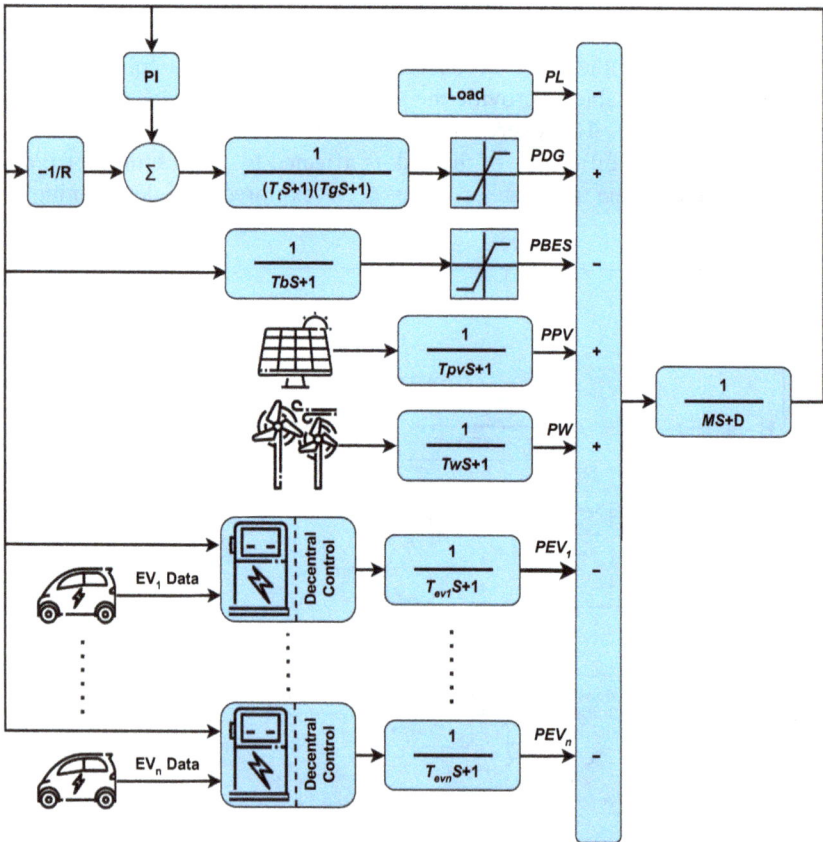

Figure 7.7 Decentralized control structure

challenging to guarantee an optimal solution and complicating the provision of ancillary services [65]. Nevertheless, the decentralized approach exhibits high scalability and is well-suited for large-scale EV fleets. By distributing the computational burden across all EVs, this approach ensures that each vehicle autonomously manages its own charge and discharge tasks.

- **Hierarchical control structure**: In terms of computational burden and network requirements, hierarchical control structure has the advantages of both centralized and decentralized control. The hierarchical control framework usually consists of two layers. As shown in Figure 7.8, at the upper level, a central controller, such as the system operator, manages all EV aggregators. At the lower layer, aggregators are responsible for scheduling the charging and discharging of several EVs [66]. By integrating the ancillary services provided with individual EVs, these aggregators can create a single source of power customized for a specific application. Upon receiving requests for ancillary services from grid operators, the aggregators dispatch commands to the vehicles that are available and eager to offer those services. The aggregator can then bid on providing the requested ancillary services after estimating the EV participation. Then, EVs are paid for their contribution, for example, according to the number of minutes they have assigned. However, this framework is vulnerable to disintegration if the central controller fails. In this context, an alternative approach can be presented in which the central controller is removed, the aggregators make schedules through communication among themselves, and then each aggregator controls its own EV in accordance with the schedules established. Using this model, where an aggregator fails, only the units under the control of that aggregator are affected, and the rest of the system continues to function.

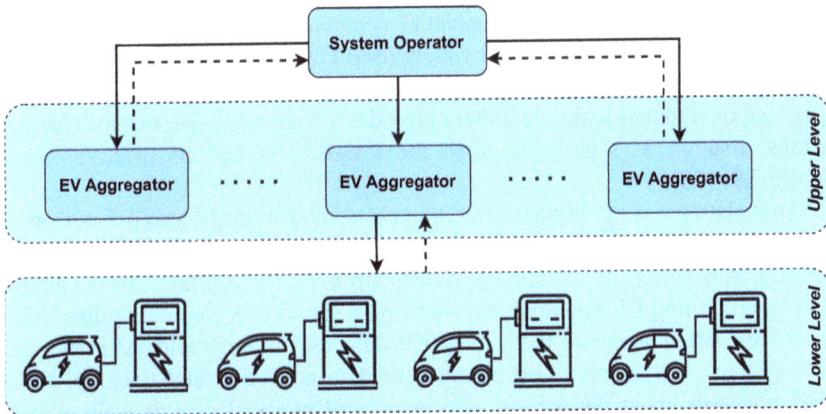

Figure 7.8 Hierarchical control structure

7.5.2.2 Control mechanisms

There are two different approaches to implementing charging/discharging control strategies for EVs, including the direct approach and the market-based mechanism. By using the direct method, EV charging and discharging are directly controlled. As part of this method, EV owners delegate control of their EV batteries to aggregators or system operators through contractual agreements [67]. This allows the aggregator to manage electricity demand and ensure that the EV charging/ discharging process is optimized for the grid. It also ensures that EV owners receive economic benefits from participating in the process. This method also reduces costs for grid operators by allowing them to better manage demand. Also, it enables a more rapid response to changes in demand and supply. The market-based mechanism, on the other hand, relies on the market forces of supply and demand to incentivize EVs to charge and discharge when necessary. This method considers consumers' decisions and allows EVs to optimize their charging and discharging behaviors based on electricity prices to maximize cost savings [68]. This mechanism is also more flexible and adaptive than traditional control schemes, allowing for better optimization of the electricity system. In recent years, market-based strategies have become increasingly popular due to the growth of electricity markets with adjustable price incentives. By utilizing this approach, real-time management of EVs can be realized to ensure regulation of demand and frequency, considering time of use and real-time pricing.

7.5.3 Data collection, security, and privacy for EVs

The availability of EV data plays a critical role in network and power system management. Collecting information on the quantity and rate of charging and discharging of existing EVs, as well as data related to their involvement in V2G services, can assist network operators in designing future network expansion plans and support other stakeholders like aggregators in developing incentive programs to increase EV participation. Hence, establishing a comprehensive database, implementing a robust data acquisition system for EVs, and establishing clear protocols for data collection are essential components of effective management [69]. PlugShare is a prominent and widely used EV database and charging station locator platform, known for its user-friendly interface and extensive coverage. Operating on a global scale, PlugShare provides a comprehensive map of charging stations, making it a go-to resource for EV owners and prospective buyers (Figure 7.9) [70].

This platform is not limited to finding charging stations; it also offers features like route planning, station reservation, and integration with various payment methods, enhancing the overall EV ownership experience. PlugShare's commitment to promoting EV adoption and fostering a supportive community has solidified its reputation as one of the leading EV databases in the world. As mentioned, V2G systems will involve vast amounts of data collection and tracking, which raises concerns about the security and privacy of personal data as well as about maintaining transparency with respect to the electrical grid [71]. The data

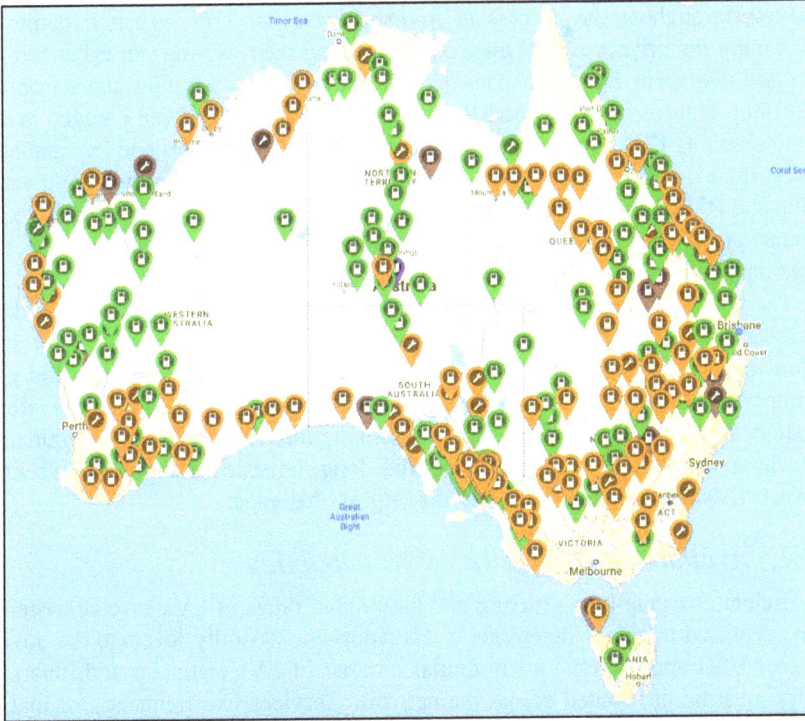

Figure 7.9   Distribution of charging station in Australia [70]

associated with V2G systems, including the timing and location of connections, could be considered sensitive. Using this data, a third party could, for example, predict when EV owners participating in V2G would be or would not be at home or determine the places they have traveled to and from. Thus, to ensure the EV owner's privacy, a variety of security measures must be established for the V2G system's data. Such concerns regarding privacy and information security are not new in the electric industry, especially as the power grids become increasingly modernized. As the smart grid involves a variety of data, a great deal of which is sensitive, such as statistics relating to household energy use. There is a variety of other potential security issues associated with V2G systems that must be addressed in addition to third parties accessing data about EV owners.

Moreover, there are several aspects of V2G that are unique, further complicating the risks to the EV owner, the aggregator, and the grid. As an example, unlike other smart grid technologies, V2G systems dynamically participate in ancillary services, which requires the system to communicate every few seconds. A further problem is that EVs are highly mobile and will move among various charging locations, making identification more difficult. Moreover, V2G systems are distinguished by their bidirectionality and decentralization to an extent greater than that of other smart grid technologies. The aggregator is therefore required to implement several key security measures throughout the process to mitigate these concerns. When it comes to maintaining the privacy of EV owners and limiting their exposure to cyber threats, there are two main factors to consider. To prevent impersonation attacks on the aggregator, authentication of the EV is undertaken once it has been plugged in and started charging [72]. For this purpose, digital certificate systems and key authentication can be used. A second aspect involves preserving the anonymity of EV owners during the collection and analysis of data. Therefore, in addition to securing communication between the EV and the aggregator, the aggregator must ensure that stored data is protected from external attacks [73].

The potential scale of V2G further complicates these security issues. As there may be millions of EVs in the future, using a key authentication system for each region may be particularly challenging. In addition, gathering, storing, and protecting petabytes of information has become increasingly challenging as well. In the short term, V2G system algorithms must take into account battery degradation and charger efficiency. However, in the long term, communication security between EVs and chargers will pose the largest challenge.

7.5.4   Availability of EVs with V2G capability

Insufficient charging infrastructure and the limited range of EVs have consistently been identified as major deterrents to EV adoption. To fully leverage the advantages of V2G applications, a substantial number of EVs equipped with this capability must be distributed across the network. Services like frequency regulation often require tapping into the battery capacity of thousands of EVs to be effectively provided [74]. In simpler terms, the extent to which V2G technologies are adopted will determine the opportunities derived from integrating EVs. With a higher number of EVs capable of V2G, there is a greater potential to offer services because the battery capacity of a single EV is limited compared to a fleet of EVs. Therefore, a high penetration of V2G technologies enhances the ability of EVs to provide ancillary services, leading to increased demand for EVs and associated technologies. This, in turn, may drive investment in EV infrastructure.

The communication and control infrastructure needed for V2G aggregation are still in the early stages of development. Moreover, there are currently limited commercial aggregators available in the market. Moreover, V2G technology is not fully appreciated by customers, and the value proposition is not well understood by end users [75]. These barriers restrict the adoption of EVs with V2G capability. As this infrastructure is further developed and improved, the cost and complexity of V2G aggregation will decrease, allowing more EV owners to participate in ancillary services and increasing the potential for EVs to provide grid services. Additionally, the increased demand for EVs and their associated technologies will create incentives for EV manufacturers to develop more efficient and cost-effective vehicles. This will create a positive feedback loop, driving the development of the V2G infrastructure and improving the economic prospects of EV owners. Ultimately, this will lead to greater adoption of EVs and increased participation in grid services, leading to a more sustainable energy system.

7.5.5   Establishing key players for effective EV integration

The potential benefits of V2G technology can only be realized if a sufficient number of EVs become involved. Thus, the presence of active players, such as EV aggregators and interested consumers, is essential for integrating EVs and unifying battery capacities. EV aggregators serve as a link between EVs and grid operators. They manage the costs and profits associated with the operation of a large fleet of EVs participating in ancillary services, such as frequency regulation. According to their strategy, EV aggregators intend to maximize their revenue based on their knowledge of the grid and EVs. EV aggregators collect fees from the EVs for their services, and they benefit from price arbitrage opportunities, allowing them to purchase energy when prices are low and sell it to the vehicles when prices are higher. They can also receive incentives from utility companies for providing grid services. Moreover, EV owners should be able to meet their mobility requirements, as well as maximize their profits. EV owners with the proper equipment, such as a V2G charger, can participate in DR programs and sell excess energy back to the grid [76]. This helps to balance the grid, and utilities are willing to pay for it. This creates an additional revenue stream for EV owners.

Hence, it is crucial to define and delineate the roles of these stakeholders, as well as the extent of their responsibilities, within the power system industry and the electricity market environment. This can be done through the development of regulations and standards that support the integration of these players into the grid [77]. In addition, policies should be established to ensure that these players are able to compete fairly and efficiently within the power system. Furthermore, grid codes in numerous countries do not acknowledge V2G or EVSE as DERs capable of injecting power into the grid. Consequently, technical and regulatory frameworks are necessary to facilitate the operation and profitability of EVs equipped with V2G capabilities in electricity markets.

7.5.6   Power grid hosting capacity for V2G integration

Existing electric networks are not designed to accommodate the high penetration of EVs and to deal with the adverse effects that may result from charging and discharging them frequently. Increased EV integration may result in overloaded distribution transformers, overhead lines, and cables if the network hosting capacity is insufficient. It is due to the fact that the current infrastructure is not equipped to handle the additional load associated with charging EVs, which would require more power than is typically demanded by the existing grid. As a result, the grid could become overloaded, leading to voltage dips, power outages, and increased maintenance costs. Therefore, limited network capacity to host EVs also affects the realization of V2G technology [78,79]. Network configuration, voltage levels, equipment, load consumption, component utilization ratios, and EV owner behavior all contribute to the hosting capacity of the system [80].

In the short term, the coordinated management of EVs may enhance network hosting capacity. This coordination helps to even out the peaks and troughs in electricity demand, ensuring that the network is not overloaded and can accommodate as many EVs as possible. By intelligently managing the charging process of EVs and spreading out the peak electricity demand, the strain on the electricity network is reduced. This allows the network to accommodate more EVs without the need for costly upgrades. In the longer term, however, increasing EV penetration levels may require network upgrades and modifications to accommodate the increased loads and peak power requirements for V2G. This could include expanding the transmission networks, as well as increasing the capacity of local distribution networks [81]. Additionally, the strategic placement of charging stations can help reduce the load on any particular segment of the grid. With the proper incentives and infrastructure, V2G technology can be realized, and network hosting capacity can be increased. Thus, designing network backbones and developing grid expansion plans that consider V2G is essential in preparing for the high penetration of EVs and enabling V2G capabilities in power systems.

7.5.7   Adaptive and intelligent protection systems

The traditional distribution network is passive and radial in nature, and the process of information collection and scheduling is relatively straightforward. However, in the future, the battery of an EV will act as a mobile energy storage unit, which will make it more difficult for dispatchers to accurately predict load requirements, increasing the difficulty of monitoring and protecting. It is expected that many EVs will be connected to the distribution network for charging and discharging as EVs become more popular [31]. This will have significant effects on the original power supply, load, and network structure. In turn, it will change the direction of power flow in the distribution network, which will have a substantial impact on the protection system.

Overcurrent protection is one of the most critical aspects of protecting any distribution system. Proper coordination between overcurrent relays is necessary to prevent damage to power devices arising from fault currents. Substantial EV integration may significantly increase the fault current magnitudes in the local network. Besides, increasing the integration of EVs into power grids could have other detrimental effects on the coordination and protection of relays [82,83]. The malfunction of the relays' coordination is one of the most significant effects of EVs. In this context, traditional techniques for configuring overcurrent relays may present serious problems due to improper and non-optimal settings. Additionally, EVs can deliver energy to the power system during the discharging process, employing V2G ability, which corresponds to the EVs behaving in a DGs mode. Operation in V2G mode will exacerbate the situation for the protection system [84]. Also, charging and discharging behaviors are unpredictable in time and space, which presents further difficulties for the coordination and configuration of the protection system. It should be considered how the fault characteristics of the grid might be influenced by the discharging process of EVs and their unpredictable charging features. Thus, implementing adaptive and smart protection system may alleviate the challenges regarding the change of power flow direction and provide a feasible foundation for enabling V2G services.

7.6   Economic and financial considerations

The economic considerations for V2G technology are multifaceted, encompassing infrastructure costs, revenue opportunities for participants, grid stability benefits, market dynamics and business models, and environmental impacts. Numerous studies in the literature provide valuable insights into the potential economic viability and advantages of V2G technology. As technology continues to evolve, ongoing research and real-world implementations will further refine our understanding of the economic and financial implications of this transformative technology, ultimately shaping the future of sustainable energy systems. In this section, we delve into the economic related considerations for V2G technology.

7.6.1   Infrastructure investment and implementation costs

One crucial factor influencing the widespread adoption of V2G technology is the initial infrastructure investment required. Deploying the necessary charging infrastructure, bidirectional inverters, and communication systems comes with a substantial cost. A notable real-world example is the California Energy Commission's investment in V2G infrastructure. In 2021, California allocated $436 million for EV charging infrastructure projects, including those incorporating bidirectional capabilities [85]. The state's commitment reflects the recognition of V2G technology as a key component of its strategy to promote sustainable transportation and grid resilience. Furthermore, the EU, a pioneer in renewable energy integration, has been actively investing in V2G infrastructure. Projects like the European Project INVADE (integrated electric vehicles and batteries to empower distributed and centralized storage in distribution grids) have received substantial funding [16]. This collaborative initiative, involving multiple European countries and industry partners, aims to demonstrate the technical and economic feasibility of V2G technologies. The investment from both public and private sectors in such initiatives showcases the commitment to addressing infrastructure challenges and fostering the development of V2G technology. Despite these investments, challenges persist, and understanding the nuances of infrastructure costs remains critical. Studies, such as the International Energy Agency's analysis of global V2G infrastructure costs, provide valuable insights into the economic considerations. The mentioned study [86] found that the global average cost of implementing V2G infrastructure is approximately $500 per kW of bidirectional power capacity. The study also highlights regional variations, with North America experiencing slightly higher implementation costs ($550 per kW) compared to Europe ($480 per kW). These costs encompass the installation of bidirectional charging stations, grid connection upgrades, and communication systems necessary for effective V2G operation.

The findings underscore the need for continued research, collaborative efforts, and strategic investments to propel V2G technology from a promising concept to a commercially viable and widespread solution for sustainable energy systems.

7.6.2   Revenue streams for V2G participants

One of the key attractions for EV owners to participate in V2G programs is the potential for additional revenue streams. A comprehensive study conducted by the National Renewable Energy Laboratory (NREL) in the United States illustrates the financial benefits for V2G participants [8]. According to the findings, EV owners engaging in V2G programs can earn substantial income, with an average annual revenue potential of up to $500. This income is derived from participating in grid services such as frequency regulation, where the EVs discharge stored energy back into the grid during periods of high demand. The study illustrated that this potential income is contingent on factors like electricity market dynamics, regulatory frameworks, and the availability of grid services programs. The ability of V2G-enabled EVs to contribute to grid stability becomes not only an environmental asset but also a lucrative opportunity for participants seeking additional income.

Further evidence of the revenue potential for V2G participants comes from a study conducted by the Rocky Mountain Institute (RMI). The research delved into the income-generating possibilities in the United States and revealed that optimizing the timing of energy discharge could significantly enhance the financial returns for EV owners [87]. By strategically aligning energy discharge with peak demand periods, V2G participants could unlock greater revenue, potentially exceeding the average annual estimate. These numerical findings underscore the importance of thoughtful participation in V2G programs and highlight the considerable financial advantages available to proactive EV owners.

Moreover, the evolving landscape of electricity markets presents additional opportunities for V2G revenue generation. A pilot project in Denmark, led by the energy company Ørsted and the EV manufacturer Nissan, demonstrated the potential of V2G to provide grid services [88]. The project involved a fleet of Nissan LEAF EVs that collectively delivered 1 MW of power back to the grid. This not only showcased the feasibility of V2G at scale but also hinted at the income-generating possibilities in markets actively seeking flexible and responsive energy resources. As the energy transition progresses, V2G technology is poised to become a key player in the dynamic and evolving energy landscape, offering tangible economic benefits to participants.

In the United Kingdom, the E-Flex project [89] demonstrated that V2G technology could generate revenue for EV owners by participating in grid services. The project showcased how a fleet of V2G-enabled EVs collectively earned over £180 000 in a year by providing frequency regulation services to the grid. This not only emphasizes the potential for revenue generation but also establishes the economic viability of V2G participation for individual EV owners.

7.6.3   Grid reliability and stability benefits

V2G technology has the potential to contribute significantly to grid reliability and stability through various mechanisms, ultimately resulting in economic benefits:

- **DR and peak shaving**: V2G allows EVs to serve as flexible energy resources by charging during periods of low demand and discharging electricity back to the grid during peak demand or grid instability events. This DR capability helps smooth out fluctuations in electricity demand, reducing the strain on the grid during peak hours and enhancing overall grid stability. By mitigating peak demand spikes, V2G helps avoid grid overloads and potential blackouts, thus improving reliability.

- **Frequency regulation**: EV batteries can respond rapidly to changes in grid frequency, providing ancillary services such as frequency regulation. When integrated into V2G systems, EVs can inject or absorb power as needed to help maintain grid frequency within acceptable limits. This real-time adjustment of power flows enhances grid stability by ensuring a balanced supply–demand equilibrium. A more stable grid reduces the likelihood of frequency deviations that could lead to equipment damage, power interruptions, or system-wide failures.
- **Grid balancing and voltage support**: V2G technology can also assist in balancing supply and demand across different parts of the grid and providing voltage support. By strategically dispatching power from EV batteries to areas experiencing voltage fluctuations or congestion, V2G systems help maintain grid reliability and minimize voltage deviations. This proactive grid balancing improves overall system resilience and reduces the risk of voltage instability-related outages or equipment failures.
- **Grid resilience and redundancy**: Integrating V2G into the grid enhances its resilience and redundancy by diversifying energy resources and providing distributed energy storage capabilities. In the event of unforeseen disruptions, such as extreme weather events or equipment failures, V2G-enabled EVs can serve as backup power sources, providing critical electricity supply to support essential services and emergency response efforts. This increased resilience helps minimize the economic impact of grid disruptions by reducing downtime costs and mitigating losses for businesses and households.

The improved reliability resulting from V2G technology translates into several economic benefits:

- **Reduced economic losses**: A more reliable grid minimizes the occurrence and duration of power outages, reducing economic losses associated with disrupted operations, lost productivity, and damaged equipment for businesses and industries. By avoiding downtime costs, businesses can maintain continuous operations and preserve revenue streams, leading to overall economic stability and growth.
- **Increased energy efficiency**: Enhanced grid reliability enables more efficient utilization of electricity resources, reducing wastage and optimizing energy flows. This increased energy efficiency translates into cost savings for consumers and businesses, as they benefit from lower energy bills and improved operational efficiency. By minimizing energy waste, V2G contributes to a more sustainable and cost-effective energy system, further supporting economic prosperity.

• **Stimulated investment and innovation**: A reliable grid infrastructure creates a conducive environment for investment and innovation in energy technologies and services. Businesses are more likely to invest in energy-intensive operations and adopt advanced technologies when they can rely on a stable and resilient power supply. This investment and innovation drive economic growth, job creation, and technological advancement, fostering a competitive and vibrant energy sector.

Thus, by leveraging the flexibility of EVs and integrating them into grid operations, V2G contributes to a more reliable, resilient, and cost-effective energy system, ultimately supporting sustainable economic development. Real-world examples reinforce this concept, such as the V2G pilot project in the United Kingdom led by EDF Energy and Nissan [90]. The project successfully demonstrated how a fleet of Nissan LEAF EVs equipped with V2G technology could support grid stability by responding to demand fluctuations. Another illustrative example comes from Japan, where the automaker Mitsubishi Motors collaborated with the utility company Chubu Electric Power to deploy V2G technology [91]. In this initiative, a fleet of EVs were utilized as mobile storage units to support the grid during peak demand periods. The successful implementation of this project showcased the adaptability of V2G systems to different types of electric vehicles and their potential to contribute to grid stability on a broader scale. Furthermore, a joint project by the University of Delaware and NRG Energy in the United States exemplifies the potential of V2G technology in enhancing grid reliability [92]. The study involved a fleet of EVs equipped with bidirectional chargers, demonstrating their ability to respond to grid signals in real time. By dynamically adjusting the charging and discharging patterns of the vehicles, the V2G system exhibited its capability to provide grid support services, reinforcing the notion that V2G technology can be a valuable tool for grid operators in maintaining stability and reliability. These examples underscore the practical benefits of V2G technology in bolstering grid resilience.

A study conducted by the NREL assessed the impact of V2G on grid stability in a simulated environment [88]. The results demonstrated that widespread adoption of V2G technology could lead to a 10% reduction in the need for additional grid infrastructure investments, translating to a saving of approximately $30 billion over a decade. Furthermore, the study found that V2G-enabled grid services, such as voltage regulation and frequency response, could enhance grid stability and reduce the occurrence of blackouts. Furthermore, a pilot project in Denmark [90] showcased the economic advantages of grid stability. By utilizing V2G capabilities, the project demonstrated a reduction in the need for grid balancing services, resulting in cost savings of approximately €230 000. In addition, Nissan's V2G pilot project in Denmark estimated that the aggregated value of frequency regulation services provided by V2G-enabled EVs could reach up to €800 per vehicle annually. This indicates substantial potential savings for grid operators compared to traditional frequency regulation methods [4]. The University of Delaware's V2G demonstration project [93] also calculated that V2G-enabled EVs could collectively provide regulation services valued at approximately $1500 per vehicle annually. This translates to significant economic benefits in terms of reduced operational costs and increased grid stability. Pacific Gas and Electric (PG&E) V2G Initiatives also suggest that V2G-enabled EVs could provide frequency control services at a fraction of the cost of traditional methods. While specific cost figures may vary, preliminary analyses indicate substantial potential savings for grid operators through the deployment of V2G technology [94]. Moreover, the California Independent System Operator also conducted a V2G pilot project to assess the economic viability of using V2G for grid stability. The pilot demonstrated that V2G-enabled EVs could provide grid stability services at a cost of around $100–$200 per vehicle annually, representing substantial savings compared to traditional ancillary services [95]. The Danish V2G initiative, "Parker," projected that EVs with V2G capabilities could deliver frequency regulation services worth around €600–€900 per vehicle annually. This project involved collaboration between several stakeholders including Danish transmission system operator Energinet and the University of Denmark [96]. The Dutch V2G project "V2G Hub" conducted by the Technical University of Eindhoven demonstrated that vehicles equipped with V2G could offer regulation services valued at approximately €500–€800 per vehicle annually [97]. This project aimed to showcase the economic benefits of V2G for grid operators and EV owners alike. A V2G pilot conducted in Portugal by EDP (Energias de Portugal) also estimated that EVs with V2G capabilities could provide frequency control services valued at approximately €300–€600 per vehicle annually [98]. The pilot aimed to assess the technical and economic feasibility of integrating V2G into the Portuguese electricity grid. A V2G trial conducted in Sweden by Vattenfall, a leading European energy company, found that V2G-enabled vehicles could provide grid stability services valued at around €400–€700 per vehicle annually [99]. This trial demonstrated the potential of V2G to support grid stability and reduce carbon emissions. Several V2G demonstrations in Japan, including projects by Tokyo Electric Power Company (TEPCO) and NEDO (New Energy and Industrial Technology Development Organization), also estimated that V2G-enabled EVs could provide grid stability services valued at approximately ¥50 000–¥100 000 per vehicle annually. These demonstrations aimed to assess the scalability and economic viability of V2G in the Japanese context [100].

Although the precise advantages of V2G technology for the power grid may fluctuate due to factors such as the scale of the V2G fleet, grid conditions, and regulatory frameworks, compiling estimates from the cases discussed above can underscore the substantial economic benefits of V2G technology for power grids.

7.6.4   *V2G business models*

Establishing appropriate business models plays a pivotal role in shaping the landscape of V2G technology, influencing its adoption and success in the broader energy ecosystem. Understanding the intricate interactions within the market is essential for stakeholders, including automakers, utilities, and technology providers, as they navigate the evolving terrain of V2G implementation. One key aspect of market dynamics for V2G technology is the emergence of new business models that capitalize on the synergies between EVs and the energy grid. Traditional models, such as pay-per-charge or subscription-based services, are evolving to incorporate V2G capabilities [101]. Utilities are exploring innovative pricing structures and service agreements that reward EV owners for participating in grid services, creating a symbiotic relationship between the transportation and energy sectors. For instance, a subscription-based model could offer EV owners discounted charging rates in exchange for allowing the grid to access their vehicle's battery during peak demand periods. This dynamic interplay between market forces and business models is crucial for establishing the economic viability of V2G technology.

Moreover, some utilities are adopting a platform-based approach, creating digital marketplaces where EV owners can offer their vehicles' storage capacity to the grid. These platforms facilitate transactions between grid operators and EV owners, creating a decentralized and dynamic marketplace for V2G services. Additionally, partnerships between automakers and energy companies are exploring integrated business models, where the purchase of an EV comes bundled with incentives for V2G participation, fostering a holistic approach to sustainable transportation and energy management. Moreover, market dynamics for V2G technology are influenced by regulatory frameworks and policy incentives. Governments worldwide are recognizing the importance of V2G in achieving energy and environmental goals, leading to supportive policies such as tax incentives, grants, and mandates. In California, for example, regulations encourage utilities to invest in V2G infrastructure and offer financial incentives to EV owners participating in grid services [12]. These policy measures create a conducive environment for market growth, stimulating both the supply and demand sides of the V2G ecosystem. Thus, the market dynamics and business models surrounding V2G technology are inextricably linked, shaping the trajectory of this transformative technology. The evolution of innovative business models, coupled with supportive regulatory frameworks, is essential for unlocking the full potential of V2G and establishing it as a viable and sustainable component of the energy landscape. As the market continues to mature, collaboration among stakeholders and ongoing adaptation of business strategies will be crucial in realizing the economic and financial benefits of V2G technology.

7.6.5 Environmental-related cost savings

Beyond the economic and financial aspects, V2G technology plays a crucial role in advancing environmental sustainability. A life-cycle analysis conducted by the European Commission [102] compared the environmental impact of traditional internal combustion engine vehicles with EVs equipped with V2G capabilities. The study revealed that over the lifespan of a vehicle, including manufacturing, use, and disposal, V2G-enabled EVs resulted in a 30% reduction in greenhouse gas emissions compared to conventional vehicles. This reduction in emissions not only aligns with global climate goals but also carries economic implications. The associated health benefits from reduced air pollution, estimated to be around $5 billion annually, contribute to the overall economic case for V2G technology adoption.

7.7 Conclusions

V2G technology holds immense importance in shaping the future of sustainable transportation and energy management. This technology facilitates the efficient use of RES, grid stabilization, and peak load management, all of which are critical for a greener and more resilient energy infrastructure. By allowing EVs to feed surplus energy back into the grid when not in use, V2G promotes a symbiotic relationship between the transportation and energy sectors, ultimately reducing greenhouse gas emissions and enhancing grid reliability.

Supportive policies and regulations play a pivotal role in fostering the development and adoption of V2G technology. Governments and regulatory bodies need to create a conducive environment that encourages investment in V2G infrastructure, ensures fair compensation for vehicle owners, and establishes interoperability standards. These policies can provide incentives for both consumers and businesses to embrace V2G, thereby accelerating its integration into mainstream transportation and energy systems. The success of V2G hinges on a collaborative effort between industry stakeholders and policymakers to create a regulatory framework that promotes innovation, safeguards consumer interests, and aligns with broader sustainability goals. While V2G holds great promise, addressing technical challenges is paramount for its widespread adoption. Key technical issues include standardizing communication protocols, developing robust cybersecurity measures to protect V2G systems from cyber threats, and improving battery technology for increased energy storage and discharge efficiency. Additionally, grid integration and management algorithms must be refined to maximize the benefits of V2G while minimizing the strain on the grid infrastructure. Overcoming these technical hurdles will be essential in ensuring the reliability, safety, and scalability of V2G technology, ultimately paving the way for a more sustainable and resilient energy future.

References

[1] Mwasilu F, Justo JJ, Kim EK, Do TD, and Jung JW. Electric vehicles and smart grid interaction: A review on vehicle to grid and renewable energy sources integration. *Renewable and Sustainable Energy Reviews*. 2014; 34:501–16.

[2] Liu C, Chau KT, Wu D, and Gao S. Opportunities and challenges of vehicle-to-home, vehicle-to-vehicle, and vehicle-to-grid technologies. *Proceedings of the IEEE*. 2013;101(11):2409–27.

[3] Yilmaz M and Krein PT. Review of the impact of vehicle-to-grid technologies on distribution systems and utility interfaces. *IEEE Transactions on Power Electronics*. 2012;28(12):5673–89.

[4]   Kempton W and Tomić J. Vehicle-to-grid power implementation: From stabilizing the grid to supporting large-scale renewable energy. *Journal of Power Sources*. 2005;144(1):280–94.

[5]   Precedence Research. Vehicle-to-grid technology market. Available online: https://www.precedenceresearch.com/vehicle-to-grid-technology-market (accessed on 05 August 2023).

[6]   Fortune Business Insights. Vehicle-to-grid market. Available online: https://www.fortunebusinessinsights.com/vehicle-to-grid-v2g-market-107673 (accessed on 16 July 2023).

[7]   Rajper SZ and Albrecht J. Prospects of electric vehicles in the developing countries: A literature review. *Sustainability*. 2020;12(5):1906.

[8]   Steward DM. *Critical Elements of Vehicle-to-Grid (V2G) Economics*. National Renewable Energy Lab. (NREL), Golden, CO, 2017.

[9]   Lopez, L., Warichet, J., Hernandez Alva, C.A., *et al. Grid Integration of Electric Vehicles: A Manual for Policy Makers*. International Energy Agency (IEA), Paris, 2022.

[10] Rahimi F and Ipakchi A. Demand response as a market resource under the smart grid paradigm. *IEEE Transactions on Smart Grid*. 2010;1(1):82–8.

[11] Cartwright ED. FERC order 2222 gives boost to DERs. *Climate and Energy*. 2020;37(5):22.

[12] Collantes G and Sperling D. The origin of California's zero emission vehicle mandate. *Transportation Research Part A: Policy and Practice*. 2008;42(10):1302–13.

[13] Bowen T, Koebrich S, McCabe K, and Sigrin B. The locational value of distributed energy resources: A parcel-level evaluation of solar and wind potential in New York state. *Energy Policy*. 2022;166:112744.

[14] Ofuji K and Tatsumi N. Wholesale and retail electricity markets in Japan: Results of market revitalization measures and prospects for the current. *Economics of Energy & Environmental Policy*. 2016;5(1):31–50.

[15] Taljegard M. Impact of vehicle-to-grid on the European electricity system-the electric vehicle battery as a storage option. In *2019 IEEE Transportation Electrification Conference and Expo (ITEC)* 2019 (pp. 1–5). Piscataway, NJ: IEEE.

[16] Deb S, Al Ammar EA, AlRajhi H, Alsaidan I, and Shariff SM. V2G pilot projects: Review and lessons learnt. *Developing Charging Infrastructure and Technologies for Electric Vehicles*. 2022 (pp. 252–67). New York: IGI Global.

[17] Wei Y, Huang H, Han X, *et al.* Whole-system potential and benefit of energy storage by vehicle-to-grid (V2G) under carbon neutrality target in China. In *2022 IEEE 5th International Electrical and Energy Conference (CIEEC)* 2022 (pp. 4006–4012). Piscataway, NJ: IEEE.

[18] Lucas-Healey K, Sturmberg BC, Ransan-Cooper H, and Jones L. Examining the vehicle-to-grid niche in Australia through the lens of a trial project. *Environmental Innovation and Societal Transitions*. 2022;42:442–56.

[19] Ustun TS, Zayegh A, and Ozansoy C. Electric vehicle potential in Australia: Its impact on smartgrids. *IEEE Industrial Electronics Magazine*. 2013;7(4):15–25.

[20] Bakker S, Maat K, and Van Wee B. Stakeholders interests, expectations, and strategies regarding the development and implementation of electric vehicles: The case of the Netherlands. *Transportation Research Part A: Policy and Practice*. 2014;66:52–64.

[21] van Heuveln K, Ghotge R, Annema JA, van Bergen E, van Wee B, and Pesch U. Factors influencing consumer acceptance of vehicle-to-grid by electric vehicle drivers in the Netherlands. *Travel Behaviour and Society*. 2021;24:34–45.

[22] Kester J, Noel L, de Rubens GZ, and Sovacool BK. Promoting vehicle to grid (V2G) in the Nordic region: Expert advice on policy mechanisms for accelerated diffusion. *Energy Policy*. 2018;116:422–32.

[23] Chen CF, de Rubens GZ, Noel L, Kester J, and Sovacool BK. Assessing the socio-demographic, technical, economic and behavioral factors of Nordic electric vehicle adoption and the influence of vehicle-to-grid preferences. *Renewable and Sustainable Energy Reviews*. 2020;121:109692.

[24] Meelen T, Doody B, and Schwanen T. Vehicle-to-grid in the UK fleet market: An analysis of upscaling potential in a changing environment. *Journal of Cleaner Production*. 2021;290:125203.

[25] Begley J and Berkeley N. UK policy and the low carbon vehicle sector. *Local Economy*. 2012;27(7):705–21.

[26] Pillai JR and Bak-Jensen B. Integration of vehicle-to-grid in the western Danish power system. *IEEE transactions on sustainable energy*. 2010;2 (1):12–9.

[27] Noel L, De Rubens GZ, and Sovacool BK. Optimizing innovation, carbon and health in transport: Assessing socially optimal electric mobility and vehicle-to-grid pathways in Denmark. *Energy*. 2018;153:628–37.

[28] Loisel R, Pasaoglu G, and Thiel C. Large-scale deployment of electric vehicles in Germany by 2030: An analysis of grid-to-vehicle and vehicle-to-grid concepts. *Energy Policy*. 2014;65:432–43.

[29] Wang D, Locment F, and Sechilariu M. Modelling, simulation, and management strategy of an electric vehicle charging station based on a DC microgrid. *Applied Sciences*. 2020;10(6):2053.

[30] Das HS, Rahman MM, Li S, and Tan CW. Electric vehicles standards, charging infrastructure, and impact on grid integration: A technological review. *Renewable and Sustainable Energy Reviews*. 2020;120:109618.

[31] Habib S, Kamran M, and Rashid U. Impact analysis of vehicle-to-grid technology and charging strategies of electric vehicles on distribution networks – A review. *Journal of Power Sources*. 2015;277:205–14.

[32] Australian Energy Market Operator (AEMO). *Vehicle-Grid Integration Standards Taskforce – Key Findings*. Distributed Energy Integration Program, 2021.

[33] Schmutzler J, Wietfeld C, and Andersen CA. Distributed energy resource management for electric vehicles using IEC 61850 and ISO/IEC 15118. In *2012 IEEE Vehicle Power and Propulsion Conference* 2012 (pp. 1457–1462). Piscataway, NJ: IEEE.

[34] Jang Y, Sun Z, Ji S, *et al.* Grid-connected inverter for a PV-powered electric vehicle charging station to enhance the stability of a microgrid. *Sustainability.* 2021;13(24):14022.

[35] DeBlasio R and Tom C. Standards for the smart grid. In *2008 IEEE Energy 2030 Conference* 2008 (pp. 1–7). Piscataway, NJ: IEEE.

[36] Sachan S, Deb S, Singh PP, Alam MS, and Shariff SM. A comprehensive review of standards and best practices for utility grid integration with electric vehicle charging stations. *Wiley Interdisciplinary Reviews: Energy and Environment.* 2022;11(3):e424.

[37] Gopstein A, Nguyen C, O'Fallon C, Hastings N, and Wollman D. *NIST Framework and Roadmap for Smart Grid Interoperability Standards, Release 4.0.* Gaithersburg, MD: Department of Commerce. National Institute of Standards and Technology, 2021.

[38] Corchero, C., Sanmartí, M., González-Villafranca, S., and Chapman, N. *V2X Roadmap 2019: Task 28 on Home Grids and V2X Technologies.* International Energy Agency (IEA), Paris, 2019.

[39] Venegas FG, Petit M, and Perez Y. Active integration of electric vehicles into distribution grids: Barriers and frameworks for flexibility services. *Renewable and Sustainable Energy Reviews.* 2021;145:111060.

[40] https://energy-rules.aemc.gov.au/ner/379.

[41] Lu MH and Jen MU. Safety design of electric vehicle charging equipment. *World Electric Vehicle Journal.* 2012;5(4):1017–24.

[42] Kuljeet K, Sahil G, Georges K, *et al.* A secure, lightweight, and privacy-preserving authentication scheme for V2G connections in smart grid. In *IEEE Conference on Computer Communications Workshops.* Paris, France. Piscataway, NJ: IEEE, 2019.

[43] LaMonaca S and Ryan L. The state of play in electric vehicle charging services: A review of infrastructure provision, players, and policies. *Renewable and Sustainable Energy Reviews.* 2022;154:111733.

[44] Gong X and Rangaraju J. *Taking Charge of Electric Vehicles: Both in the Vehicle and on The Grid.* Dallas, TX: Texas Instruments, 2018:1–3.

[45] Evenergi Pty Ltd and SD Planning. *Charging Gippsland for Future Transport*, prepared for South Gippsland Shire Council and Gippsland Regional Councils, Victoria, Australia, 2019. [Online]. Available: https://www.localgovernment.vic.gov.au/__data/assets/pdf_file/0021/168015/South-Gippsland-Charging-Gippsland-for-Future-Transport-Final-Report.pdf.

[46] Electric Vehicle Council. State of Electric Vehicles, March 2022. [Online]. Available: https://electricvehiclecouncil.com.au/wp-content/uploads/2022/03/EVC-State-of-EVs-2022.pdf.

[47] Hemavathi S and Shinisha A. A study on trends and developments in electric vehicle charging technologies. *Journal of Energy Storage.* 2022;52:105013.

[48] Pearre NS and Ribberink H. Review of research on V2X technologies, strategies, and operations. *Renewable and Sustainable Energy Reviews.* 2019;105:61–70.

[49] Hannan MA, Mollik MS, Al-Shetwi AQ, *et al.* Vehicle to grid connected technologies and charging strategies: Operation, control, issues and recommendations. *Journal of Cleaner Production.* 2022;339:130587.

[50] https://www.evenergi.com/vehicle-to-grid-an-overview.

[51] https://www.racv.com.au/royalauto/transport/electric-vehicles/bidirectional-charging-explained.html.

[52] Sharma A and Sharma S. Review of power electronics in vehicle-to-grid systems. *Journal of Energy Storage.* 2019;21:337–61.

[53] https://www.abc.net.au/news/science/2022-02-14/electric-vehicle-first-ev-chargers-v2g-v2h-to-arrive-australia/100811130.

[54] Vadi S, Bayindir R, Colak AM, and Hossain E. A review on communication standards and charging topologies of V2G and V2H operation strategies. *Energies.* 2019;12(19):3748.

[55] Quinn C, Zimmerle D, and Bradley TH. The effect of communication architecture on the availability, reliability, and economics of plug-in hybrid electric vehicle-to-grid ancillary services. *Journal of Power Sources.* 2010; 195(5):1500–9.

[56] Qiao L, Liu X, and Jiang B. Design and implementation of the smart meter in vehicle-to-grid. In *2011 4th International Conference on Electric Utility Deregulation and Restructuring and Power Technologies (DRPT)* 2011 (pp. 618–621). Piscataway, NJ: IEEE.

[57] Noel L, De Rubens GZ, Kester J, and Sovacool BK. *Vehicle-to-Grid.* Cham: Springer. 2019.

[58] Patarau TM, Petreus DM, Ferencz I, and Orban Z. Comparison between LLC and phase-shift converter with synchronous rectification for high power, high current applications. In *2020 IEEE 26th International Symposium for Design and Technology in Electronic Packaging (SIITME)* 2020 (pp. 398–403). Piscataway, NJ: IEEE.

[59] https://frenetic.ai/magnetic-notes/comparison-of-isolated-dc-dc-converter-topologies-for-high-power-applications.

[60] https://www.energy.gov/sites/default/files/2021-06/elt238_lukic_2021_o_5-14_535pm_KF_TM.pdf.

[61] https://www.freedm.ncsu.edu/2022/07/06/xfc-development-continues/.

[62] Ma Y, Zhang B, Zhou X, *et al.* An overview on V2G strategies to impacts from EV integration into power system. In *2016 Chinese Control and Decision Conference (CCDC)* 2016 (pp. 2895–2900). Piscataway, NJ: IEEE.

[63] Tan KM, Ramachandaramurthy VK, and Yong JY. Integration of electric vehicles in smart grid: A review on vehicle to grid technologies and optimization techniques. *Renewable and Sustainable Energy Reviews.* 2016;53:720–32.

[64] Marinelli M, Martinenas S, Knezović K, and Andersen PB. Validating a centralized approach to primary frequency control with series-produced electric vehicles. *Journal of Energy Storage.* 2016;7:63–73.

[65] Liu H, Hu Z, Song Y, and Lin J. Decentralized vehicle-to-grid control for primary frequency regulation considering charging demands. *IEEE Transactions on Power Systems.* 2013;28(3):3480–9.

[66] Shao C, Wang X, Wang X, Du C, and Wang B. Hierarchical charge control of large populations of EVs. *IEEE Transactions on Smart Grid*. 2015;7 (2):1147–55.

[67] Saunders E, Butler T, Quiros-Tortos J, Ochoa LF, and Hartshorn R. Direct control of EV charging on feeders with EV clusters. In *23rd International Conference on Electricity Distribution* 2015 (pp. 1–5).

[68] Heussen K, You S, Biegel B, Hansen LH, and Andersen KB. Indirect control for demand side management: A conceptual introduction. In *2012 3rd IEEE PES Innovative Smart Grid Technologies Europe (ISGT Europe)* 2012 (pp. 1–8). Piscataway, NJ: IEEE.

[69] Coignard J, MacDougall P, Stadtmueller F, and Vrettos E. Will electric vehicles drive distribution grid upgrades?: The case of California. *IEEE Electrification Magazine*. 2019;7(2):46–56.

[70] PlugShare. Available on: https://www.plugshare.com/.

[71] Ghosal A and Conti M. Security issues and challenges in V2X: A survey. *Computer Networks*. 2020;169:107093.

[72] Kalogridis G, Sooriyabandara M, Fan Z, and Mustafa MA. Toward unified security and privacy protection for smart meter networks. *IEEE Systems Journal*. 2013;8(2):641–54.

[73] Ferrag MA, Maglaras LA, Janicke H, Jiang J, and Shu L. A systematic review of data protection and privacy preservation schemes for smart grid communications. *Sustainable Cities and Society*. 2018;38:806–35.

[74] Zecchino A, Prostejovsky AM, Ziras C, and Marinelli M. Large-scale provision of frequency control via V2G: The Bornholm power system case. *Electric Power Systems Research*. 2019;170:25–34.

[75] Noel L, Zarazua de Rubens G, Kester J, *et al.* The technical challenges to V2G. *Vehicle-to-Grid: A Sociotechnical Transition Beyond Electric Mobility*. 2019 (pp. 65–89). Berlin: Springer.

[76] Zagrajek K, Paska J, Sosnowski Ł, Gobosz K, and Wróblewski K. Framework for the introduction of vehicle-to-grid technology into the Polish electricity market. *Energies*. 2021;14(12):3673.

[77] Shafie-Khah M, Neyestani N, Damavandi MY, Gil FA, and Catalão JP. Economic and technical aspects of plug-in electric vehicles in electricity markets. *Renewable and Sustainable Energy Reviews*. 2016;53:1168–77.

[78] Lamedica R, Geri A, Gatta FM, Sangiovanni S, and Maccioni M, Ruvio A. Integrating electric vehicles in microgrids: Overview on hosting capacity and new controls. *IEEE Transactions on Industry Applications*. 2019;55 (6):7338–46.

[79] Paudyal P, Ghosh S, Veda S, Tiwari D, and Desai J. EV hosting capacity analysis on distribution grids. In *2021 IEEE Power & Energy Society General Meeting (PESGM)* 2021 (pp. 1–5). Piscataway, NJ: IEEE.

[80] Zhao J, Wang J, Xu Z, Wang C, Wan C, and Chen C. Distribution network electric vehicle hosting capacity maximization: A chargeable region optimization model. *IEEE Transactions on Power Systems*. 2017;32(5):4119–30.

[81] Lin X, Sun J, Ai S, Xiong X, Wan Y, and Yang D. Distribution network planning integrating charging stations of electric vehicle with V2G. *International Journal of Electrical Power & Energy Systems*. 2014;63:507–12.

[82] Goodarzi M, Sadat Nouprvar B, Safaei A, and Mozaffari M. A novel algorithm for improving the overcurrent relay coordination with the consideration of EV charging station. In *2019 International Power System Conference (PSC)* 2019 (pp. 280–286). Piscataway, NJ: IEEE.

[83] Saldarriaga-Zuluaga SD, López-Lezama JM, Zuluaga Ríos CD, and Villa Jaramillo A. Effects of the incorporation of electric vehicles on protection coordination in microgrids. *World Electric Vehicle Journal*. 2022;13 (9):163.

[84] Yadav PK, Yadav RP, Sah S, and Jha S. Analysis of the impact of EV penetration on protection coordination. Institute of Engineering, Tribhuvan University, Nepal, 2023.

[85] https://www.cpuc.ca.gov/industries-and-topics/electrical-energy/infrastructure/transportation-electrification/charging-infrastructure-deployment-and-incentives.

[86] IEA, *Global EV Outlook 2023*, IEA. Paris. 2023. https://www.iea.org/reports/global-ev-outlook-2023, License: CC BY 4.0.

[87] Crow A, Mullaney D, Liu Y, and Wang Z. *A New EV Horizon: Insights from Shenzhen's Path to Global Leadership in Electric Logistics Vehicles*. Boulder, CO, Rocky Mountain Institute, 2019. [Online]. Available: https://www.rmi.org/insight/a-new-ev-horizon.

[88] International Renewable Energy Agency (IRENA). *Innovation Landscape for a Renewable-Powered Future: Solutions to Integrate Variable Renewables*, Abu Dhabi, 2019. [Online]. Available: https://www.irena.org/-/media/Files/IRENA/Agency/Publication/2019/Feb/IRENA_Innovation_Landscape_2019_report.pdf.

[89] https://nuvve.com/projects/eflex-united-kingdom/.

[90] https://www.edfenergy.com/media-centre/news-releases/edf-and-nissan-launch-new-commercial-v2g-service-ev-fleets.

[91] https://www.v2g-hub.com/projects/chubu-electric-power-toyota-v2g-pilot/.

[92] https://www.udel.edu/udaily/2021/march/ud-developed-vehicle-to-grid-technology-gains-traction/.

[93] Kempton W and Letendre SE. Electric vehicles as a new power source for electric utilities. *Transportation Research Part D: Transport and Environment*. 1997;2(3):157–75.

[94] Pacific Gas and Electric Company. Electric vehicles and grid modernization. 2020. Available online at https://www.pge.com/en_US/sustainability/clean-energy-future/zero-emission-vehicles/ev-grid-modernization.

[95] California ISO. *Vehicle-to-Grid (V2G) Pilot*. 2017.

[96] Energinet, "Parker project," available at: https://en.energinet.dk/.

[97] V2G Hub project, available at: https://www.tue.nl/en/research/research-groups/sustainable-innovation-and-entrepreneurship/research/v2g-hub/.

[98] EDP. Vehicle-to-grid (V2G). available at: https://www.edp.com/en/innovation/vehicle-grid-v2g.

[99]   Vattenfall, Vehicle-to-grid, available at: https://group.vattenfall.com/se/
        our-business/projects/vehicle-to-grid.

[100]   TEPCO, V2G (vehicle-to-grid) demonstration project, available at: https://
        www.tepco.co.jp/en/.

[101]   https://energy5.com/garage-and-parking-lot-ev-charging-business-models-
        for-vehicle-owners-and-operators.

[102]   Hill N, Raugei M, Pons A, Vasileiadis N, Ong H, and Casullo L. *Envir-
        onmental Challenges Through the Life Cycle of Battery Electric Vehicles*,
        Directorate-General for Internal Policies, European Parliament (Committee
        on Transport and Tourism), Brussels, 2023. [Online]. Available: https://
        www.europarl.europa.eu/RegData/etudes/STUD/2023/733112/IPOL_STU
        (2023)733112_EN.pdf

Chapter 8

Emerging EV charging technologies and V2G applications

Bingkun Song[1], Weihao Dong[1] and Udaya K. Madawala[1]

8.1 Introduction

This chapter introduces emerging electric vehicle (EV) charging technologies, including wired EV charging and wireless EV charging. In addition, vehicle-to-everything (V2X), including vehicle-to-grid (V2G) systems and applications, are presented. First, the EV charging systems are classified and summarised in Section 8.2. It introduces wired EV charging technologies, EV charging standards, wireless charging based on inductive power transfer (IPT) technology and commercial advancements. Next, grid impacts and demand-side issues related to EV charging are analysed in Section 8.3. Afterwards, in Section 8.4, strategies to mitigate grid impacts and demand-side issues are introduced. These strategies include optimal planning and operation of EV charging infrastructure. Then, V2G opportunities are discussed in Section 8.5. The opportunities involve V2G for demand-side management and wireless V2G systems. Finally, Section 8.6 concludes this chapter.

8.2 EV charging systems

An EV charging system refers to the equipment and infrastructure that charge the batteries of EVs to provide electric energy. The batteries of EVs need to be charged by DC power, but the power supplied by the grid is generally AC power. Thus, EV charging systems need to convert AC power to DC power. As shown in Figure 8.1, EV charging systems can be classified as wired and wireless types, depending on whether there is physical contact between EVs and the charging equipment. Depending on the direction of power flow to and from EVs, the charging system can be further classified into unidirectional and bidirectional charging.

[1]Department of Electrical, Computer, and Software Engineering, University of Auckland, New Zealand

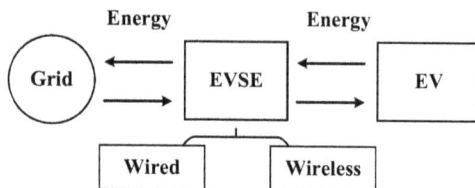

Figure 8.1   EV charging system

8.2.1   EV supply equipment

Electric vehicle supply equipment (EVSE) is the infrastructure that provides the power from the local distribution network (LDN) to an EV [1]. The main components of the EVSE include electrical conductors, control units, metres, communication devices, software systems, payment systems and other devices [1]. EVSEs are commonly known as charging stations or charging points, which are typically located at EV car parks [1]. Some EVSEs can convert AC power to DC power. Then, they provide DC power to EVs directly. These EVSEs are referred to as fast EV chargers as well. Some other EVSEs provide AC power to EVs. Then, EVs can be charged using their on-board chargers to convert AC power to DC power. Depending on whether there is physical contact between an EV and an EVSE, EVSEs can be classified into wired and wireless types, as shown in Figure 8.1.

8.2.2   Wired EV charging systems

Wired EV charging systems require physical contact between EVSEs and EVs, and the connection is made by charging cables. Depending on the form of power supplied to EVs, there are on-board and off-board charging systems. When the power supplied to an EV is in the form of AC, the EV battery must be charged by an on-board charger to convert AC power to DC power. When the power supplied to an EV is in the form of DC, an off-board charger must be used to convert AC power to DC power.

8.2.2.1   On-board EV charging

On-board charging systems are integrated inside EVs, providing convenient charging at private locations such as personal garages. An example of charging at a personal garage using the on-board charging system is shown in Figure 8.2(a) [2]. This type of charging can be realised by simply plugging the charging cable into a household power outlet. However, due to the power limitation of the power outlets, the charging power of this method is the lowest. Generally, this charging method is referred to as Level-1 charging. To increase the charging power and speed, the on-board charging systems can be connected to dedicated charging equipment. These types of equipment can be installed in private and public places, such as personal garages and public car parks. An example of EV charging at a parking place using the on-board charging system with dedicated charging equipment is shown in Figure 8.2(b) [3]. Generally, this charging method is referred to as Level-2 charging.

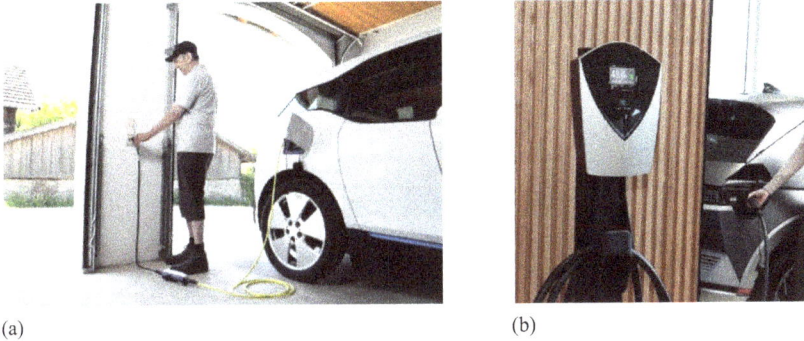

Figure 8.2   (a) An example of Level-1 charging [2] and (b) an example of Level-2 charging [3]

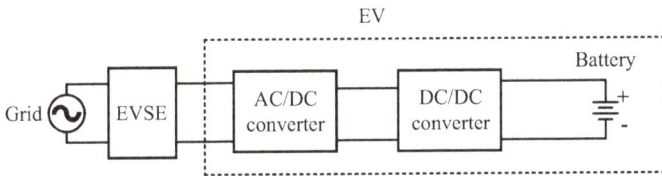

Figure 8.3   Block diagram of the on-board EV charging system

Considering the cost, size and weight of EVs, on-board charging systems are designed only for low-power applications. Consequently, on-board charging systems draw power from the grid through single-phase power supply because of the low charging power. Generally, EV charging using on-board systems is referred to as slow charging. The block diagram of the EV on-board charging system is shown in Figure 8.3. Most on-board charging systems can provide unidirectional power flow, where the power flows from the grid to the EV battery. Generally, a unidirectional on-board charging system comprises a rectifier for AC/DC conversion and a DC/DC converter. In addition to the unidirectional on-board charging system, the on-board charging system can be upgraded by replacing the rectifier and the unidirectional DC/DC converter with a bidirectional AC/DC converter and DC/DC converter [4]. With the bidirectional on-board charging systems, EVs can provide V2G services to support the grid. The galvanic isolation can be achieved using either an LF transformer on the grid side or a high-frequency transformer integrated into the DC/DC converter.

8.2.2.2   Off-board EV charging

To facilitate fast EV charging, higher-power off-board charging systems are required [4–8]. This charging refers to Level-3 and extreme-fast charging. An example of fast EV charging is shown in Figure 8.4.

Figure 8.4   An example of fast EV charging [9]

These charging systems are located outside the EVs and are not affected by the cost, size and weight of EVs. Therefore, the off-board charging systems can be customised with higher power capacity to provide fast charging services for EVs. Typically, these charging systems are installed at public car parks. Generally, these charging systems, known as DC fast chargers, convert AC power to DC outside the EV and supply the DC power directly to EV batteries.

As a result of the high charging power of off-board charging systems, they are integrated into a three-phase power network. Based on the connection setup between off-board charging systems and the power grid, the charging systems can be classified into a common AC bus and a common DC bus, as demonstrated in Figures 8.5 and 8.6, respectively [10].

The two types of fast charging systems have distinct features in terms of their configurations and the ways they connect charging infrastructure and LDN. The common AC bus charging system requires relatively fewer upgrades to the distribution network. The fast chargers can be integrated to the point of common coupling of the LDN. The AC power is converted to DC power inside each fast charger, resulting in a relatively higher fault tolerance for the power converters. The common DC bus charging system requires the construction of the DC network. Therefore, it involves complexity and a higher investment in the initial construction. In addition, as the capacity of the DC network is limited by the central AC/DC converter, the number of fast chargers that can be connected should be determined according to the capacity of the AC/DC converter. Moreover, the proper functioning of the entire charging system relies heavily on the central AC/DC converter; thus, its fault tolerance is relatively lower compared to the common AC bus. However, the common DC bus system does have some advantages. First, the local energy storage devices such as batteries and renewable energy sources (RESs) can be easily integrated into the common DC bus. In addition, the common DC bus can

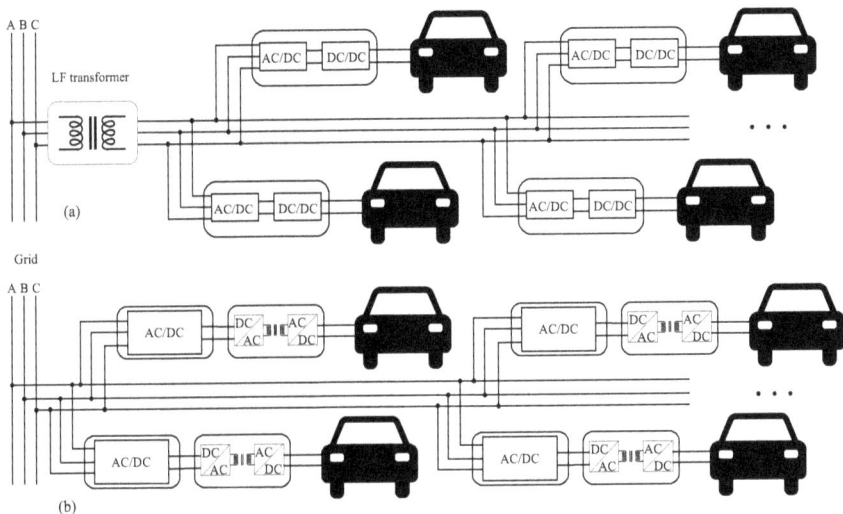

Figure 8.5   Common AC bus off-board EV charging system (a) with a central LF transformer and (b) with individual high-frequency transformers

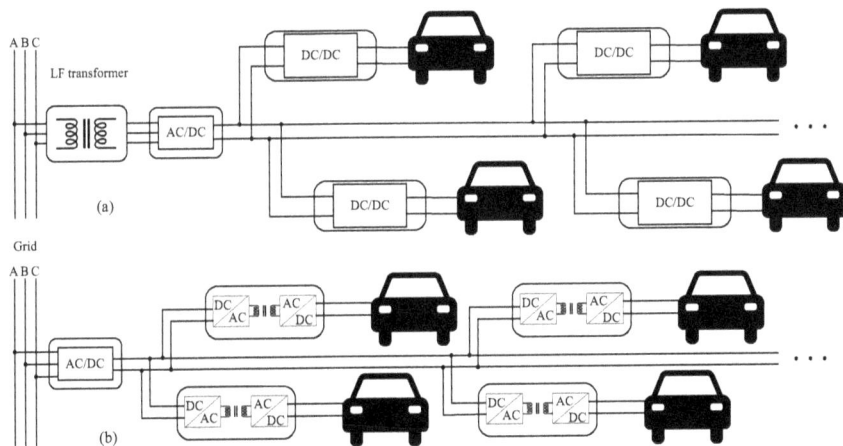

Figure 8.6   Common DC bus off-board EV charging system (a) with a central LF transformer and (b) with individual high-frequency transformers

be lengthened to connect more fast chargers if the capacity of the central AC/DC converter is sufficient. Moreover, the common DC bus system has a higher overall efficiency, as fewer conversion stages are required [11].

Depending on the typology of AC/DC and DC/DC converters, the off-board charging system can provide unidirectional or bidirectional charging. The bidirectional off-board charging system can provide V2G services to support the grid. The establishment of galvanic isolation between the EV batteries and the grid involves either the installation of a central LF transformer or the integration of individual high-frequency transformers inside DC/DC converters [10].

8.2.2.3 Summary of wired EV charging levels

Table 8.1 summarises the features associated with on-board and off-board charging systems, according to the SAE J1772 standard [4–8, 10].

8.2.3 Emerging wired EV charging infrastructure

This sub-section introduces five emerging charging technologies: pop-up charging stations, lamppost charging stations, street cabinet charging stations, reconfigurable EV chargers (REVCs) and movable charging stations.

Pop-up charging stations are charging infrastructure installed on the pavements. This charging infrastructure charges EVs next to pavements, as shown in Figure 8.7(a) [12]. The benefits of this charging infrastructure are that it reduces the footprint, and EV users do not need to find particular EV parking spaces. The key feature of this charging infrastructure is that it pops into the ground when there are no EVs, not taking up space on the pavement, as shown in Figure 8.7(b) [12].

Lamppost charging stations are charging infrastructure integrated into the street lamppost. This charging infrastructure charges EVs next to street lampposts, as shown in Figure 8.8(a) [13]. The benefits of this charging infrastructure are similar to the pop-up charging stations. In addition, there is no need to construct new infrastructure. Moreover, it makes the existing infrastructure becoming multi-functional. Another advantage of this charging infrastructure is that it can provide convenient charging services to EV users who do not have private charging capability.

Street cabinet charging stations are charging infrastructure transformed by street cabinets, as shown in Figure 8.8(b) [14]. These cabinets used to be telecommunication servers. However, some telecommunication infrastructure has been upgraded. Therefore, those old cabinets can be repurposed as charging infrastructure without destruction.

The increasing demand for fast charging has prompted the emergence of modular fast chargers, which enhance charging capacity by incorporating additional power modules. These chargers are composed of several identical power modules in parallel within an individual unit, improving overall charging capability. Unlike conventional single-module chargers, modular fast chargers offer improved redundancy and scalability. However, extra modules result in larger physical dimensions and higher costs. Moreover, modular fast chargers are always underutilised, as most EVs do not need the highest charging power. Therefore, the underutilisation adversely impacts the cost-effectiveness of these chargers.

In recent years, an innovative modular fast charger has been developed by ChargePoint [15]. This type of charger can dynamically allocate interior power modules of a power cabinet to enable different combinations of charging modules. These charging modules deliver an amount of power to a distributor, and several distributors are connected to a power cabinet, allowing the connection to be switched to diverse power modules. The control unit manages the connection status of power

Table 8.1 EV charging power levels

Voltage level	Charger location	Typical usage	Energy supply interface	Power level	Charging time	EV battery capacity
Level 1 120 V AC	On-board 1-Phase	At private places	Convenience outlet	1.4 kW 1.9 kW	14–36 h 11–26 h	EVs (20–50 kWh)
Level 2240 V AC	On-board 1- or 3-Phase	At private places or public places	Dedicated EV charging equipment	4 kW 8 kW 19.2 kW	5–12.5 h 2.5–6 h 1–2.5 h	EVs (20–50 kWh)
Level 3 (Fast) 200–600 V DC	Off-board 3-Phase	At public places	Dedicated EV charging equipment	50 kW 100 kW	0.4–1 h 0.2–0.5 h	EVs (20–50 kWh)
Extreme–Fast 800 V DC	Off-board 3-Phase	At public places	Dedicated EV charging equipment	400 kW	<10 min	EVs (20–50 kWh)

(a)

(b)

Figure 8.7 *(a) Photo of a working pop-up charging station and (b) photo of an idle pop-up charging station [12]*

(a)

(b)

Figure 8.8 *(a) Photo of a lamppost charging station [13] and (b) photo of a street cabinet charging station [14]*

modules from distributors, providing high flexibility in operating charging modules. For example, if an EV is charged with only a fraction of the rated power, other EVs can utilise idle modules, facilitating the concurrent charging of several EVs.

Furthermore, the self-contained nature of these modules streamlines their installation to chargers. The scalable design enables the modified rated power by increasing modules, allowing charger operators to expand the initial units as required. In addition, the high-power range makes chargers appropriate for diverse EVs and scenarios. These benefits, incorporating adaptability and scalability, have encouraged many manufacturers to develop products with the same functionalities. These chargers are called REVCs and have been studied in [16]. A comparison of regular, modular and REVCs is given in Table 8.2.

REVCs come in two types, with the first being the integrated-type REVCs. In this configuration, power modules and ancillary devices are housed within a single cabinet, and the charging cables are directly connected to this cabinet. An example of the integrated-type REVC is shown in Figure 8.9(a). To increase the maximum charging power, two integrated-type REVCs can be interconnected, enabling them to share power modules [17]. The alternative type is the split-type REVCs,

Table 8.2   Comparison of regular fast charger, modular fast charger and REVCs

Charger type	Regular charger	Modular charger	REVC
Modular design	No	Yes	Yes
No. of EVs charged simultaneously	1	1	>1
Flexibility	Low	Low	High
Redundancy	Low	High	High
Power range	Limited	Wide	Wide
Scalability	Low	High	High
Power module utilisation	Low	Low	High

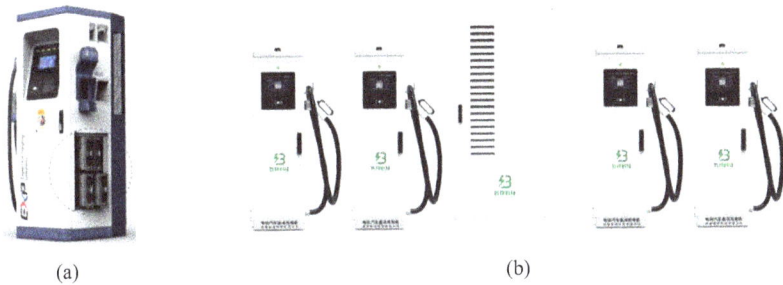

(a)   (b)

Figure 8.9   Reconfigurable EV chargers: (a) integrated-type [19] and (b) split-type [18]

characterised by a division between the power cabinet and charging posts. Several identical power modules, control units and ancillary devices are accommodated inside the power cabinet. Outside the power cabinet, several charging posts are connected to this power cabinet, as shown in Figure 8.9(b) [18]. The power cabinet regulates the power provided to a charging post by changing the operational modules. The concurrent charging capacity is determined by the number of charging posts. Given that multiple charging posts can be connected to a power cabinet, the split-type charger is more appropriate for concurrent charging multiple EVs. Similar to integrated-type REVCs, the maximum power can be expanded by interconnecting split-type chargers.

Dynamic power management (DPM) is the key technology realising the operational capabilities of REVCs. DPM plays an important role in enabling chargers to adapt to varied charging powers and support several EVs without extensive upgrades to the LDN. DPM dynamically regulates the existing power across several charging spaces instead of incorporating extra electrical components such as new circuits and transformers. This innovative approach allows for charging more EVs within the current power capacity limits [20]. For example, when numerous EVs are parked at a place for a long time, DPM efficiently manages REVCs to charge several EVs concurrently at a reduced power level. Moreover, if

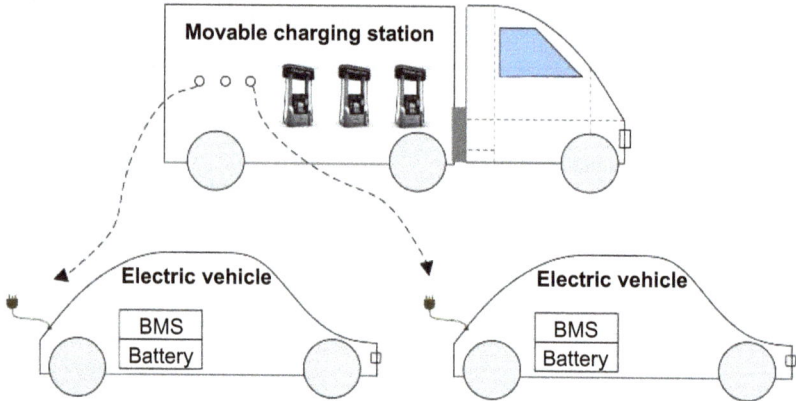

Figure 8.10   Schematic of a movable charging station [21]

there is a failure, the failsafe mechanism integrated into DPM will automatically diminish the charging power to ensure charging.

Movable charging stations are integrated into trucks or vans. These charging stations can be moved anywhere with trucks or vans to charge EVs, as shown in Figure 8.10 [21]. This breaks the conventional concept of EV charging, where the charging infrastructure is fixed at specific locations, and EVs must go to specific locations for charging. This charging station can significantly improve the travel satisfaction of EV users, as there is no need to plan a particular driving route for charging.

However, this charging station also has several problems that need to be solved. First, arranging appropriate recharging time for itself and planning routes to charge EVs to improve operational effectiveness and economy is crucial. This charging station is essentially a large-scale mobile energy storage device equipped with chargers. Therefore, it also needs to be recharged. Second, how to determine the capacity of this charging station also determines its operating rationality and economy. If the capacity is too small, it may not be able to satisfy EV charging demands. If the capacity is too large, it may cause the truck to be too large and waste too much energy during transportation. Third, how many mobile charging stations should be arranged in a region is crucial to meeting the EV charging demands and improving the economics of operations. Fourth, it is also important to set the charging price. If the price is too high, EV users may not choose this form of charging. If the price is too low, it can be unprofitable, especially for long-distance conditions. Some studies have investigated the movable charging stations, as detailed in [21–26].

8.2.4   Wireless EV charging systems

Wireless EV charging technology, based on IPT technology, offers a solution to the limitations of wired charging systems. IPT provides numerous advantages that address the issues associated with wired charging, including enhanced safety, increased convenience, reduced maintenance and cost-effectiveness. This section highlights the benefits of IPT technology.

8.2.4.1 Challenges of wired EV charging

Although wired EV charging has been successfully commercialised, it faces several significant issues. One major concern is the trip hazard it poses to pedestrians. For example, charging cables laid across pavements, as depicted in Figure 8.11(a) and (b) [27], can block walkways and endanger individuals, particularly those with disabilities or pushchairs. Additionally, there is an electric shock risk in adverse weather conditions; for instance, Figure 8.11(c) shows a charging port blocked by snow, which may increase the potential risk of electric shock. Furthermore, the US Department of Energy [28] points out that high-power public charging stations require structures like roofs to address users' concerns about charging in rainy weather. Additionally, as power levels increase, the weight of charging cables may exceed Occupational Safety and Health Administration regulations [28], becoming too heavy (>22.7kg) for one person to manage. These shortcomings significantly hinder the promotion of EVs. To address these issues, IPT technology has emerged as a research focus in academia and industry.

IPT technology presents a revolutionary approach to EV charging, providing many benefits over traditional wired methods [30–32]. By utilising magnetic coils to transform high-frequency current into an alternating magnetic field, IPT facilitates wireless power transmission, eliminating the necessity for cumbersome plugs and cables, as shown in Figure 8.12. This method enhances appeal and convenience and increases reliability by removing the potential for wear and tear due to physical connections. IPT's safety advantages are particularly notable in adverse weather conditions such as heavy rain or snow, where traditional plug-in systems may present hazards, providing a secure and reliable charging solution.

The IPT system's unique feature of 'snack charging' offers reassuring convenience, allowing the vehicle's battery to be charged during short stops at various locations, such as homes, workplaces, shopping centres or even traffic lights. This approach significantly alleviates range anxiety and diminishes the demand for larger battery capacities. Moreover, IPT's compatibility with automated charging processes, including automated parking and charging features, aligns with the anticipated needs of future autonomous EVs.

(a) (b) (c)

Figure 8.11   EV wired charging posing trip hazard (a) BWM i3 in London, (b) Nissan Leaf in Paris [27], and (c) exposed port of wired EV charging in the snow [29]

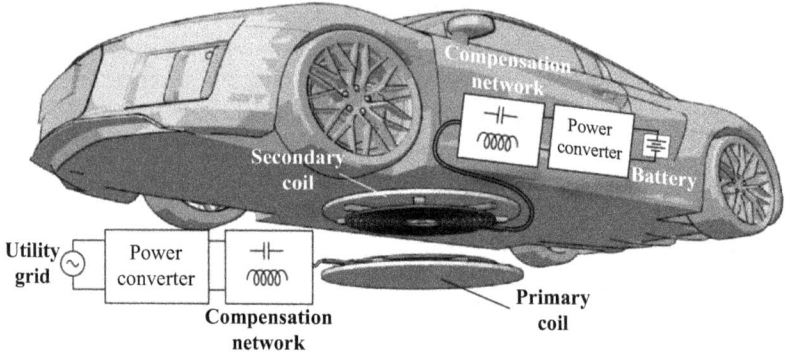

Figure 8.12   Wireless EV charging system

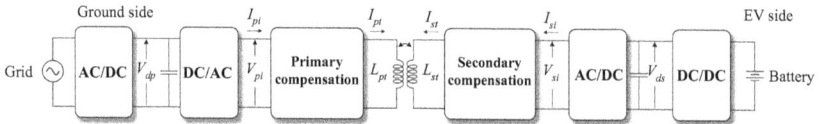

Figure 8.13   IPT-based wireless charging system

On the technical side, the components on the vehicle side convert the received power back into DC, powering the vehicle's equipment and battery without the high maintenance costs and complexity associated with wired systems. The possibility of integrating an active rectifier in place of a traditional passive one further broadens IPT's applicability, enabling V2G or V2X functionalities through various control strategies. This flexibility and the host of advantages underscore IPT's crucial role in advancing EV technology and adoption in the future.

8.2.4.2   IPT-based wireless EV charging system

Figure 8.13 illustrates a typical wireless EV charging system configuration. It comprises interconnected primary and secondary sides, linked by high-frequency magnetic flux across an air gap, represented by the mutual inductance M. Each side includes a converter, a compensation network and a magnetic coil. The primary side AC–DC converter features a power factor correction (PFC) unit connected to the electrical mains, serving as an interface, producing a stable DC bus for the system. A DC–AC inverter converts the stable DC into high-frequency AC. This is followed by a volt-ampere reactive (VAR) compensation network and a primary coil that generate magnetic flux. Near the primary coil, the secondary coil links this magnetic flux, inducing an electromotive force. Power transfer on the secondary side is facilitated by a VAR compensation network, a high-frequency AC–DC rectifier, and, optionally, a DC–DC converter.

8.2.4.3 Standards

Given the importance of IPT technology in promoting EVs, the industry actively pursues the development of higher power IPT charging systems to accelerate stationary wireless EV charging and enable in-motion EV charging [31,33,34]. In response, industry standards have been updated accordingly. For instance, the SAE J2954 standard, setting guidelines for wireless charging of light-duty and plug-in EVs [35], recently introduced higher power classes, such as WPT3 (11 kW) and WPT4 (22 kW). The SAE J2954 standard specifies a wireless EV charging system as including a ground assembly (GA) and a vehicle assembly (VA). Table 8.3 details the WPT power class classifications and their target efficiencies. Target efficiency measures the input power from the utility grid on the GA side against the output power delivered to the EV battery on the VA side. Power classes WPT1, WPT2 and WPT3 have been defined, with WPT4 classification currently in progress.

While international standards like GB28775 closely align with SAE J2954, the SAE Task Force is developing SAE J2954-2 to standardise wireless power transfer systems for heavy-duty electric vehicles [36,37]. Although still under development, SAE J2954-2's initial power classes, expected to reach up to 1 MW, are detailed in Table 8.4.

8.2.4.4 Commercialised IPT systems

This section summarises the commercialisation of IPT technology, distinguishing static wireless charging systems that require stationary vehicles for charging from dynamic systems enabling vehicle charging in motion.

In 1998, the University of Auckland introduced an IPT battery charging system for electric buses in Whakarewarewa Geothermal National Park, as shown in Figure 8.14(a). This was the first commercial IPT system for such an application, delivering 20 kW to onboard batteries with over 80% efficiency [38]. In 2000,

Table 8.3 WPT power classification for light-duty vehicles [36]

WPT power class	WPT1	WPT2	WPT3	WPT4
Max. input volt-amp (kVA)	3.7	7.7	11.1	22
Min. target efficiency (%)	>85	>85	>85	To be determined
Min. target efficiency at offset position	>80	>80	>80	To be determined
Frequency	85 kHz within the international frequency band of 81.38 and 90 kHz			

Table 8.4 Preliminary WPT power classification for heavy-duty vehicles [37]

WPT power class	WPT5	WPT6	WPT7	WPT8	WPT9
Maximum input volt-amp (kVA)	60	180	295	590	1175

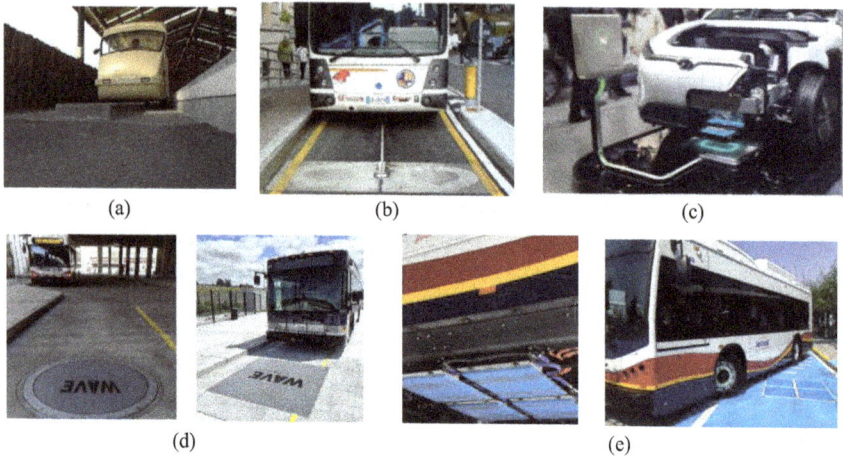

Figure 8.14 (a) 1998: University's pioneering IPT battery charging system for electric buses in Whakarewarewa Geothermal National Park [38]. (b) Conductix Wampfler's IPT-based electric bus chargers in Genoa and Turin, featuring underground primary coils, between 2000 and 2003 [39,40]. (c) WiTricity's DRIVE 11 evaluation system: a static EV wireless charger transmitting 11 kW at 94% efficiency [41]. (d) WAVE's advancements in WPT since 2011: offering up to 250 kW charging capabilities [43–46]. (e) Momentum dynamics' wireless charging solutions: ranging from 50 to 450 kW, implemented in Washington State's Link Transit by 2018 [19,47–50].

Conductix Wampfler promoted IPT chargers, installing the first system in Genoa, Italy, in 2002, followed by Turin in 2003. This system, shown in Figure 8.14(b), featured two underground circular primary coils with surface markings to guide buses, delivering 60 kW [39,40]. In 2007, WiTricity Corporation, founded by an MIT team, became a leader in static EV wireless charging, achieving 11 kW with 94% efficiency with their DRIVE 11 system, depicted in Figure 8.14(c) [41,42]. From 2011, WAVE, a Utah State University spin-off offering up to 250 kW, as shown in Figure 8.14(d) [43–46], was rebranded as Wave Charging by Ideanomics in 2021. Momentum Dynamics also leads in this field, offering solutions from 50 to 450 kW, implemented in Washington state's Link Transit by 2018, as shown in Figure 8.14(e) [19,47–50].

Dynamic wireless charging transfer (DWPT), while still in its early stages, holds the promise of greatly extending EV driving ranges and advancing electrified transportation. KAIST's on-line electric vehicle (OLEV) technology, launched in 2009, demonstrated power ranges from 3 to 17 kW, with the first 62 kW tram loop at Seoul Zoo in 2010 and a 100 kW OLEV bus at Expo 2012, as shown in Figure 8.15(a) [51–53]. Bombardier's PRIMOVE project, beginning in 2010, tested wireless EV charging up to 200 kW in Europe, as shown in Figure 8.15(b) [54–57].

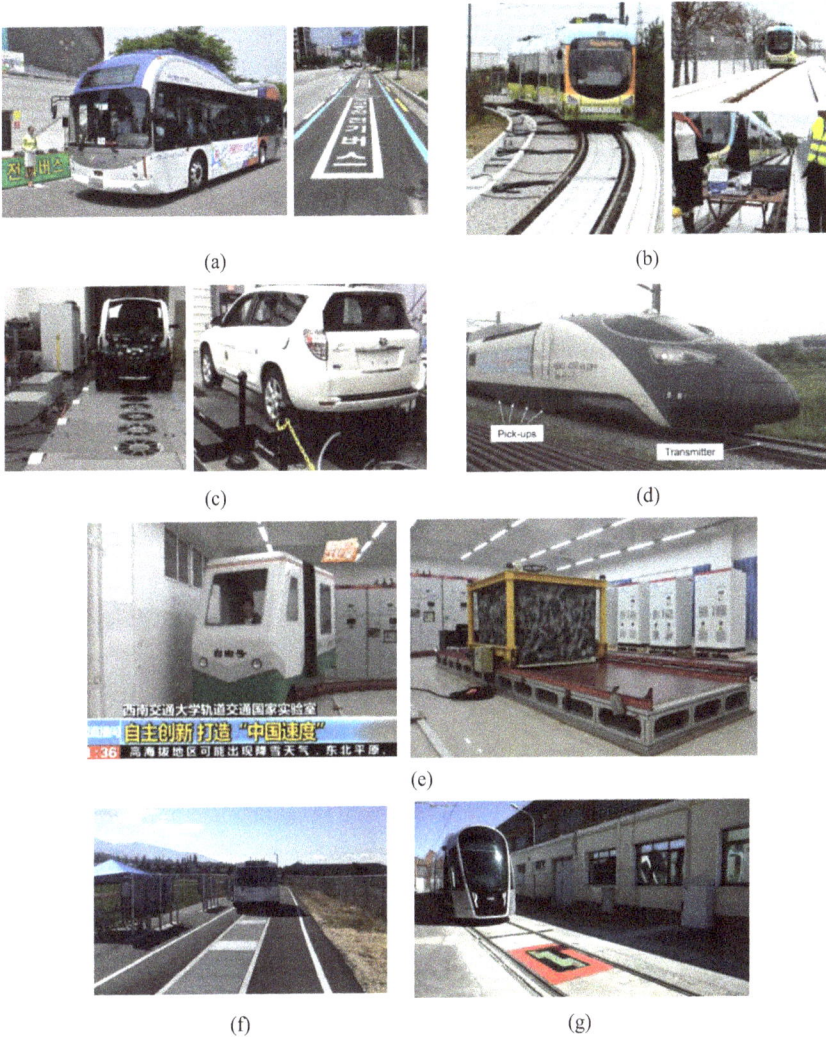

Figure 8.15   *(a) KAIST's OLEV system deployment in South Korea since 2009: advancing DWPT commercialisation with power levels up to 27 kW [51–53]. (b) Bombardier's PRIMOVE project since 2010: testing and deploying WPT technology in European cities with outputs up to 200 kW [54–57]. (c) ORNL's dynamic wireless charging technology progress since 2011: 12 kW system on Toyota RAV4 in 2016 with 95% efficiency [58–61]. (d) Korea Railroad Research Institute's 2015 achievement: enhancing high-speed railways with a 1-MW IPT system, 128 m transmitter, 82.7% efficiency and 10 km/h acceleration [62]. (e) Southwest Jiaotong University's 40 kW DWPT-powered rail test vehicle achieving 100 kW transfer across 12–15 cm air gap at 85% efficiency in 2016 [63]. (f) Utah State University's 2018 Breakthrough: 25 kW IPT system with dual coils achieves 86% efficiency, overcoming ±15 cm misalignments [64]. (g) IK4-IKERLAN's 2018 Power Electronics Innovation: 50 kW IPT system for railways with over 88% efficiency and up to ± 1,000 mm tolerance [65].*

Since 2011, ORNL has developed DWPT systems, with a 120 kW system achieving 97% efficiency by 2018, as shown in Figure 8.15(c) [58–61]. In 2015, a 1-MW wireless EV charging system for high-speed railways demonstrated 82.7% efficiency at an 818 kW output, as shown in Figure 8.15(d) [62]. Southwest Jiaotong University developed a 40 kW DWPT system in 2016 and achieved 100 kW transfer at 85% efficiency, as shown in Figure 8.15(e) [63]. In 2018, a 25 kW DWPT system with dual primary coils achieved 86% efficiency, as shown in Figure 8.15(f) [64]. That same year, IK4-IKERLAN's 50 kW DWPT system for railway traction demonstrated over 88% efficiency and tolerance for misalignment up to ±1,000 mm, as shown in Figure 8.15(g) [65].

In summary, IPT technology effectively addresses the limitations of wired EV charging, enhancing system safety and maintainability without notably compromising performance. It serves as a robust alternative to wired charging. Crucially, IPT has been validated for V2G and V2X applications, with further details in Section 8.5.

8.3   Grid impacts and demand-side issues of EV charging

As a grid load and transportation tool, EV poses challenges and opportunities for both the power grid and EV-related entities. As introduced in the previous section, the charging power of an EV is generally very high, especially when a large number of EVs are connected to the power grid for charging. This can significantly impact the power grid, thus affecting its safe and reliable operation. Therefore, the power grid requires upgrades to accommodate these EV loads. However, improper upgrading will affect the economic operation of the power grid and even increase the impacts on the power grid.

In addition, the demand-side issues of EVs need to be addressed. From the perspectives of charging infrastructure operators and EV users, considering their demand-side issues, and from the perspective of grid operators, considering the impacts of EVs on the grid, some effective technologies and concepts that solve the problems of both parties should be significantly developed.

In this section, the grid impacts from EV charging and the demand-side issues are summarised.

8.3.1   Grid impacts of EV charging

Large-scale EV charging can seriously impact the security and stable operation of the power grid. The grid impacts can be categorised into six types: increased power load, grid instability, asset overloading, increased power losses, harmonics and phase imbalance. These grid impacts are summarised and shown in Figure 8.16 [66].

When a large number of EVs are integrated into the grid for charging, the first and immediate impact is the increased power load. A study investigated the increased power load with the increase of EV penetration. A case study was conducted to show the increased load when the EV penetration is 11.3%, 35% and 45%, respectively. The results were compared with no EV scenario, as shown in

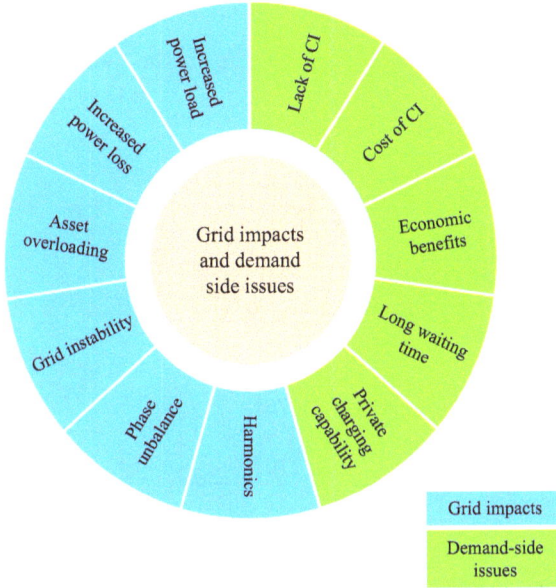

Figure 8.16   Types of grid impacts and demand-side issues of EV charging [66]

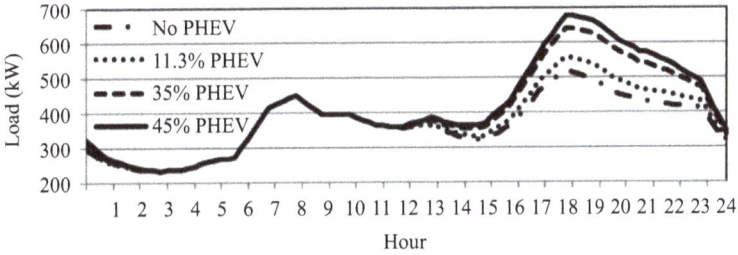

Figure 8.17   Increased power load with the increased EV penetration studied in [67]

Figure 8.17 [67]. In this case study, there were 166 households, and the EV penetration represented the percentage of 166 households. The lower and upper limits of charging power were set up as 1.56 and 5.52 kW, respectively. Based on the results, the peak load increased by around 40%, and the peak-valley difference increased by 170%, comparing 45% penetration with 0% penetration.

Another study was conducted based on the case in Germany. The study investigated the increased power load with the increased EV number in the country. Two scenarios with 1 million and 42 million EVs were simulated and compared with no EV scenario, as shown in Figure 8.18 [68]. From the results, the load fluctuation increased by only 1.5% compared to no EVs under the scenario of

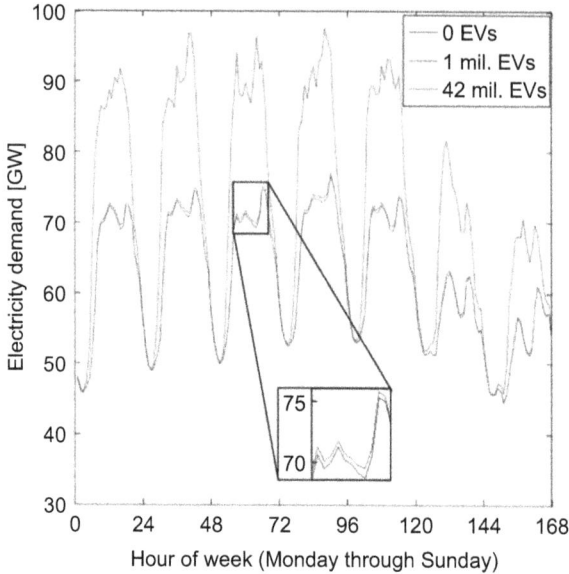

Figure 8.18   Increased power load with the increased EV number studied in [68]

1 million EVs. However, the load fluctuation increased by 92% compared to no EVs with the scenario of 42 million EVs. Therefore, large-scale EV charging can significantly increase the power load on the grid.

When a large number of EVs are connected to the grid for charging, the power load increases dramatically, which may result in the grid not being able to supply enough power, leading to an imbalance between supply and demand in the grid and thus affecting frequency stability. A study has shown that EV charging without proper control can lead to fluctuations in grid frequency, as shown in Figure 8.19 [69]. An effective method to alleviate grid frequency fluctuations caused by EV charging is to charge EVs in a coordinated manner, the blue curve as shown in Figure 8.19 [69]. By using coordinated charging, the grid frequency fluctuations can be reduced significantly. The detailed method is explained in the following section.

In addition, large-scale EV charging can cause voltage instability in the power grid. A study has demonstrated that uncontrolled charging of large-scale EVs can cause huge deviations in grid voltage, as shown in Figure 8.20 [70]. Voltage deviations worsen when the EV penetration increases. The voltage fluctuations can be mitigated by coordinated charging. The detailed method is introduced in the following section.

The increased power load of the grid can cause asset overloading of the grid. Long-term overloading will cause ageing and damage to the grid asset, consequently affecting the safe operation of the power grid. A study has been conducted on asset overloading under different EV penetrations, as shown in Figure 8.21 [71].

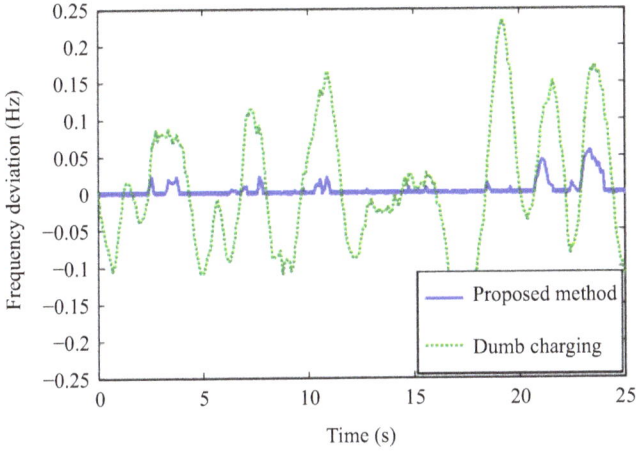

Figure 8.19   Frequency deviation with the uncontrolled and controlled EV charging studied in [69]

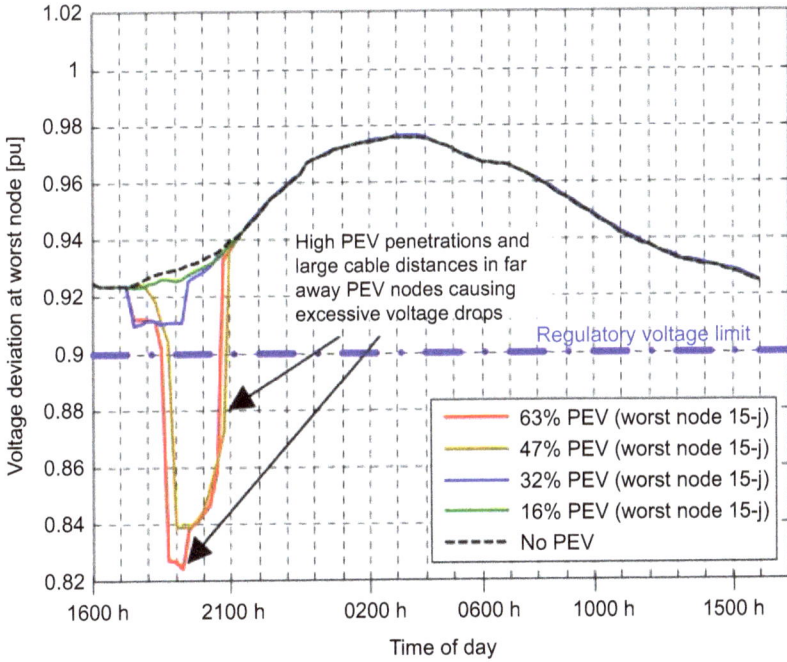

Figure 8.20   Voltage deviation with the uncontrolled EV charging studied in [70]

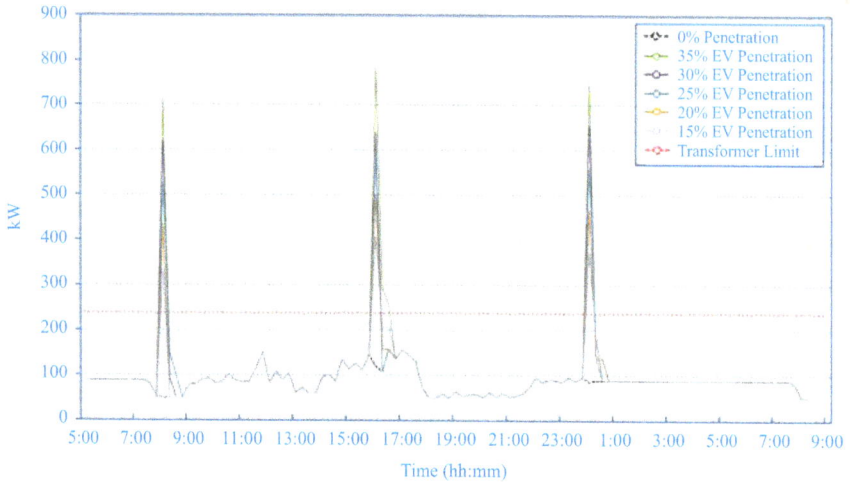

Figure 8.21   Overloading on the grid asset with different EV penetrations studied in [71]

It can be seen from the results that when EV penetration increases, the overloading of the transformer becomes worse, and these overloading periods are gathered in specific periods, while there is no overloading in other periods. This result shows that EV charging is usually clustered at particular periods. Based on this feature, some effective mechanisms and control methods should be applied to guide and manage EV charging to avoid overloading. Moreover, the grid asset should be upgraded under serious conditions to accommodate additional charging loads. Detailed grid upgrades and EV charging management methods will be introduced in the following section.

When the power load of the grid increases, the power loss also increases. A study verified that as EV penetration increases, the power losses on the grid increase, as shown in Figure 8.22 [70]. It can be seen from the results that when the EV penetration is 63%, the peak power loss increases by more than four times compared to the scenario without EVs.

EV charging can inject harmonics into the grid due to the nonlinear power electronics components in charging equipment [1]. As shown in Figure 8.23, currents are distorted when the passive diode rectifier works. The study recommended using active converters instead of passive rectifiers to eliminate harmonics and realise the unity power factor [72].

Single-phase EV charging may lead to phase imbalance in the LDN [73,74]. If single-phase charging loads are connected to the LDN in an unbalanced manner, it will cause load imbalance, thus, phase imbalance. The phase imbalance will cause voltage drops on heavily loaded phases, neutral line power losses, wrong operation of protection relays and low power factor.

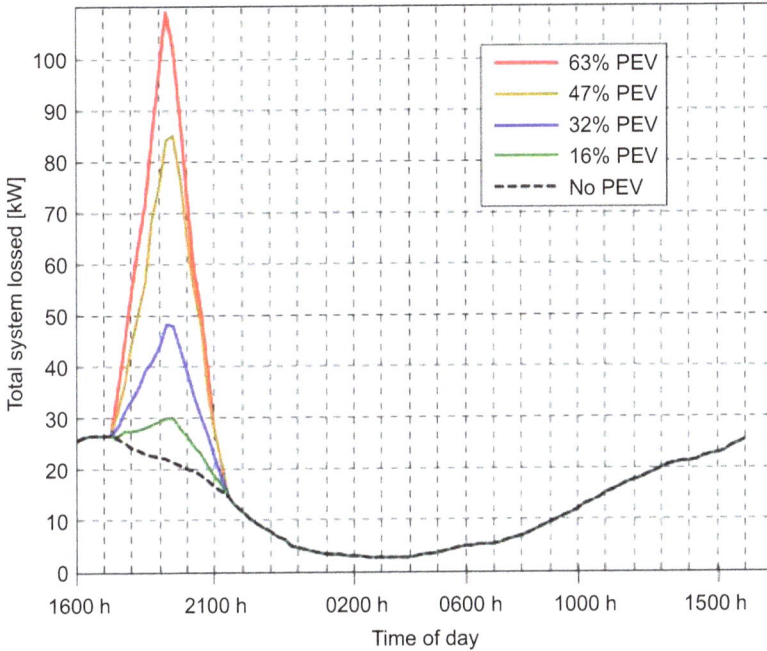

Figure 8.22   Power losses with the uncontrolled EV charging under different EV penetration rates studied in [70]

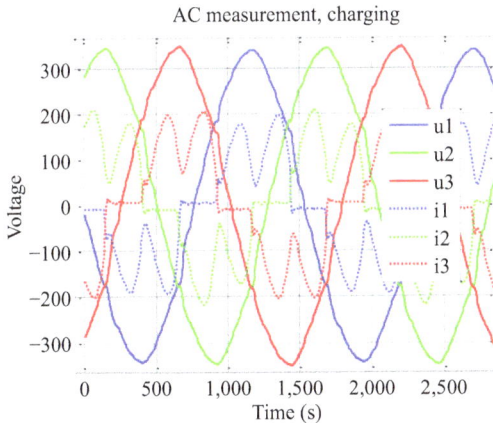

Figure 8.23   Current harmonics studied in [72]

8.3.2   Demand-side issues of EV charging

To increase the EV uptake, it is crucial to realise the demand-side issues of EV charging. These issues directly affect whether people choose to use EVs and the quality of charging services that EV charging infrastructure operators provide. These issues can be classified into five types: lack of charging infrastructure, high cost of deploying charging infrastructure, economic benefits of charging infrastructure operators, long waiting time of EV users and private charging capability of EV users. These demand-side issues are summarised in Figure 8.16 [66].

Due to the battery capacity limitation, range anxiety is the key issue preventing people from buying EVs. Therefore, the charging infrastructure is crucial in extending the driving range. However, the lack of charging infrastructure is currently the most significant demand-side issue. This is evident from comparing the number of EVs and public chargers worldwide. As shown in Figure 8.24, the number of EVs increased by over ten million worldwide in the past decade [75]. Nevertheless, as shown in Figure 8.25, the number of fast and slow EV chargers only increased to less than 400 thousand and 800 thousand, respectively, from 2015 to 2020 [75].

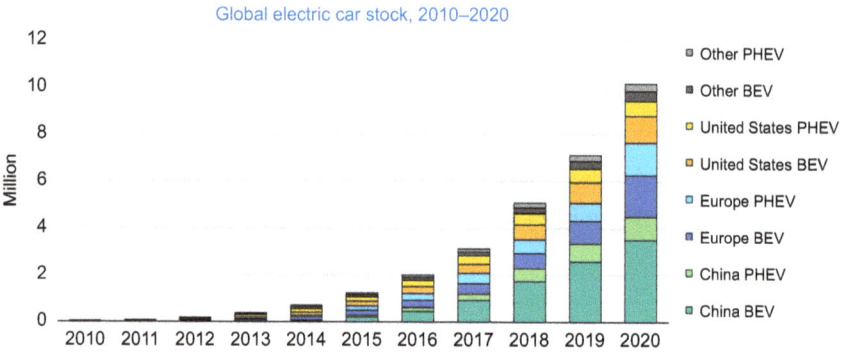

Figure 8.24   The number of EVs worldwide from 2010 to 2020 [75]

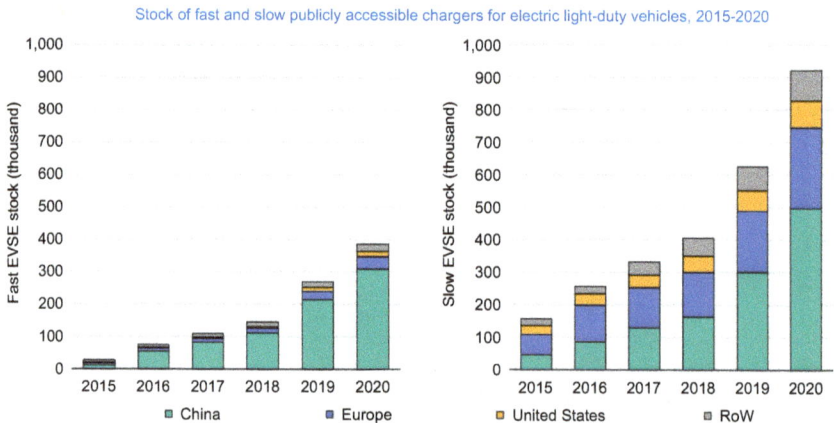

Figure 8.25   The number of EV public chargers worldwide from 2015 to 2020 [75]

To solve this issue, more charging infrastructure should be deployed. However, the cost of deploying charging infrastructure is high. As shown in Figures 8.26 and 8.27, charging infrastructure costs and related expenses are given [76,77]. It can be seen from the figures that chargers are costly, especially DC fast chargers. Therefore, for charging infrastructure operators, proper planning of the location and number of chargers before installation is critical to future charger operations. In addition, properly operating and managing charging services is crucial to obtaining considerable economic benefits. To make charging infrastructure profitable in the long term, operators need to consider the payback period, rationalise pricing and manage the power and time of EV charging.

As shown in Table 8.1, the EV charging time generally exceeds an hour, and long waiting times will significantly affect the utility of EV users. Therefore, to solve this problem, EV fast-charging technology needs to be developed. Moreover, a large amount of fast-charging infrastructure needs to be deployed to alleviate the time wasted by EVs on the road when looking for chargers. In addition, some

COST ELEMENT	LOWEST COST	HIGHEST COST
Level 2 residential charger	$380 (2.9 kW)	$689 (7.7 kW)
Level 2 commercial charger	$2,500 (7.7 kW)	$4,900 (16.8 kW); outlier: $7,210 (14.4 kW)
DCFC (50 kW)	$20,000	$35,800
DCFC (150 kW)	$75,600	$100,000
DCFC (350 kW)	$128,000	$150,000
Transformer (150–300 kVA)	$35,000	$53,000
Transformer (500–750 kVA)	$44,000	$69,600
Transformer (1,000+ kVA)	$66,000	$173,000
Data contracts	$84/year/charger	$240/year/charger
Network contracts	$200/year/charger	$250/year/charger
Credit card reader	$325	$1,000
Cable cost	$1,500	$3,500

Note: DCFC denotes direct-current fast chargers.

Figure 8.26   The ranges of charging infrastructure-related costs [76]

	50 kW				150 kW				350 kW			
	1 charger per site	2 chargers per site	3–5 charger per site	6–50 chargers per site	1 charger per site	2 chargers per site	3–5 chargers per site	6–20 chargers per site	1 charger per site	2 chargers per site	3–5 chargers per site	6–10 chargers per site
Labor	$19,200	$15,200	$11,200	$7,200	$20,160	$15,960	$11,760	$7,560	$27,840	$22,040	$16,240	$10,440
Materials	$26,000	$20,800	$15,600	$10,400	$27,300	$21,840	$16,380	$10,920	$37,700	$30,160	$22,620	$15,080
Permit	$200	$150	$100	$50	$210	$158	$105	$53	$290	$218	$145	$73
Taxes	$106	$85	$64	$42	$111	$89	$67	$45	$154	$123	$92	$62
Total	$45,506	$36,235	$26,964	$17,692	$47,781	$38,047	$28,312	$18,577	$65,984	$52,541	$39,097	$25,654

Figure 8.27   Installation costs of DC fast chargers per site [77]

mechanisms should be designed to manage the queuing and overstaying problems of EVs at charging [78–82].

Another issue is the private charging capability. Some EV users do not have private places to charge their EVs, such as people living in rented flats. Therefore, they can only charge at public charging infrastructure. Thus, the development of public charging infrastructure is crucial.

8.4   Strategies to mitigate grid impacts and demand-side issues

This section discusses some strategies to mitigate grid impacts and demand-side issues. The strategies include the optimal planning and operation of EV charging infrastructure.

8.4.1   *Planning strategies for EV charging infrastructure*

This subsection introduces some planning strategies: collaborative planning of charging infrastructure, grid expansion planning with EV integration, joint planning of charging infrastructure with distributed energy resources (DERs) and planning of battery swapping stations (BSSs).

8.4.1.1   Collaborative planning of charging infrastructure

EVs are different from conventional ICE vehicles in that they are not only transportation tools, but also power loads. Therefore, collaboration between the transport and power sectors is required when planning EV charging infrastructure in cities or on highways. Thus, the optimal planning must incorporate the traffic flow and power grid conditions.

The planning objects should be the locations, the number of charging infrastructure, and the size of the charging infrastructure at each location. From the transportation network operators' perspective, selecting optimal locations, numbers and sizes of charging infrastructure in a transportation network can improve utilisation, boost profitability, prevent traffic congestion and avoid long waiting. From the power grid operators' perspective, selecting optimal locations, numbers and sizes of charging infrastructure can prevent grid overloading and avoid large-scale grid upgrades. From EV users' perspective, selecting optimal locations, numbers and sizes of charging infrastructure can reduce the routing time, improve satisfaction and increase individual utility.

Various strategies have been developed for collaborative planning of charging infrastructure. Typically, some studies have developed planning models based on multi-objective optimisation [83,84]. Optimisation models can be established from the perspectives of the transportation network and power grid operators. Typically, the coupled network of the transportation and the power grid are obtained for the planning, as shown in Figure 8.28 [83].

The objective function of the power grid is to minimise the annual cost of charging infrastructure investment and energy losses. The key constraints of constructing the charging infrastructure in the power grid include the power balance

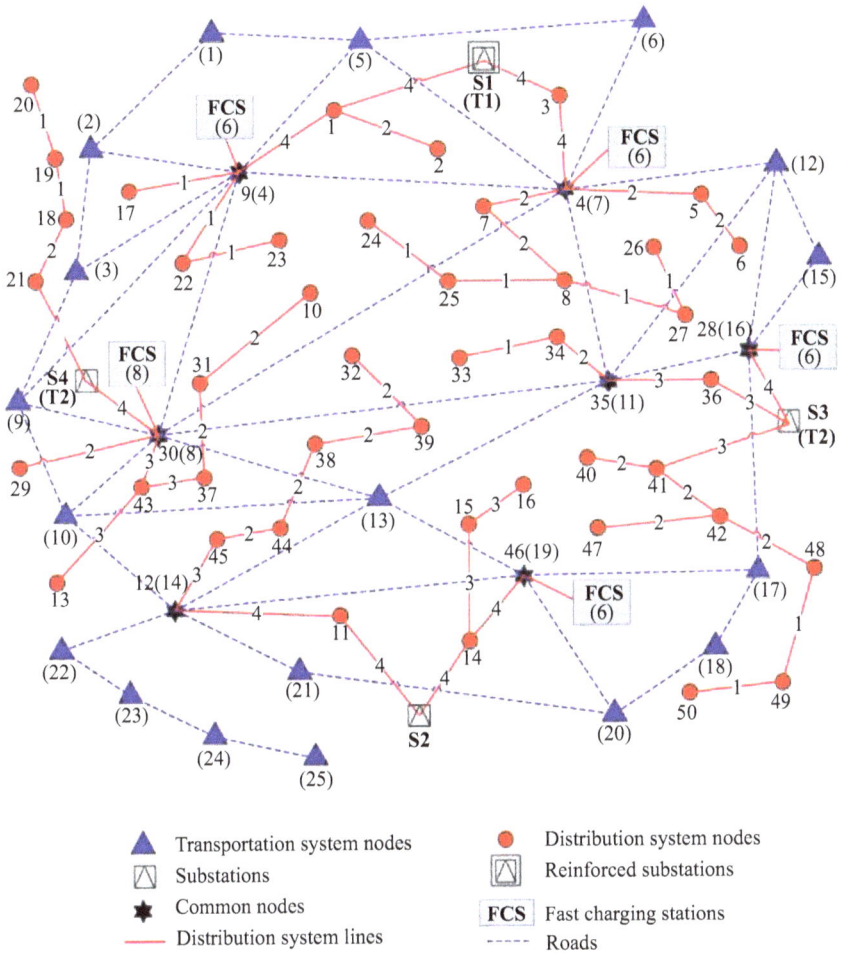

Figure 8.28   An example of the coupled network of transportation and power grid [83]

constraints, capacity constraints of substations, voltage limit constraints of buses and capacity constraints of charging infrastructure.

Some other studies have developed the planning model for deploying charging infrastructure on the highway networks [85–87]. Typically, the traffic flow refuelling location model in a highway network is first developed to formulate the captured EV charging demands under the driving range limits. Then, an optimisation model is developed to formulate the investment cost of charging infrastructure considering the grid and transportation constraints based on the traffic flow refuelling location model.

The investment cost of charging stations can be divided into the fixed cost of constructing charging stations and the variable cost of charger numbers. The investment cost of grid expansion incorporates the distribution lines and the substation expansion. In addition, the cost of unsatisfied charging demands is considered as well. The constraints of the optimisation model are transportation, power grid and charging demand constraints.

8.4.1.2   Grid expansion planning with EV integration

The high EV uptake poses significant challenges to the existing LDN due to the substantial charging demand. Consequently, to address the challenges associated with the increased load from EV charging, it might be necessary to expand the current LDN to prevent asset overloading. Some studies have investigated expansion planning issues, exploring the most economical and sustainable ways to construct new components or expand existing assets within the LDN [88,89]. Some other studies have proposed grid expansion planning methods considering EV charging, RESs and energy storage systems (ESSs) [90,91].

The planning objects include the locations and number of charging infrastructure to be deployed and the locations and the number of substations and feeders to be constructed or expanded. An example of the expansion of the LDN for EV charging is shown in Figure 8.29 [88].

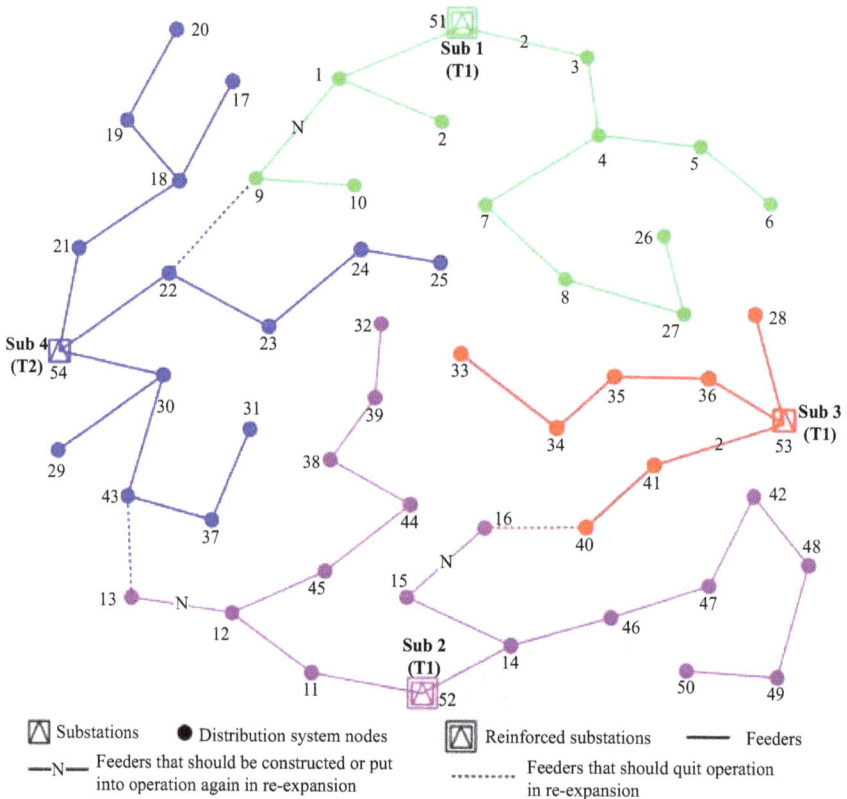

Figure 8.29   An example of the expansion of the LDN for EV charging [88]

Typically, the objective function of the optimisation model for expansion planning is to minimise the total cost. The total cost includes two parts: the investment and the operation costs. The investment cost incorporates the construction of charging infrastructure, substations and feeders. The operation cost involves charging infrastructure and the LDN operations. The key constraints of the optimisation model include the LDN constraints and charging demand constraints.

8.4.1.3   Joint planning of charging infrastructure with DERs

RESs can increase power generation capacity sustainably, reducing the impact on grid stability caused by growing charging demands. However, RES generation faces challenges of intermittency and low stability. Integrating ESSs is crucial to address these issues. Because ESSs can store the energy generated by RESs during high-generation periods or conventional power plants during off-peak periods. Therefore, with the strategic planning of RESs and ESSs, they can work together to provide stable, sustainable and economical electricity to the EVs. RESs and ESSs work as DERs. They can be jointly considered with EV charging loads to achieve a local balance of power generation and consumption to reduce energy transmission losses in the grid.

Typically, the planning objects include the optimal locations and the number of charging infrastructure and DERs, including PV, wind turbines and ESSs. An example of the joint planning of charging infrastructure with DERs is shown in Figure 8.30 [88]. Generally, the objective function of the optimisation model for joint planning is to minimise the total annual cost, including the investment cost and the operation cost. The investment cost incorporates the construction of

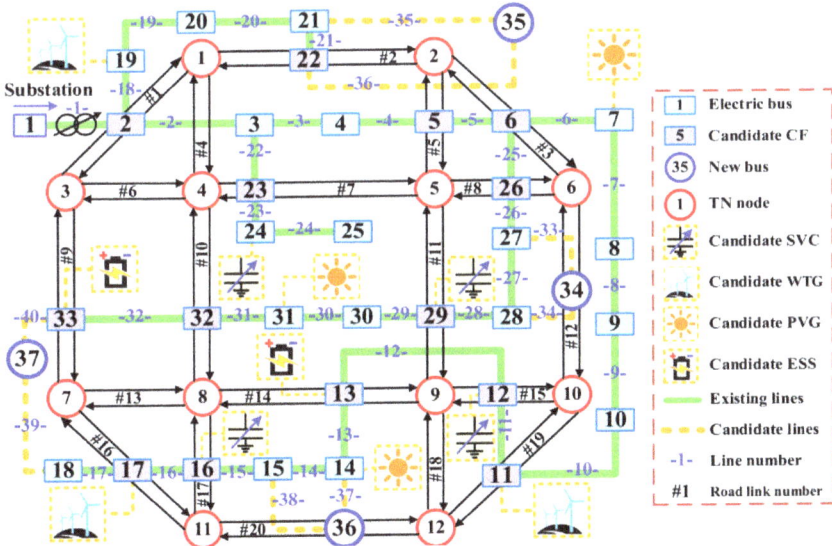

Figure 8.30   An example of the joint planning of charging infrastructure with DERs [88]

charging infrastructure, RESs and ESSs. The operation cost can be divided into the operation of charging infrastructure, RESs, ESSs, grid power losses and EV charging dispatches. The key constraints of the optimisation model include power flow constraints, voltage magnitude constraints, current constraints, DER size constraints, EV charging demand constraints and EV charging dispatch constraints.

8.4.1.4 Planning of BSSs

As a complementary way to EV charging stations, EV BSSs are more commonly used for public EVs, such as electric buses and taxis. Most public EVs use BSSs because replacing a battery takes less time than recharging it and increases the utilisation of public EVs. In addition, using BSSs can extend the battery lifespan. Because BSSs only need to replace batteries and then charge them during idle time, there is no need to apply high power to charge batteries like fast charging infrastructure. Thus reduce the loss of battery lifespan.

According to the business models, there are two modes of BSSs. The first mode is that the operator manages a single BSS. A schematic of the mode is shown in Figure 8.31 [26]. In the beginning, EV users send requests to the BSS when they need to swap batteries. Next, if the BSS operator accepts the requests, the EVs can be allocated to the waiting queue. Then, the EV batteries are swapped at the BSS and collected into the swapped battery groups. Afterwards, the EVs leave the BSS, and the swapped batteries are assigned to chargers for recharging. Finally, the fully charged batteries are collected and are waiting to be swapped to incoming EVs.

There are several issues worth studying for BSS with the first mode. First, a smart control system for managing battery swapping requests should be designed to decide whether to accept or reject a request according to the battery swapping demands, EV queuing status, number of fully charged batteries and recharging batteries. In addition, incentive mechanisms should be designed to incentivise EVs to submit swapping requests in advance and reduce the number of EVs walking in. Advance swapping requests can help BSSs perform optimal management, improving operational efficiency and profitability. Furthermore, charging management of the swapped batteries is also crucial to proper operation. The content of charging management includes the charging schedule, charging power and charging sequence of each swapped battery. Additionally, dynamically optimising the energy management of swapped batteries can improve operational efficiency and profitability. When a BSS is busy, the swapped batteries may not be fully charged, which enhances the efficiency of battery swapping and reduces the EV waiting time. Moreover, determining the number of batteries in a BSS is also crucial for economic operation. Having too many batteries in stock can increase investment and operating costs. If the number of batteries in stock is insufficient, the service experience of EVs will be affected, thereby affecting the revenue of the BSS. Apart from these, the pricing for the battery swapping processes is another crucial issue that determines operational efficiency and profitability.

The second business mode of BSS is that the operator manages multiple BSSs at different locations. A schematic of the operation mode is shown in Figure 8.32 [26]. To coordinate and manage numerous BSSs, there is a control centre that

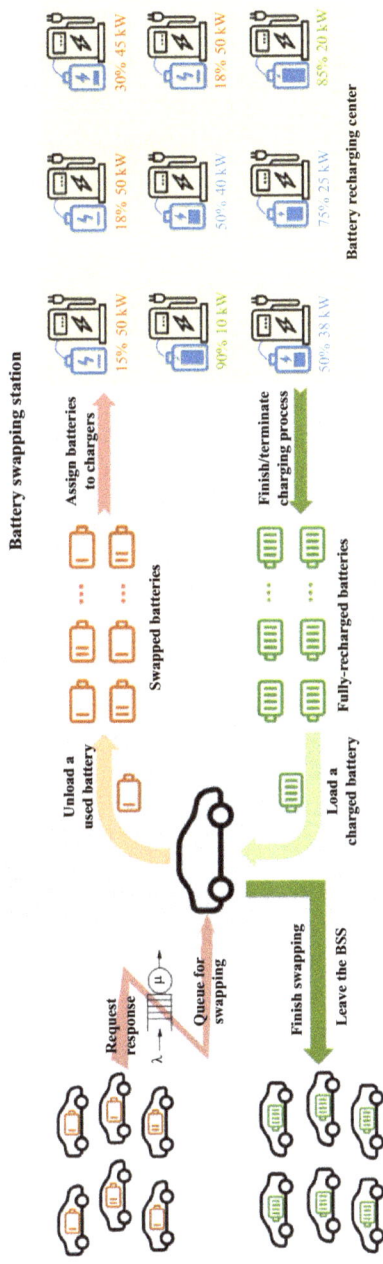

Battery swapping station

Assign batteries to chargers

Swapped batteries

Unload a used battery

Finish/terminate charging process

Fully-recharged batteries

Load a charged battery

Request response

Queue for swapping

Finish swapping

Leave the BSS

15% 50 kW
90% 10 kW
50% 38 kW

18% 50 kW
50% 40 kW
75% 25 kW

30% 45 kW
18% 50 kW
85% 20 kW

Battery recharging center

Figure 8.31 Schematic of BSS business mode one [26]

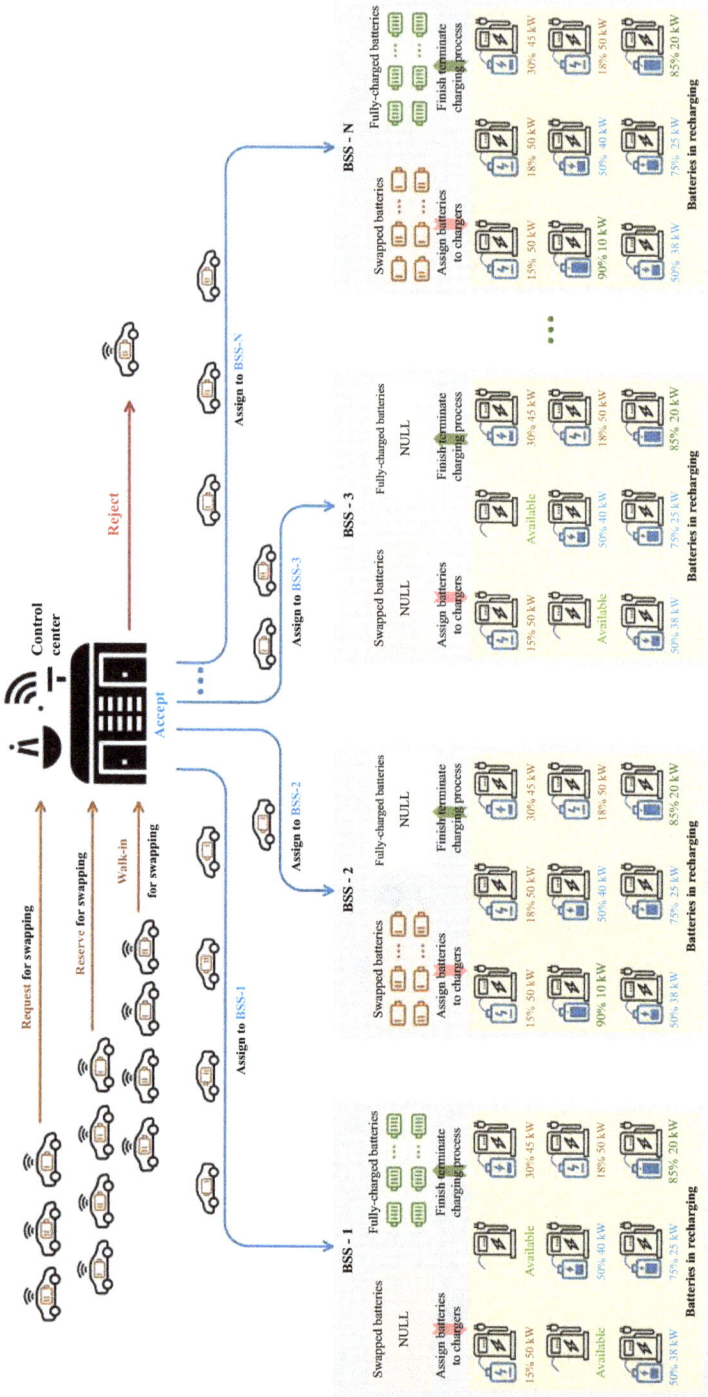

Figure 8.32 Schematic of BSS business mode two [26]

receives EV requests and allocates EVs to appropriate BSSs for battery swapping. Depending on the way of starting swapping orders, the control centre arranges EVs for different BSSs with different swapping fees. Some EVs reserve the swapping every day, which can help the BSSs make the optimal scheduling in day-ahead conditions. Some other EVs request the swapping in advance (normally one hour before), and these EVs can help the BSSs schedule intraday. The rest of the EVs walk in for swapping without advanced booking. These EVs normally pay the highest swapping fees and have a relatively higher probability of rejection under busy periods. Once the control centre receives these swapping requests, the EVs are allocated to the candidate BSS to maximise the economic benefits and utility without breaking the constraints. Next, EV users accept the allocations and go to the candidate BSSs or reject the allocations and leave away. Then, candidate BSSs swap the batteries for the accepted EVs. In the meantime, the control centre communicates with the BSSs and EVs in real-time to update the information.

There are several issues worth studying for BSSs with the second mode. First, when planning BSSs, the number and the locations should be considered carefully to facilitate the swapping of most EVs. Generally, the number and locations of BSSs are determined by the battery swapping demands, EV population, EV travel behaviours and grid capacity. In addition, real-time scheduling of EV swapping requests and recharging the swapped batteries is another crucial issue in increasing operational efficiency and profitability. Moreover, properly arranging the routes for EVs to candidate BSSs is critical to service satisfaction. For example, EVs must arrive at candidate BSSs without finishing the remaining energy. Furthermore, some incentive mechanisms should be designed to attract EVs to less busy BSSs.

A comparative study of planning EV charging infrastructure and BSSs has been investigated, and results show that BSSs are more suitable for public EVs [92]. Various studies have investigated different planning methods and models for BSSs. Generally, two different approaches can be applied depending on the operation modes for BSSs.

One approach is to plan the locations of BSSs and battery charging stations separately [93,94]. The BSSs only have battery inventory and battery swapping functions, and the swapped batteries must be transported to specific battery charging stations. Therefore, the battery charging stations are responsible for charging the swapped batteries and then transporting the fully charged batteries to the BSSs. The concept schematic for this operation mode is shown in Figure 8.33 [93]. In the figure, SEV, BDS and CBCS represent EVs going to swap batteries, BSSs and battery charging stations, respectively.

The spatial and temporal traffic and grid conditions are required to locate the BSSs and battery charging stations optimally. Then, based on the conditions of traffic and grid, the planning model is developed to determine the locations and size of BSSs. Typically, the planning model is formulated into two levels. The objective of the upper level is to minimise the investment and operation costs of BSSs and battery charging stations. The key constraints in the upper level incorporate power flow constraints, grid security constraints, transformer capacity constraints, station construction constraints and swapping demand constraints. Then, the objective of the lower level is to minimise the annual transportation cost of battery logistics.

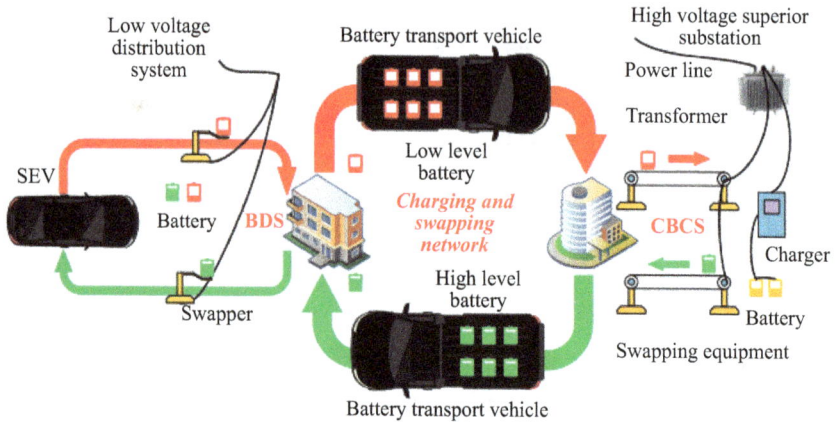

Figure 8.33   *The schematic for separate BSSs and battery charging station locations [93]*

The key constraints at the lower level include transportation constraints, logistics capacity constraints and battery demand constraints.

Another approach is to plan BSSs and battery charging stations at the same locations [95–97]. In other words, the swapped batteries can be charged at the same location with BSSs. Generally, the objective of the planning model is to minimise the investment and operation costs.

8.4.2   Operation strategies of EV charging infrastructure

As introduced in the previous sections, substantial EV charging demands can significantly increase grid loading, impacting the stable and cost-effective operation of the power grid. Consequently, some operation strategies for charging infrastructure have been implemented, involving the coordinated control of charging demands. This aims to mitigate impacts on the grid while ensuring the satisfaction of EV users without compromising their utility.

Generally, the concept of coordinated control of charging demands is to schedule and arrange the EV charging demands in the most appropriate periods and locations. Typically, in a scheduling scheme, the charging sessions of EV fleets are strategically scheduled to the most appropriate periods, preventing adverse impacts on the grid while satisfying the EV charging demands. Especially for EV fleets undergoing overnight charging, sessions can be scheduled during off-peak hours at night to avoid overloading assets and reduce power losses. This improves economic benefits for charging service operators, taking advantage of low tariff periods, and contributes to overall grid stability. Scheduling schemes are generally categorised into three types: centralised, decentralised and hierarchical, as shown in Figure 8.34 [98].

Typically, the scheduling of EV fleet charging can be realised from two control objects. The first object is to flatten the load profile. In other words, minimise the peak-valley difference of the power load. This is more appropriate to be applied to

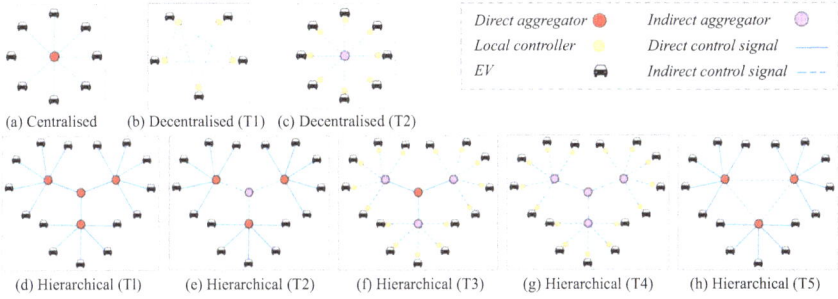

Figure 8.34   Scheduling schemes of charging EV fleet [98]

overnight EVs. This is because the valley period of electricity consumption is usually at night, while the peak period is usually after work and before people go to bed. Therefore, charging sessions can be postponed to the evening to move the charging loads to the valley period. Another objective is to minimise the charging costs of EV fleets. This can be realised using the time-of-use (TOU) electricity tariff to incentivise EVs to be charged during low-tariff periods. This is more suitable for EV charging at public charging stations.

8.4.2.1   Centralised scheduling of EV charging demands

In centralised scheduling schemes, a central operator aggregates all charging information from individual EVs. Leveraging the information and considering grid constraints, the central operator determines the optimised charging schedule for each EV to achieve an overall optimum. Although this approach allows for optimal solutions given complete charging information, it has certain drawbacks. The scalability of the communication system and the computation complexity become potential concerns as EV penetration increases. Advanced and costly communication and control infrastructure is necessary to handle extensive charging data effectively and accurately. Furthermore, managing large-scale data may require significant storage space and lead to data congestion. Privacy concerns from EV owners also arise due to the extensive information collected, and reliability issues of communication systems may emerge, given the massive scale of data collection.

8.4.2.2   Decentralised scheduling of EV charging demands

In decentralised scheduling schemes, each EV independently determines its charging schedule depending on its local optimum, functioning as an individual decision-maker. This approach eliminates the necessity for deploying an advanced communication infrastructure. Although decentralised schemes exhibit improved reliability and scalability compared to centralised scheduling schemes, they do not ensure a global optimum due to the individualised decision-making process of each EV.

8.4.2.3   Hierarchical scheduling of EV charging demands

A hierarchical scheduling scheme comprises multiple levels, incorporating central and lower-level operators interconnected in a tree-like structure. Lower-level

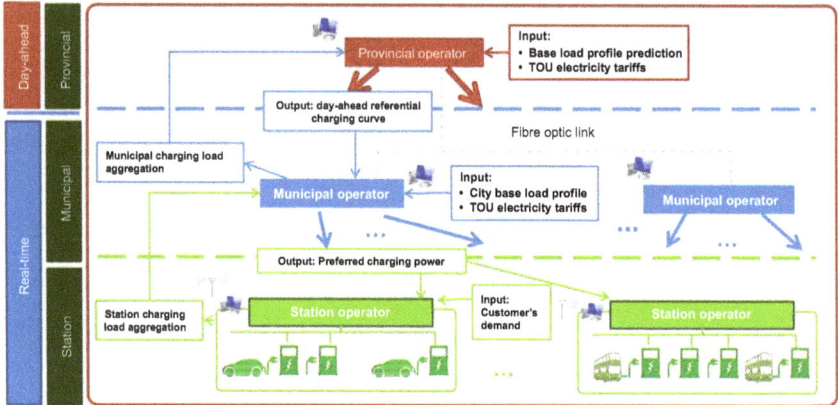

Figure 8.35   A typical hierarchical scheduling framework [95]

operators collect overall power demand information rather than specific details from individual EVs. The central operator then uses this aggregated power demand to schedule the amount of power to be supplied to the lower-level operators. Then, the lower-level operators arrange individual charging schedules for each EV. Central and lower-level operators aggregate information and make decisions within their own levels. The schematic of a typical hierarchical scheduling framework is shown in Figure 8.35 [99]. In this particular framework, the individual EV charging demands are aggregated by the station-level operators and sent to the municipal-level operators. Next, the municipal-level operators aggregate multiple charging demands from charging stations and send them to the provincial-level operator. Then, provincial-level operators make scheduling decisions based on the aggregated demand and dispatch power to municipal-level operators. Afterwords, municipal-level operators dispatch power to each charging station. Finally, each station operator arranges individual charging schedules for each EV.

8.5   V2G opportunities

Traditional EV charging systems use passive rectifiers to charge an EV battery, which cannot achieve bidirectional power flow. Replacing the passive rectifier with an active rectifier enables bidirectional power transfer. This system can also support functions such as V2G and V2X. Systems featuring V2G enable EVs to interact with the power grid, allowing bidirectional power flow for both charging and discharging. These systems emphasise convenience, high efficiency and flexibility and have become the main focus of current research in both industrial and academic communities. A typical wired V2G system is shown in Figure 8.36(a). Key components include isolation transformers for electrical safety and voltage management, power converters for converting AC to DC and vice versa, and control systems for coordinating power transfer.

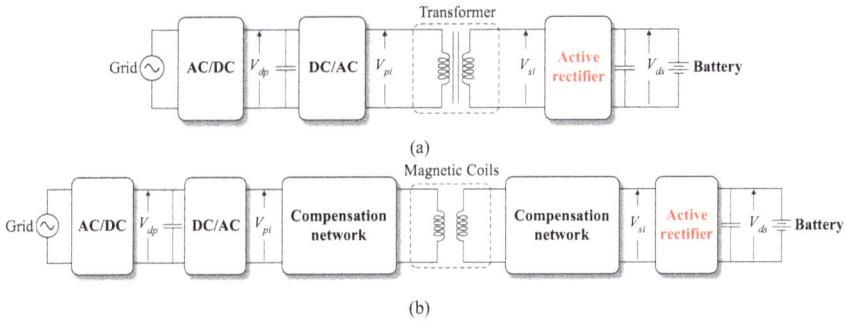

Figure 8.36   (a) A typical wired V2G system and (b) A typical wireless V2G system

Replacing tightly coupled isolation transformers with loosely coupled IPT coils suitable for large air gaps and using VAR compensation networks expands the V2G concept into a wireless approach, as shown in Figure 8.36 (b). Wireless charging systems offer higher flexibility, maintenance-free operation, safety and cost-effectiveness, making wireless V2G techniques a viable choice that meets most of the above attributes.

From the grid operators' perspective, V2G technology can use the energy of idle EVs to provide auxiliary services to the grid, improving its stability, reliability and economy. Furthermore, V2G plays a crucial role in integrating RESs. Because of the intermittency and instability of RESs, it is difficult for the power grid to control and accommodate their generation in a timely manner. However, idle EVs can solve this problem. Using idle EVs to absorb the power generation from RESs, they can discharge energy to the grid when needed. Therefore, the V2G technology can reduce the waste of electricity generated by renewable energy and improve the sustainability of the power grid. In addition, economic incentives are provided to EV owners and charging infrastructure operators participating in V2G ancillary services, reducing their charging costs and increasing their economic benefits. Therefore, V2G technology has the potential to assist the power grid in transforming into a low-carbon, clean, resilient and smart grid.

8.5.1   V2G for demand-side management

With the development of bidirectional chargers, EVs can participate in V2G to support the grid by providing ancillary services. In the V2G scenario, EVs are not only transportation tools for EV users but also controllable loads and distributed energy sources for the grid. Apart from the V2G concept, discharging EVs can also provide energy to buildings (V2B), homes (V2H) and other EVs (V2V). In general, these applications are called V2X. An illustration of the V2X framework is shown in Figure 8.37 [100].

In the framework, there are four types of V2X applications. First, EVs can be connected to personal homes (V2H). In the V2H scenarios, EVs can store energy during low-tariff periods and supply energy to home appliances during high-tariff

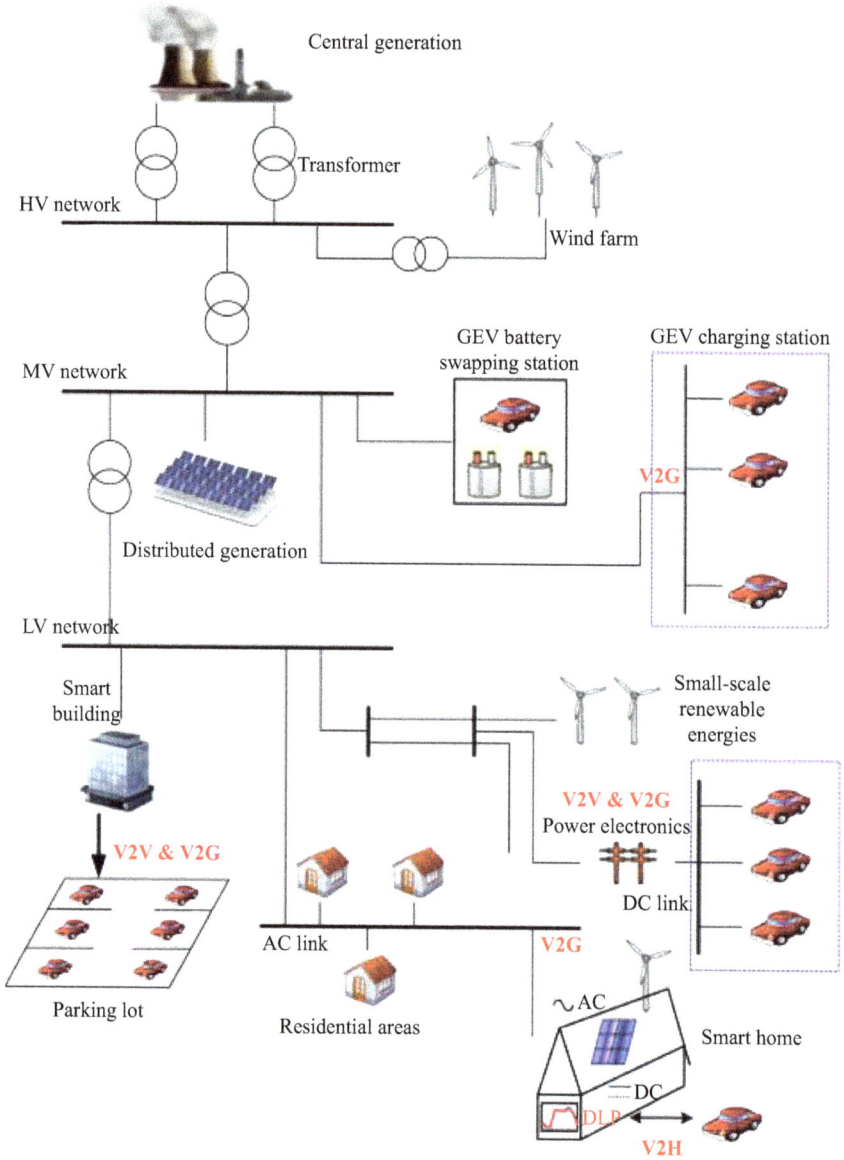

Figure 8.37   A V2X framework [100]

periods. Generally, EVs can adjust the active power exchange between the grid and the homes, operating economically. In addition, EVs can also provide reactive power to homes [100]. The basic devices for establishing V2H systems are bidirectional EV chargers and energy controllers.

Second, EVs can be aggregated in car parks and participate in the V2V operation. Using V2V, the excessive energy in idle EVs can be supplied to those requiring urgent charging. The V2V operation can significantly reduce economic and power losses during peak-loading periods. Because during the peak-loading periods, the grid suffers high power demands, which may aggravate the aging and failure probability of grid assets. Moreover, tariffs during peak loading periods are usually high. Therefore, by applying the V2V concept, both the grid operators and EV users can benefit. The establishment of the V2V system is relatively more complicated than the V2H system. The V2V system requires not only bidirectional chargers and energy controllers but also EV aggregators, privacy and security systems, technical standards and incentive mechanisms for user participation.

Third, EVs can provide energy to nearby buildings (V2B). This application is similar to the V2H system, but the power load of buildings is larger and more complex. Therefore, more advanced energy controllers and smart metering infrastructure are required.

Fourth, EVs supply energy to the grid directly. The support of a single EV on the grid is negligible; thus, V2G requires the participation of a large number of EVs. Therefore, EV aggregators are required to incentivise a large number of EVs to participate in V2G. In addition, EV aggregators act like brokers connecting individual EVs to the grid. Due to the flexible features of EVs, they can be used as controllable loads to participate in the demand-side response (DSR) of the power grid for peak loading and frequency regulation, improving grid stability and resilience. Moreover, by utilising the energy storage capabilities of EV batteries, they can be discharged when the grid requires additional power, thus functioning as a virtual power plant. In addition, by leveraging V2G technology, the grid can manage load demands more efficiently, thereby reducing the need for upgrades in grid infrastructure. Therefore, it helps the grid reduce overall operating costs. Furthermore, the V2G technology attracts more EV users to participate in and benefit from the electricity market.

In summary, in the V2G scenario, EVs can be used as loads, energy storage equipment and power generation equipment, which benefits all the related entities. Therefore, various V2G applications, such as frequency regulation, voltage regulation and spinning reserve, have been developed to support the power grid.

The applications of V2G typically rely on DSR strategies and mechanisms to incentivise adjustments in EV users' charging behaviours, energy consumption and energy release. Through DSR from EVs and V2G ancillary services, improvements can be achieved in the operational efficiency, stability and economics of the grid. Furthermore, the active participation of EV users in DSR and V2G ancillary services can result in economic benefits. Therefore, strategically guiding EVs to participate in DSR and V2G ancillary services can create a win-win situation for both grid and EV users.

8.5.1.1 Frequency regulation

Normally, aggregators are required to aggregate individual EVs altogether, providing MW-scale active power to the grid for frequency regulation [101–104], as shown in Figure 8.38. Technically, frequency regulation can be regarded as active power balancing between supply and demand. A typical objective function of the

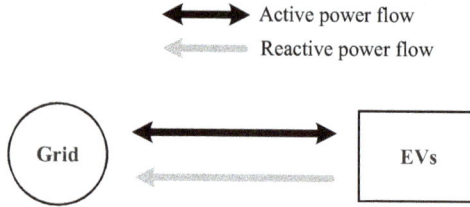

Figure 8.38   *Block diagram for frequency and voltage regulation provided by EVs*

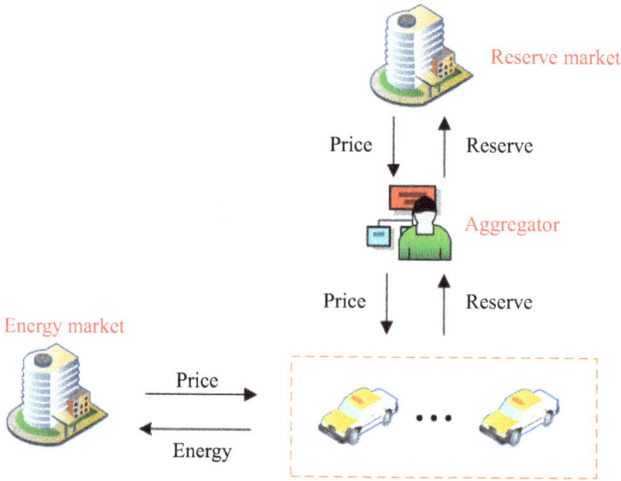

Figure 8.39   *A simplified spinning reserve framework [106]*

optimisation model for frequency regulation is to minimise the deviation between active power provided by EVs and the required active power supplied by EVs.

8.5.1.2   Voltage regulation

Technically, voltage regulation can be regarded as reactive power balancing between supply and demand. Conventionally, capacitors such as shunt capacitors and static VAR compensators are applied to compensate reactive power in the grid. Under the scenario of high EV uptake, using EVs to provide reactive power becomes an alternative way to conventional capacitors, as shown in Figure 8.38.

8.5.1.3   Spinning reserve

Spinning reserve in the power system refers to the reserved power capacity synchronised with the grid, and it is ready to supply immediate response to changes in demands or supplies due to unexpected disruptions. It is a backup to keep the grid operation stable and reliable [105]. Considering that EVs are idle for most of the day, they have the potential to provide the spinning reserve. A simplified framework of spinning reserve using EVs is shown in Figure 8.39, where an aggregator aggregates EVs and participates in the reserve market [106]. Optimisation models are required to calculate the optimal amount of power supplied by EVs to provide a sufficient spinning reserve. Typically, the objective function of the optimisation model for spinning reserve is to minimise the total operation cost for spinning reserve in a time slot. The key constraints incorporate power balance constraints, upper and lower limits of provided spinning reserve by each unit, power generation constraints, and upper limits of ramp-up and ramp-down for each unit.

8.5.2　Wireless V2G systems

Traditional V2G is implemented through wired charging techniques. However, combining V2G with IPT technology to form wireless V2G makes the system more flexible. This section introduces the basic operational principles of bidirectional IPT systems, common power conversion topologies and control strategies.

8.5.2.1　Basic of bidirectional IPT (BD-IPT) system

Traditional IPT systems employ passive rectifiers on the secondary side, limiting bidirectional power flow and failing to meet advanced application needs like V2G or V2X. To overcome this limitation and expand the IPT technology's applications, particularly for bidirectional energy flow, Professor Udaya Madawala and his team at the University of Auckland introduced a revolutionary concept in 2010: BD-IPT technology [107]. By substituting the secondary side's AC–DC converter with an active rectifier, they successfully adapted IPT technology for bidirectional operation. The primary distinction between wireless and traditional wired EV charging systems lies in the implementation of VAR compensation networks and control strategies, as illustrated in Figure 8.36. BD-IPT technology, known for its enhanced control flexibility and efficiency, has garnered widespread academic attention and emerged as a research hotspot.

Figure 8.40 shows a typical BD-IPT system. The grid inverter converts the power from the grid into DC link voltage V_{dp} to supply the primary inverter. The primary inverter then generates high-frequency AC voltage V_{pi} to drive the primary compensation network. The secondary circuit is implemented with similar devices, including an inverter, compensation network and controller. For simplicity, the active load on the pickup side is represented by the voltage source V_{si}. In practice, the DC voltage source V_{ds} can be the battery pack of an EV, utilised for storage or energy regeneration. Typically, the primary and secondary inverters are composed of half-bridge, full-bridge or push-pull inverters and are controlled to produce pulse-width modulated AC voltages V_{pi} and V_{si} at a fundamental frequency f.

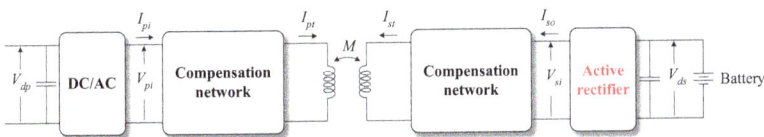

Figure 8.40　Bidirectional IPT system

Therefore, active rectifiers in the BD-IPT system introduce control variables, including (1) Relative phase angle between V_{pi} and V_{si}, (2) Magnitude of V_{pi} and V_{si}, (3) Operating frequency (f). By adjusting these control variables, the system can operate more effectively.

A typical BD-IPT system can be equivalently represented as a gyrator [108]. This equivalence arises because the secondary converter current (I_{si}) is determined by the primary voltage (V_{pi}). Therefore, by manipulating the relative phase angle between V_{pi} and V_{si}, it is feasible to effectively modify the phase relationship between V_{pi} and V_{si}, consequently varying the power output. The amplitudes of V_{pi} and V_{si} are typically used to control the magnitude of power flow. IPT systems with various compensation networks can be equivalently represented as different types of gyrators, with the distinction lying in the phase difference between V_{pi} and V_{si}. This difference results in varying phase angle ranges for controlling power flow. LCL compensation and SS compensation are two common structures in BD-IPT systems, and the output power P_{out} as phase shift (θ) between V_{pi} and V_{si} vary is depicted in Figure 8.41. It can be seen that, actively adjusting the phase angle between the primary and secondary voltages can easily reverse the direction of power flow, enabling functions like V2G.

In practical applications, misalignment between the primary and secondary coils is inevitable. Such physical displacement can lead to self-inductance and M variations, affecting the BD-IPT systems' efficacy in power transfer. This problem can be solved by adopting an optimised converter topology or by using advanced control strategy implementation.

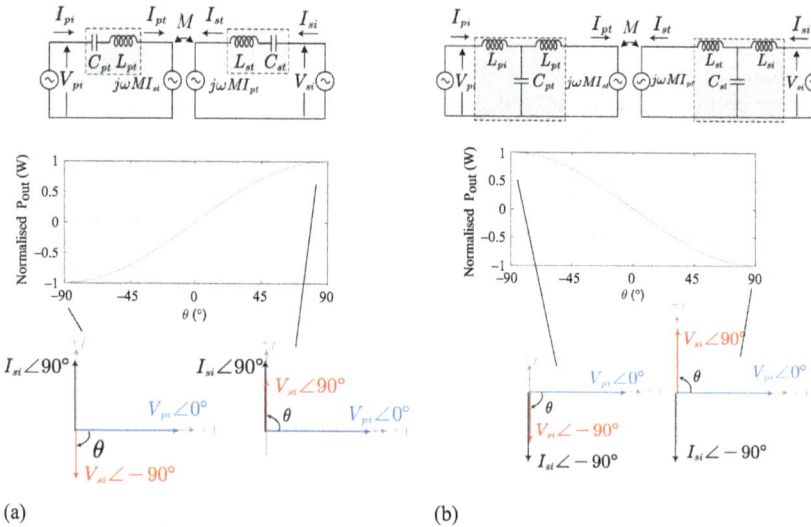

Figure 8.41 P_{out} of (a) SS-compensated BD-IPT system and (b) LCL-compensated BD-IPT system as θ vary

8.5.2.2   Power converters

Dual active bridge (DAB) converter

The traditional DAB system is shown in Figure 8.42(a), where electrical isolation between the primary and secondary sides is achieved by a tightly coupled transformer. Implementing a full-bridge converter on both the primary and secondary sides forms a DAB-based IPT system, as illustrated in Figure 8.42 (b). This structure offers significant power capacity and extensive control flexibility, enabling zero-voltage switching (ZVS), impedance matching for optimal AC efficiency and phase control, among others. Due to its adaptability to significant parameter changes and its ability to maintain high system performance, it is one of the most widely used structures in BD-IPT systems. For BD-IPT systems using an LCL configuration, as shown in Figure 8.41(b), adjusting the phase angle between the AC voltage of the primary and secondary converters allows the direction of power flow to be controlled, thereby facilitating the realisation of V2G. As shown in the waveforms in Figure 8.43, when the phase angle (θ) is $-90°$, power is transmitted from the primary side to the secondary side. Conversely, when θ is $-90°$, power flows from the EV side to the grid side, thereby realising the V2G function [109].

Multi-level converter

The diode-clamped multi-level converter, as shown in Figure 8.44, can produce a low-harmonic waveform with its switching devices operating at the fundamental frequency. It employs cascaded capacitors and a unique arrangement of diodes to

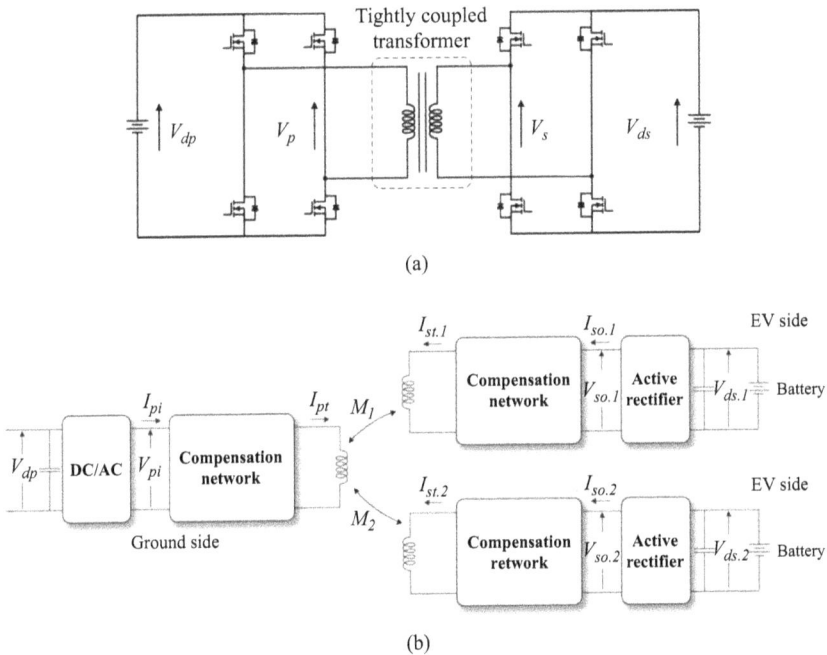

(a)

(b)

Figure 8.42   (a) Traditional DAB system and (b) IPT-based DAB converter

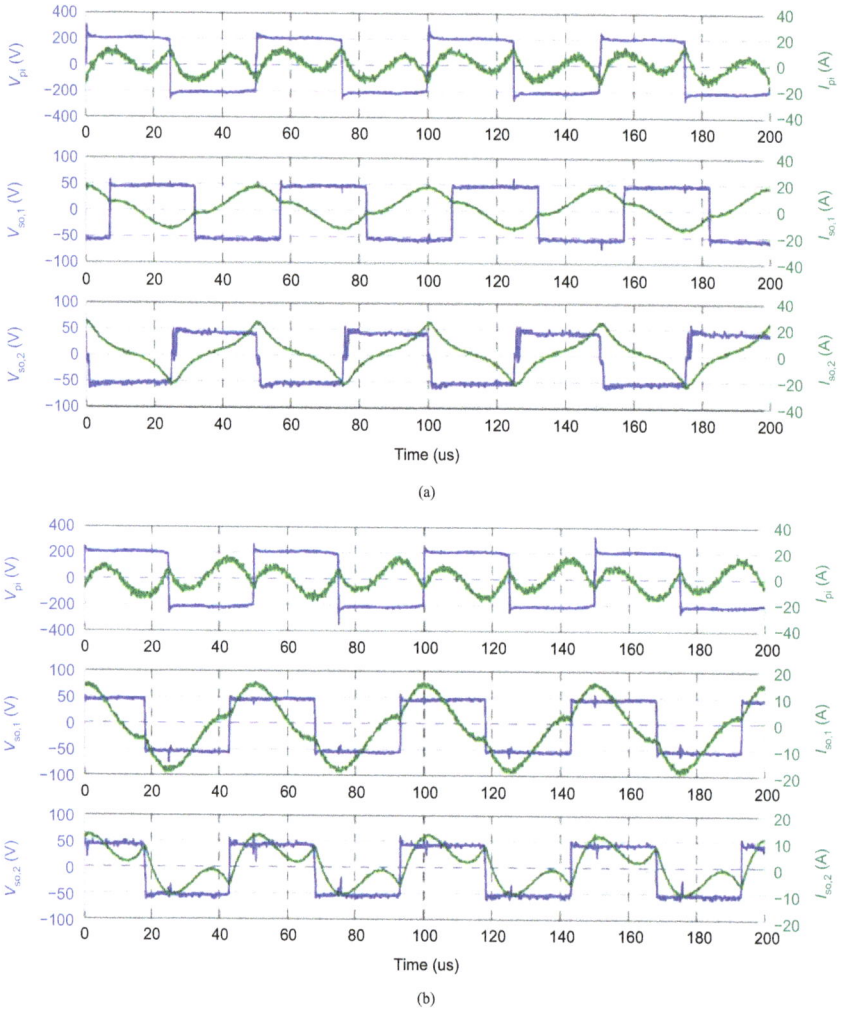

Figure 8.43   Converter voltage and currents waveforms of LCL BD-IPT system with (a) forward power transfer and (b) reverse power transfer [109]

create an output voltage waveform with multiple levels. This capability reduces switching losses in the systems. However, the drawbacks include the complexity of arranging clamping diodes for numerous voltage levels and the challenge of balancing the capacitor voltages. Moreover, the flying capacitor multi-level converter uses a unique arrangement of capacitors to generate a pulse amplitude-modulated output voltage. The output point is connected to a clamping capacitor branch, either individually or in series, with the DC source. This topology's drawbacks include the requirement for numerous bulky capacitors and the significant circulating current caused by imbalanced capacitor voltages.

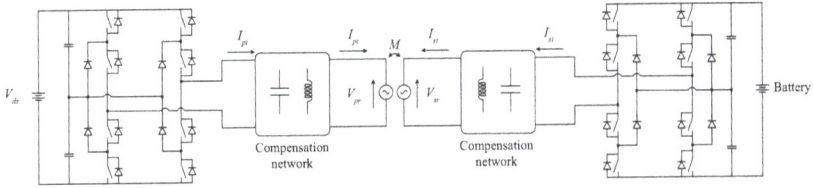

Figure 8.44   Diode-clamped multi-level converter

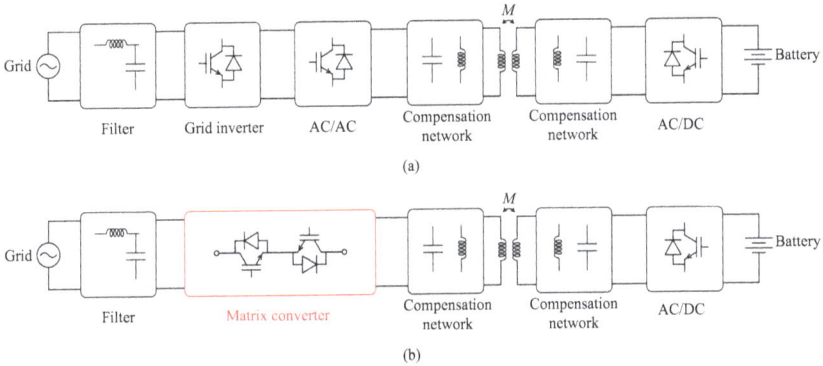

(a)

(b)

Figure 8.45   Grid interface converter (a) traditional grid interface and (b) grid interface based on matrix converter [112]

In addition, the cascaded H-bridge multi-level converter is composed of several series-connected full-bridge modules, each with its own storage capacitors. However, this topology requires isolated DC sources, leading to increased losses and costs. Other topology types include Z-source inverter-based IPT systems, which are also capable of bidirectional operation as rectifiers. These systems have great potential for applications in transportation electrification, such as V2G chargers [110].

Matrix converter
The traditional inductive coupled grid interface converter, shown in Figure 8.45(a), is gaining popularity as an attractive solution for V2G and G2V systems. However, these systems typically employ a large electrolytic DC-link capacitor and a substantial input inductor, resulting in expensive, bulky and less reliable [111]. The matrix converter topology presents an attractive solution due to its high-power density and small volume, as shown in Figure 8.45(b). This design eliminates the DC-link capacitor, enhancing reliability and power density. It comprises two back-to-back converters capable of regulating both the magnitude and direction of power flow. Unlike traditional systems, the grid converter operates at a much lower switching frequency with a simpler switching strategy, maintaining steady efficiency, reasonably low THD, and unity power factor [111]. Therefore, this

topology is a promising solution [112,113]. As an example, a single-phase bidirectional grid interface based on IPT technology is shown in Figure 8.46.

Figure 8.47 shows the current and voltage waveforms of the converter and grid under forward and reverse power transmission of a wireless matrix converter-based system. Similar to the system based on the DAB converter, bidirectional power flow can be achieved by adjusting the phase shift angle between the primary and secondary converter voltages.

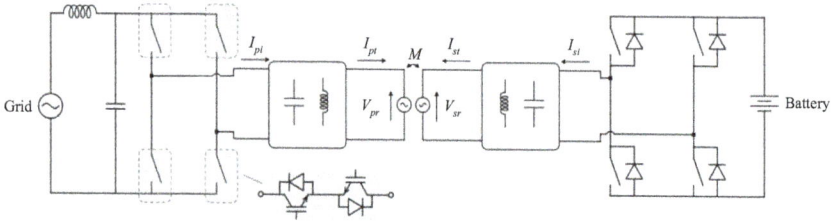

Figure 8.46   *A single-phase matrix converter [114]*

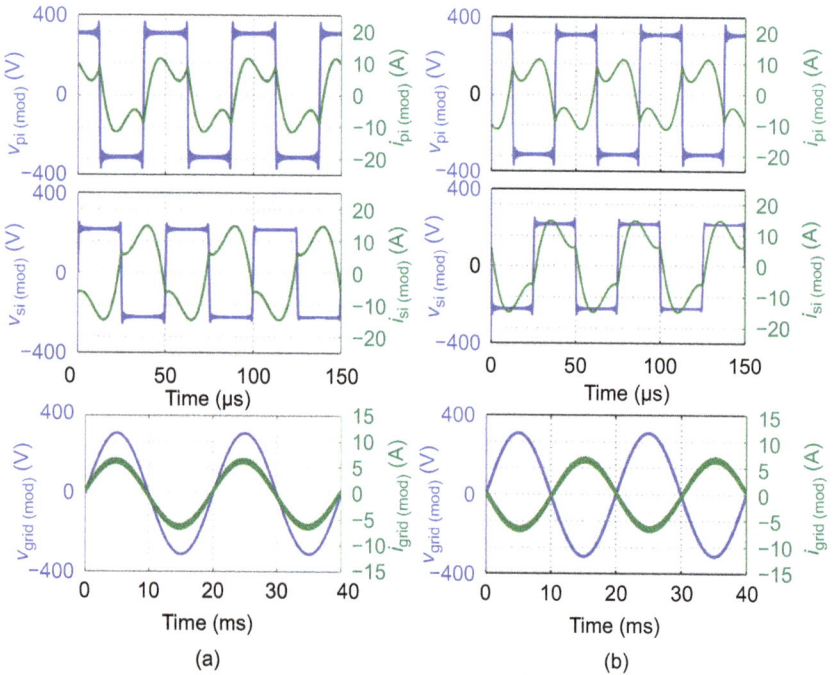

Figure 8.47   *Waveforms of converter voltage, current, and grid voltage and current in a wireless matrix converter system for (a) forward direction and (b) reverse direction [114]*

In addition, the multi-port T-type matrix converter is another feasible type [115]. This topology offers advantages such as bidirectional power flow capability and reduced output capacitance due to ripple cancellation. However, it also has drawbacks, including a high switch count and a complicated control strategy.

8.5.2.3 Control strategy

Active power converters, such as the full-bridge converter of a bidirectional IPT system [116,117], and the semi-bridgeless converters [118,119] or the secondary DC–DC converter [120,121] of a unidirectional IPT system, offer freedom to either directly or indirectly control the IPT systems. Control strategies can be classified according to the degrees of freedom adopted.

The single-variable control method is simple and robust but achieving efficient power transfer in the case of coil misalignment, power variation and load variation is challenging. It can be divided into primary-side control and secondary-side control. The former avoids the secondary-side converter, making it particularly suitable for applications that limit the volume of the secondary side [122,123], such as biomedical implantable or distributed sensors. Similar control strategies include [124], a power flow control method was proposed to control the primary current's magnitude based on the energy injection and regeneration concept, and in [125], based on the integrated boost multilevel converter, the primary coil was driven by a staircase-modulated boosted AC voltage, thereby improving the converter efficiency under a wide range of operating conditions.

Dual variable control includes two types: dual-side control and single-side control combined with varying frequency. For the former, impedance matching on the load side can be implemented to balance the loss distribution on the primary and secondary sides to achieve optimal efficiency. The latter achieves zero phase angle (ZPA) or soft switching while meeting power requirements. Therefore, compared with single-variable control, dual-variable control can satisfy multiple control goals at the same time. However, it often leads to hard switching when significant changes in system parameters occur. In [126], a dual side control for SS tuned system was proposed to increase the efficiency, especially for systems with large M variations. The ratio of currents on both sides was kept optimal to track the highest efficiency. Similar strategies were also presented in [127]. In [128], a control method that can counteract fluctuations of M and battery voltage was proposed. This approach adopted a battery management converter on the secondary side. However, soft switching was not taken into consideration. Other similar control strategies include [129–132].

In contrast, control methods with three variables can improve system efficiency by achieving a wider range of soft switching, given the large misalignment between primary and secondary pads. In [133], a triple phase shift (TPS) control strategy was proposed to achieve impedance matching while realising ZVS for all power switches within the entire power range. Other similar control strategies include [134,135]. Table 8.5 summarises typical control strategies for BD-IPT systems.

Table 8.5   Control strategies for IPT systems

Reference	Control variables	Compensation network	Control objectives			
			Power	Impedance matching	Soft switching	Minimum secondary reactive power
[122,124,125]	α	SS	√	–	–	–
[126–128]	α, β	SS	√	√	–	–
[129]	α, f	SS	√	√	–	–
[130–132]	α, f	SS	√	–	√	–
[135]	α, β, θ	LCL	√	√	√	–
[133]	α, β, θ	SS	√	√	√	–
[136]	α, β, θ, f	SS	√	√	√	√

8.6   Conclusion

This chapter has offered a detailed introduction to EV charging technologies, covering both wired and wireless approaches, and has discussed the integration of V2G systems. Section 8.2 has initiated the discussion by categorising EV charging systems, presenting both wired and wireless EV charging systems. It has introduced wired EV charging technologies, EV charging standards, IPT principles and commercial advancements. In Section 8.3, grid impacts and demand-side issues of EV charging have been introduced. Section 8.4 has discussed strategies to mitigate the demand-side issues and grid impacts. These strategies include optimal planning and operation of EV charging infrastructure. Moreover, V2G opportunities have been discussed in Section 8.5. These incorporate V2G for demand-side management and wireless V2G systems. The chapter aims to equip readers and researchers with a deep understanding of the challenges, essential aspects and prospective research pathways in EV charging and V2G technology.

References

[1]   M. R. Khalid, I. A. Khan, S. Hameed, M. S. J. Asghar, and J.-S. Ro, "A comprehensive review on structural topologies, power levels, energy storage systems, and standards for electric vehicle charging stations and their impacts on grid," *IEEE Access*, vol. 9, pp. 128069–128094, 2021.

[2]   H. Johnson, "Easy as 1 2 3: Everything you need to know about installing an EV charger at home and charging on the go," 2023, https://www.carparts.com/blog/   easy-as-1-2-3-everything-you-need-to-know-about-installing-an-ev-charger-at-home-and-charging-on-the-go/, Accessed: 18 June 2024.

[3]　LECTRON, "Lectron EV chargers," 2023, https://ev-lectron.com/collections/ev-chargers, Accessed: 18 June 2024.

[4]　M. Yilmaz and P. T. Krein, "Review of battery charger topologies, charging power levels, and infrastructure for plug-in electric and hybrid vehicles," *IEEE Transactions on Power Electronics*, vol. 28, no. 5, pp. 2151–2169, 2013.

[5]　D. Ronanki, A. Kelkar, and S. S. Williamson, "Extreme fast charging technology-prospects to enhance sustainable electric transportation," *Energies*, vol. 12, no. 19, p. 3721, 2019.

[6]　S. Habib, M. M. Khan, K. Hashmi, M. Ali, and H. Tang, "A comparative study of electric vehicles concerning charging infrastructure and power levels," in *2017 International Conference on Frontiers of Information Technology (FIT)*, pp. 327–332, 2017.

[7]　S. Habib, M. M. Khan, J. Huawei, K. Hashmi, M. T. Faiz, and H. Tang, "A study of implemented international standards and infrastructural system for electric vehicles," *in 2018 IEEE International Conference on Industrial Technology (ICIT)*, pp. 1783–1788, Lyon, 2018.

[8]　"Plugs, socket-outlets, vehicle connectors and vehicle inlets conductive charging of electric vehicles – part 3: Dimensional compatibility and interchangeability requirements for d.c. and a.c./d.c. pin and contact-tube vehicle couplers," IEC 62196-3:2014, p. 1, 2014.

[9]　EVBOX, "How to use a fast charging station," 2024, https://blog.evbox.com/how-to-use-fast-charging-stations, Accessed: 18 June 2024.

[10]　H. Tu, H. Feng, S. Srdic, and S. Lukic, "Extreme fast charging of electric vehicles: A technology overview," *IEEE Transactions on Transportation Electrification*, vol. 5, no. 4, pp. 861–878, 2019.

[11]　L. Tan, B. Wu, S. Rivera, and V. Yaramasu, "Comprehensive dc power balance management in high-power three-level dc-dc converter for electric vehicle fast charging," *IEEE Transactions on Power Electronics*, vol. 31, no. 1, pp. 89–100, 2016.

[12]　L. Alter, "Pop-up charging stations are less offensive to pedestrians," 2021, https://www.treehugger.com/pop-up-charging-stations-less-offensive-pedestrians-5195501, Accessed: 18 June 2024.

[13]　Evannex, "How about charging your electric car from a lamppost?" 2022, https://insideevs.com/news/340222/how-about-charging-your-electric-car-from-a-lamppost/, Accessed: 18 June 2024.

[14]　M. Sailsbury, "Street cabinets to EV chargers," 2023, https://www.fleetpoint.org/electric-vehicles-2/charging/street-caninets-to-ev-chargers/, Accessed: 18 June 2024.

[15]　P. Vaughan, D. Baxter, C. F. Hagenmaier Jr, *et al.*, "Dynamic allocation of power modules for charging electric vehicles," 2018, US Patent 10150380B2.

[16]　B. Song, U. K. Madawala, and C. A. Baguley, "Optimal planning strategy for reconfigurable electric vehicle chargers in car parks," *Energies*, vol. 16, no. 20, p. 7204, 2023.

[17]　ChargePoint, "Express 250 specifications and ordering information," 2019, https://chargepoint.ent.box.com/v/ CPE250-DS-EN-US, Accessed: 18 June 2024.

[18] "480kw hp dc split-type ev charging station," 2023, https://www.bene-rgyevcharger.com/ 480kw-dc-split-type-ev-charger-product/, Accessed: 18 June 2024.

[19] "Link transit reveals cost to operate electric bus fleet," 2022, https://www.masstransitmag.com/bus/maintenance/power-converters-battery-chargers-and-inverters/press-release/21271633/ momentum-dynamics-link-transit-reveals-cost-to-operate-electric-bus-fleet, Accessed: 18 June 2024.

[20] A. Agrawal, "Charge more evs with power management," 2017, https://www.chargepoint.com/blog/charge-more-evs-power-management/, Accessed: 18 June 2024.

[21] L. Liu, X. Qi, Z. Xi, J. Wu, and J. Xu, "Charging-expense minimization through assignment rescheduling of movable charging stations in electric vehicle networks," *IEEE Transactions on Intelligent Transportation Systems*, vol. 23, no. 10, pp. 17212–17223, 2022.

[22] T. D. Atmaja and Amin, "Energy storage system using battery and ultra-capacitor on mobile charging station for electric vehicle," *Energy Proc.*, vol. 68, pp. 429–437, 2015.

[23] H. Chen, Z. Su, Y. Hui, and H. Hui, "Dynamic charging optimization for mobile charging stations in internet of things," *IEEE Access*, vol. 6, pp. 53509–53520, 2018.

[24] H. Saboori, S. Jadid, and M. Savaghebi, "Optimal management of mobile battery energy storage as a selfdriving, self-powered and movable charging station to promote electric vehicle adoption," *Energies*, vol. 14, no. 3, p. 736, 2021.

[25] G. Wu, M. Fan, J. Shi, and Y. Feng, "Reinforcement learning based truck-and-drone coordinated delivery," *IEEE Transactions on Artificial Intelligence*, vol. 4, no. 4, pp. 754–763, 2021.

[26] H. Wu, "A survey of battery swapping stations for electric vehicles: Operation modes and decision scenarios," *IEEE Transactions on Intelligent Transportation Systems*, vol. 23, no. 8, pp. 10163–10185, 2022.

[27] W. V. Wang, "New multilevel converter topologies for wireless power transfer systems," Ph.D. dissertation, ResearchSpace@ Auckland, 2021.

[28] C. J. Michelbacher, S. Ahmed, I. Bloom, *et al.*, "Enabling fast charging: A technology gap assessment," Tech. Rep., Oct 1, Tech. Rep., 2017, https://www.osti.gov/servlets/purl/1466683, Accessed: 18 June 2024.

[29] T. Moloughney. "BMW i3 charge port cover", 2016, https://cdn.bmwblog.com/wp-content/uploads/2016/02/bmw-i3-charge-port-cover-1.jpg, Accessed: 18 June 2024.

[30] G. A. Covic and J. T. Boys, "Inductive power transfer," *Proc. IEEE*, vol. 101, no. 6, pp. 1276–1289, 2013.

[31] D. Patil, M. K. McDonough, J. M. Miller, B. Fahimi, and P. T. Balsara, "Wireless power transfer for vehicular applications: Overview and challenges," *IEEE Transactions on Transportation Electrification*, vol. 4, no. 1, pp. 3–37, 2017.

[32] J. M. Miller, O. C. Onar, and M. Chinthavali, "Primary-side power flow control of wireless power transfer for electric vehicle charging," *IEEE*

journal of Emerging and selected topics in power electronics*, vol. 3, no. 1, pp. 147–162, 2014.

[33] G. A. Covic and J. T. Boys, "Modern trends in inductive power transfer for transportation applications," *IEEE Journal of Emerging and Selected topics in power electronics*, vol. 1, no. 1, pp. 28–41, 2013.

[34] R. Bosshard and J. W. Kolar, "Inductive power transfer for electric vehicle charging: Technical challenges and tradeoffs," *IEEE Power Electronics Magazine*, vol. 3, no. 3, pp. 22–30, 2016.

[35] SAE International. "Wireless power transfer for light-duty plug-in/electric vehicles and alignment methodology," ID: J2954, 2017, https://www.sae.org/standards/content/j2954, Accessed: 18 June 2024.

[36] M. Nasquelier, "SAE J2954/2 heavy duty wireless power transfer," 2020, https://assets.ctfassets.net/ucu418cgcnau/2AJIFzHDbnxe41giQSCZ5b/63132d 038ef07d3dc4b467920401a41f/07_WAVE_SAEJ2954HD_EPRI_TruckBus_ Masquelier_2020.03.10.pdf, Accessed: 18 June 2024.

[37] SAE International. "SAE J2954/2 wireless power transfer and alignment for heavy duty applications," 2022, https://www.sae.org/standards/content/ j2954/2/, Accessed: 18 June 2024.

[38] G. A. Covic, G. Elliott, O. H. Stielau, R. Green, and J. Boys, "The design of a contact-less energy transfer system for a people mover system," in *PowerCon 2000. 2000 International Conference on Power System Technology. Proceedings (Cat. No. 00EX409)*, vol. 1, pp. 79–84. IEEE, 2000.

[39] J. T. Boys and G. A. Covic, "The inductive power transfer story at the university of auckland," *IEEE circuits and systems magazine*, vol. 15, no. 2, pp. 6–27, 2015.

[40] A. Zaheer, "A new magnetic coupling pad topology for inductively powered vehicular systems," Ph.D. dissertation, ResearchSpace@ Auckland, 2015.

[41] WiTricity, "Drive 11," 2017, https://witricity.com/wp-content/uploads/ 2018/02/DRIVE1120170221-1.pdf, Accessed: 15 November 2023.

[42] WiTricity, "Witricity – wireless power transfer," https://witricity.com/, 2023, Accessed: 15 November 2023.

[43] R. Carlson, "Dc charger fact sheet: Wave 250kW wireless charger on 40' byd bus," 2022, https://cet.inl.gov/ArticleDocuments/WAVE 250kW Fact-Sheet.pdf, Accessed: 18 June 2024.

[44] J. Klender, "Tesla semi's fast-charging system could be wireless," *Teslarati*, 2021, https://www.teslarati.com/tesla-semi-fast-charging-system-could-be-wireless/, Accessed: 18 June 2024.

[45] J. Guzowski, "Wave announces wireless charging with antelope valley transit authority at Lancaster city park and Palmdale transportation center," *BusRide*, 2017, https://busride.com/wave-announces-wireless-charging-antelope-valley/, Accessed: 18 June 2024.

[46] C. Morris, "Wave demonstrates 250 kw wireless en route charging for e-buses," *Charged EVs*, 2017, https://chargedevs.com/newswire/wave-demonstrates-250-kw-wireless-en-route-charging-for-e-buses/, Accessed: 18 June 2024.

[47] Momentum Dynamics Corporation, "One of America's largest electric bus fleets reveals operating costs of ev buses using wireless chargers from momentum dynamics is half of a diesel-fueled bus," 2022, https://www.prnewswire.co.uk/news-releases/one-of-america-s-largest-electric-bus-fleets-reveals-operating-costs-of-ev-buses-using-wireless-chargers-from-momentum-dynamics-is-half-of-a-diesel-fueled-bus-834145054.html, Accessed: 18 June 2024.

[48] "Electric bus project – link transit," 2022, https://linktransit.com/about-us/projects-initiatives/electric-bus-project/, Accessed: 18 June 2024.

[49] "One of america's largest electric bus fleets reveals operating costs of EV buses using wireless chargers from momentum dynamics," https://www.prnewswire.com/news-releases/one-of-americas-largest-electric-bus-fleets-reveals-operating-costs-of-ev-buses-using-wireless-chargers-from-momen-tum-dynamics-is-half-of-a-diesel-fueled-bus-301570760.html, Accessed: 18 June 2024.

[50] N. Manthey, "Wireless charging halves e-bus operating cost for link transit," 2022, https://www.electrive.com/2022/06/23/wireless-chargers-help-to-halve-operating-costs-for-link-transit/, Accessed: 18 June 2024.

[51] S. Lee, J. Huh, C. Park, N.-S. Choi, G.-H. Cho, and C.-T. Rim, "On-line electric vehicle using inductive power transfer system," in *2010 IEEE Energy Conversion Congress and Exposition*, pp. 1598–1601. IEEE, 2010.

[52] S. Ahn, N. P. Suh, and D.-H. Cho, "Charging up the road," *IEEE Spectrum*, vol. 50, no. 4, pp. 48–54, 2013.

[53] J. Kim, J. Kim, S. Kong, *et al.*, "Coil design and shielding methods for a magnetic resonant wireless power transfer system," *Proc. IEEE*, vol. 101, no. 6, pp. 1332–1342, 2013.

[54] Bombardier. "World's first electric bus with bombardier's PriMove system begins revenue service," 2014, https://bombardier.com/en/media/news/worlds-first-electric-bus-bombardiers-primove-system-begins-revenue-ser-vice, Accessed: 18 June 2024.

[55] O. Olsson, "Slide-in electric road system, inductive project report, phase 1," in *Friday, October 25, 2013*. Scania CV AB, 2013.

[56] V. Cirimele, M. Diana, F. Freschi, and M. Mitolo, "Inductive power transfer for automotive applications: State-of-the-art and future trends," *IEEE Transactions on Industry Applications*, vol. 54, no. 5, pp. 4069– 4079, 2018.

[57] Bombardier. "Bombardier PriMove to provide wireless charging and battery technology to berlin," 2015, https://www.globenewswire.com/news-release/2015/03/19/1506037/0/en/BOMBARDIER-PRIMOVE-to-Provide-Wireless-Charging-and-Battery-Technology-to-Berlin.html, Accessed: 18 June 2024.

[58] O. C. Onar, J. M. Miller, S. L. Campbell, C. Coomer, C. P. White, and L. E. Seiber, "A novel wireless power transfer for in-motion EV/PHEV charging," in *2013 Twenty-Eighth Annual IEEE Applied Power Electronics Conference and Exposition (APEC)*, pp. 3073–3080. IEEE, 2013.

[59] V. P. Galigekere, J. Pries, O. C. Onar, *et al.*, "Design and implementation of an optimized 100 kw stationary wireless charging system for ev battery

recharging," *in 2018 IEEE Energy Conversion Congress and Exposition (ECCE)*, pp. 3587–3592. IEEE, 2018.

[60] J. Pries, V. P. N. Galigekere, O. C. Onar, and G.-J. Su, "A 50-kw three-phase wireless power transfer system using bipolar windings and series resonant networks for rotating magnetic fields," *IEEE Transactions on Power Electronics*, vol. 35, no. 5, pp. 4500–4517, 2019.

[61] "Ornl demonstrates 120-kilowatt wireless charging for vehicles," 2018, https://www.ornl.gov/news/ornl-demonstrates-120-kilowatt-wireless-charging-vehicles, Accessed: 18 June 2024.

[62] J. H. Kim, B.-S. Lee, J.-H. Lee, *et al.*, "Development of 1-mw inductive power transfer system for a high-speed train," *IEEE Transactions on Industrial Electronics*, vol. 62, no. 10, pp. 6242–6250, 2015.

[63] M. Ruikun, L. Yong, H. Zhengyou, *et al.*, "Wireless energy transfer technology and its research progress in rail transit," *Journal of Southwest Jiaotong University*, vol. 29, no. 3, pp. 446–461, 2016.

[64] R. Tavakoli and Z. Pantic, "Analysis, design, and demonstration of a 25-kw dynamic wireless charging system for roadway electric vehicles," *IEEE Journal of Emerging and Selected Topics in Power Electronics*, vol. 6, no. 3, pp. 1378–1393, 2017.

[65] I. Villar, A. Garcia-Bediaga, U. Iruretagoyena, R. Arregi, and P. Estevez, "Design and experimental validation of a 50kw ipt for railway traction applications," in *2018 IEEE energy conversion congress and exposition (ECCE)*, pp. 1177–1183. IEEE, 2018.

[66] B. Song, U. Madawala, and C. Baguley, "A review of grid impacts, demand side issues and planning related to electric vehicle charging," in *2021 IEEE Southern Power Electronics Conference (SPEC)*, pp. 1–6, Kigali, Rwanda, 2021.

[67] S. Shafiee, M. Fotuhi-Firuzabad, and M. Rastegar, "Investigating the impacts of plug-in hybrid electric vehicles on power distribution systems," *IEEE Transactions on Smart Grid*, vol. 4, no. 3, pp. 1351–1360, 2013.

[68] N. Hartmann and E. Ozdemir, "Impact of different utilization scenarios of electric vehicles on the german grid in 2030," *Journal of Power Sources*, vol. 196, pp. 2311–2318, 2011.

[69] S. F. Aliabadi, S. A. Taher, and M. Shahidehpour, "Smart deregulated grid frequency control in presence of renewable energy resources by EVS charging control," *IEEE Transactions on Smart Grid*, vol. 9, no. 2, pp. 1073–1085, 2018.

[70] S. Deilami, A. S. Masoum, P. S. Moses, and M. A. S. Masoum, "Real-time coordination of plug-in electric vehicle charging in smart grids to minimize power losses and improve voltage profile," *IEEE Transactions on Smart Grid*, vol. 2, no. 3, pp. 456–467, 2011.

[71] A. Dubey and S. Santoso, "Electric vehicle charging on residential distribution systems: Impacts and mitigations," *IEEE Access*, vol. 3, pp. 1871–1893, 2015.

[72] T. Thiringer and S. Haghbin, "Power quality issues of a battery fast charging station for a fully-electric public transport system in Gothenburg city," *Batteries*, vol. 1, no. 1, pp. 22–33, 2015.

[73]   L. Wang, Z. Qin, T. Slangen, P. Bauer, and T. van Wijk, "Grid impact of electric vehicle fast charging stations: Trends, standards, issues and mitigation measures – an overview," *IEEE Open Journal of Power Electronics*, vol. 2, pp. 56–74, 2021.

[74]   B. Song, L. Wang, U. Madawala, and C. Baguley, "A multi-functional fast electric vehicle charging technique," *in 2021 IEEE Southern Power Electronics Conference (SPEC)*, pp. 1–6, Kigali, Rwanda, 2021.

[75]   IEA, "Global ev outlook 2022," https://iea.blob.core.windows.net/assets/e0d2081d-487d-4818-8c59-69b638969f9e/GlobalElectricVehicleOutlook 2022.pdf, Accessed: 18 June 2024.

[76]   C. Nelder and E. Rogers,"Reducing EV charging infrastructure costs," 2020, https://rmi.org/wp-content/uploads/2020/01/RMI-EV-Charging-Infrastructure-Costs.pdf, Accessed: 18 June 2024.

[77]   M. Nicholas, "Estimating electric vehicle charging infrastructure costs across major u.s. metropolitan areas," 2019, https://theicct.org/sites/default/files/publications/ICCT EV Charging Cost 20190813.pdf, Accessed: 18 June 2024.

[78]   F. Varshosaz, M. Moazzami, B. Fani, and P. Siano, "Day-ahead capacity estimation and power management of a charging station based on queuing theory," *IEEE Transactions on Industrial Informatics*, vol. 15, no. 10, pp. 5561–5574, 2019.

[79]   S. Esmailirad, A. Ghiasian, and A. Rabiee, "An extended m/m/k/k queueing model to analyze the profit of a multiservice electric vehicle charging station," *IEEE Transactions on Vehicular Technology*, vol. 70, no. 4, pp. 3007–3016, 2021.

[80]   T. Zeng, H. Zhang, and S. Moura, "Solving overstay and stochasticity in PEV charging station planning with real data," *IEEE Transactions on Industrial Informatics*, vol. 16, no. 5, pp. 3504–3514, 2020.

[81]   T. Zeng, S. Bae, B. Travacca, and S. Moura, "Inducing human behavior to maximize operation performance at pev charging station," *IEEE Transactions on Smart Grid*, vol. 12, no. 4, pp. 3353–3363, 2021.

[82]   B. Song and U. Madawala, "Dynamic penalty for mitigating EV overstaying problem in fast charging pools," *in IECON 2023 – 49th Annual Conference of the IEEE Industrial Electronics Society*, pp. 1–6, Singapore, 2023.

[83]   W. Yao, J. Zhao, F. Wen, *et al.*, "A multi-objective collaborative planning strategy for integrated power distribution and electric vehicle charging systems," *IEEE transactions on power systems*, vol. 29, no. 4, pp. 1811–1821, 2014.

[84]   S. Wang, Z. Y. Dong, F. Luo, K. Meng, and Y. Zhang, "Stochastic collaborative planning of electric vehicle charging stations and power distribution system," *IEEE Transactions on Industrial Informatics*, vol. 14, no. 1, pp. 321–331, 2018.

[85]   H. Zhang, S. J. Moura, Z. Hu, and Y. Song, "PEV fast-charging station siting and sizing on coupled transportation and power networks," *IEEE Transactions on Smart Grid*, vol. 9, no. 4, pp. 2595–2605, 2018.

[86]   H. Zhang, S. J. Moura, Z. Hu, W. Qi, and Y. Song, "A second-order cone programming model for planning PEV fast-charging stations," *IEEE Transactions on Power Systems*, vol. 33, no. 3, pp. 2763–2777, 2018.

[87] X. Wang, M. Shahidehpour, C. Jiang, and Z. Li, "Coordinated planning strategy for electric vehicle charging stations and coupled traffic-electric networks," *IEEE Transactions on Power Systems*, vol. 34, no. 1, pp. 268–279, 2019.

[88] W. Yao, C. Y. Chung, F. Wen, M. Qin, and Y. Xue, "Scenario-based comprehensive expansion planning for distribution systems considering integration of plug-in electric vehicles," *IEEE Transactions on Power Systems*, vol. 31, no. 1, pp. 317–328, 2016.

[89] N. B. Arias, A. Tabares, J. F. Franco, M. Lavorato, and R. Romero, "Robust joint expansion planning of electrical distribution systems and ev charging stations," *IEEE Transactions on Sustainable Energy*, vol. 9, no. 2, pp. 884–894, 2018.

[90] P. M. de Quevedo, G. Munoz-Delgado, and J. Contreras, "Impact of electric vehicles on the expansion planning of distribution systems considering renewable energy, storage, and charging stations," *IEEE Transactions on Smart Grid*, vol. 10, no. 1, pp. 794–804, 2019.

[91] S. Wang, Z. Y. Dong, C. Chen, H. Fan, and F. Luo, "Expansion planning of active distribution networks with multiple distributed energy resources and ev sharing system," *IEEE Transactions on Smart Grid*, vol. 11, no. 1, pp. 602–611, 2020.

[92] Y. Zheng, Z. Y. Dong, Y. Xu, K. Meng, J. H. Zhao, and J. Qiu, "Electric vehicle battery charging/swap stations in distribution systems: comparison study and optimal planning," *IEEE transactions on Power Systems*, vol. 29, no. 1, pp. 221–229, 2013.

[93] C. He, J. Zhu, S. Li, Z. Chen, and W. Wu, "Sizing and locating planning of ev centralized-battery-chargingstation considering battery logistics system," *IEEE Transactions on Industry Applications*, vol. 58, no. 4, pp. 5184–5197, July-Aug 2022.

[94] M. Ban, M. Shahidehpour, J. Yu, and Z. Li, "A cyber-physical energy management system for optimal sizing and operation of networked nano-grids with battery swapping stations," *IEEE Transactions on Sustainable Energy*, vol. 10, no. 1, pp. 491–502, 2019.

[95] W. Infante, J. Ma, X. Han, and A. Liebman, "Optimal recourse strategy for battery swapping stations considering electric vehicle uncertainty," *IEEE Transactions on Intelligent Transportation Systems*, vol. 21, no. 4, pp. 1369–1379, 2020.

[96] W. Infante and J. Ma, "Multistakeholder planning and operational strategy for electric vehicle battery swapping stations," *IEEE Systems Journal*, vol. 16, no. 3, pp. 3543–3553, 2022.

[97] B. Sun, X. Tan, and D. H. K. Tsang, "Optimal charging operation of battery swapping and charging stations with qos guarantee," *IEEE Transactions on Smart Grid*, vol. 9, no. 5, pp. 4689–4701, 2018.

[98] N. I. Nimalsiri, C. P. Mediwaththe, E. L. Ratnam, M. Shaw, D. B. Smith, and S. K. Halgamuge, "A survey of algorithms for distributed charging control of electric vehicles in smart grid," *IEEE Transactions on Intelligent Transportation Systems*, vol. 21, no. 11, pp. 4497–4515, 2020.

[99]   Z. Xu, W. Su, Z. Hu, Y. Song, and H. Zhang, "A hierarchical framework for coordinated charging of plug-in electric vehicles in China," *IEEE Transactions on Smart Grid*, vol. 7, no. 1, pp. 428–438, 2016.

[100]   C. Liu, K. T. Chau, D. Wu, and S. Gao, "Opportunities and challenges of vehicle-to-home, vehicle-to-vehicle, and vehicle-to-grid technologies," *Proc. IEEE*, vol. 101, no. 11, pp. 2409–2427, 2013.

[101]   S. Han, S. Han, and K. Sezaki, "Development of an optimal vehicle-to-grid aggregator for frequency regulation," *IEEE Transactions on Smart Grid*, vol. 1, no. 1, pp. 65–72, 2010.

[102]   S. Han, S. Han, and K. Sezaki, "Estimation of achievable power capacity from plug-in electric vehicles for v2g frequency regulation: Case studies for market participation," *IEEE Transactions on Smart Grid*, vol. 2, no. 4, pp. 632–641, 2011.

[103]   C. Wu, H. Mohsenian-Rad, and J. Huang, "Vehicle-to-aggregator interaction game," *IEEE Transactions on Smart Grid*, vol. 3, no. 1, pp. 434–442, 2012.

[104]   C. Wu, H. Mohsenian-Rad, J. Huang, and J. Jatskevich, "PEV-based combined frequency and voltage regulation for smart grid," *in 2012 IEEE PES Innovative Smart Grid Technologies (ISGT)*, Washington, DC, USA, 2012.

[105]   J. Zhao, C. Wan, Z. Xu, and K. P. Wong, "Spinning reserve requirement optimization considering integration of plug-in electric vehicles," *IEEE Transactions on Smart Grid*, vol. 8, no. 4, pp. 2009–2021, 2017.

[106]   W. Liu, S. Chen, Y. Hou, and Z. Yang, "Optimal reserve management of electric vehicle aggregator: Discrete bilevel optimization model and exact algorithm," *IEEE Transactions on Smart Grid*, vol. 12, no. 5, pp. 4003–4015, 2021.

[107]   U. K. Madawala and D. J. Thrimawithana, "A two-way inductive power interface for single loads," in *2010 IEEE International Conference on Industrial Technology*, pp. 673–678. IEEE, 2010.

[108]   Y. H. Sohn, B. H. Choi, G.-H. Cho, and C. T. Rim, "Gyrator-based analysis of resonant circuits in inductive power transfer systems," *IEEE Transactions on Power Electronics*, vol. 31, no. 10, pp. 6824–6843, 2015.

[109]   U. K. Madawala and D. J. Thrimawithana, "A bidirectional inductive power interface for electric vehicles in v2g systems," *IEEE Transactions on Industrial Electronics*, vol. 58, no. 10, pp. 4789–4796, 2011.

[110]   W. Xu, K. W. Chan, S. W. Or, S. L. Ho, and M. Liu, "A low-harmonic control method of bidirectional three-phase z-source converters for vehicle-to-grid applications," *IEEE Transactions on Transportation Electrification*, vol. 6, no. 2, pp. 464–477, 2020.

[111]   S. Weearsinghe, D. J. Thrimawithana, and U. K. Madawala, "Modeling bidirectional contactless grid interfaces with a soft dc-link," *IEEE Transactions on Power Electronics*, vol. 30, no. 7, pp. 3528–3541, 2014.

[112]   N. D. Weise, K. Basu, and N. Mohan, "Advanced modulation strategy for a three-phase ac-dc dual active bridge for v2g," in *2011 IEEE Vehicle Power and Propulsion Conference*, pp. 1–6. IEEE, 2011.

[113] D. Das, N. Weise, K. Basu, R. Baranwal, and N. Mohan, "A bidirectional soft-switched dab-based single-stage three-phase ac–dc converter for v2g application," *IEEE Transactions on Transportation Electrification*, vol. 5, no. 1, pp. 186–199, 2018.

[114] S. Weerasinghe, U. K. Madawala, and D. J. Thrimawithana, "A matrix converter-based bidirectional contactless grid interface," *IEEE Transactions on Power Electronics*, vol. 32, no. 3, pp. 1755–1766, 2016.

[115] F. Jauch and J. Biela, "Modelling and zvs control of an isolated three-phase bidirectional ac-dc converter," *in 2013 15th European Conference on Power Electronics and Applications (EPE)*, pp. 1–11. IEEE, 2013.

[116] F. Liu, W. Lei, T. Wang, C. Nie, and Y. Wang, "A phase-shift soft-switching control strategy for dual active wireless power transfer system," *in 2017 IEEE Energy Conversion Congress and Exposition (ECCE)*, pp. 2573–2578. IEEE, 2017.

[117] C. Zhao, Z. Wang, J. Du, J. Wu, S. Zong, and X. He, "Active resonance wireless power transfer system using phase shift control strategy," in *2014 IEEE Applied Power Electronics Conference and Exposition-APEC 2014*, pp. 1336–1341. IEEE, 2014.

[118] K. Colak, E. Asa, M. Bojarski, D. Czarkowski, and O. C. Onar, "A novel phase-shift control of semibridgeless active rectifier for wireless power transfer," *IEEE Transactions on Power Electronics*, vol. 30, no. 11, pp. 6288–6297, 2015.

[119] T. Diekhans and R. W. De Doncker, "A dual-side controlled inductive power transfer system optimized for large coupling factor variations," *in 2014 IEEE Energy Conversion Congress and Exposition (ECCE)*, pp. 652–659. IEEE, 2014.

[120] W. X. Zhong and S. Hui, "Maximum energy efficiency tracking for wireless power transfer systems," *IEEE Transactions on Power Electronics*, vol. 30, no. 7, pp. 4025–4034, 2014.

[121] H. H. Wu, A. Gilchrist, K. D. Sealy, and D. Bronson, "A high efficiency 5 kw inductive charger for evs using dual side control," *IEEE Transactions on Industrial Informatics*, vol. 8, no. 3, pp. 585–595, 2012.

[122] Y. Li, R. Mai, M. Yang, and Z. He, "Cascaded multi-level inverter based ipt systems for high power applications," *Journal of Power Electronics*, vol. 15, no. 6, pp. 1508–1516, 2015.

[123] A. P. Hu and S. Hussmann, "Improved power flow control for contactless moving sensor applications," *IEEE Power Electronics Letters*, vol. 2, no. 4, pp. 135–138, 2004.

[124] H. Li, A. Hu, and G. Covic, "A power flow control method on primary side for a cpt system," *in 2010 International Power Electronics Conference-ECCE ASIA-*, pp. 1050–1055. IEEE, 2010.

[125] W. V. Wang and D. J. Thrimawithana, "A novel converter topology for a primary-side controlled wireless EV charger with a wide operation range," *IEEE Journal of Emerging and Selected Topics in Industrial Electronics*, vol. 1, no. 1, pp. 36–45, 2020.

[126] T. Diekhans and R. W. De Doncker, "A dual-side controlled inductive power transfer system optimized for large coupling factor variations and partial load," *IEEE Transactions on Power Electronics*, vol. 30, no. 11, pp. 6320–6328, 2015.

[127] Z. Huang, S.-C. Wong, and K. T. Chi, "Control design for optimizing efficiency in inductive power transfer systems," *IEEE Transactions on Power Electronics*, vol. 33, no. 5, pp. 4523–4534, 2017.

[128] M. Kim, D.-M. Joo, and B. K. Lee, "Design and control of inductive power transfer system for electric vehicles considering wide variation of output voltage and coupling coefficient," *IEEE Transactions on Power Electronics*, vol. 34, no. 2, pp. 1197–1208, 2018.

[129] Q. Chen, S. C. Wong, K. T. Chi, and X. Ruan, "Analysis, design, and control of a transcutaneous power regulator for artificial hearts," *IEEE Transactions on Biomedical Circuits and Systems*, vol. 3, no. 1, pp. 23–31, 2009.

[130] Y. Jiang, J. Liu, X. Hu, L. Wang, Y. Wang, and G. Ning, "An optimized frequency and phase shift control strategy for constant current charging and zero voltage switching operation in series-series compensated wireless power transmission," in *2017 IEEE Energy Conversion Congress and Exposition (ECCE)*, pp. 961– 966. IEEE, 2017.

[131] Z. Huang, S.-C. Wong, and K. T. Chi, "An inductive-power-transfer converter with high efficiency throughout battery-charging process," *IEEE Transactions on Power Electronics*, vol. 34, no. 10, pp. 10245–10255, 2019.

[132] Z. Huang, S.-C. Wong, and K. T. Chi, "Design of a single-stage inductive-power-transfer converter for efficient ev battery charging," *IEEE Transactions on Vehicular Technology*, vol. 66, no. 7, pp. 5808–5821, 2016.

[133] R. Mai, Y. Liu, Y. Li, P. Yue, G. Cao, and Z. He, "An active-rectifier-based maximum efficiency tracking method using an additional measurement coil for wireless power transfer," *IEEE Transactions on Power Electronics*, vol. 33, no. 1, pp. 716–728, 2017.

[134] Y. Liu, U. K. Madawala, R. Mai, and Z. He, "Zero-phase-angle controlled bidirectional wireless ev charging systems for large coil misalignments," *IEEE Transactions on Power Electronics*, vol. 35, no. 5, pp. 5343–5353, 2019.

[135] X. Zhang, T. Cai, S. Duan, H. Feng, H. Hu, J. Niu, and C. Chen, "A control strategy for efficiency optimization and wide ZVS operation range in bidirectional inductive power transfer system," *IEEE Transactions on Industrial Electronics*, vol. 66, no. 8, pp. 5958–5969, 2018.

[136] Y. Liu, U. K. Madawala, R. Mai, and Z. He, "An optimal multivariable control strategy for inductive power transfer systems to improve efficiency," *IEEE Transactions on Power Electronics*, vol. 35, no. 9, pp. 8998–9010, 2020.

Index

www.ingramcontent.com/pod-product-compliance
Lightning Source LLC
Chambersburg PA
CBHW050512190326
41458CB00005B/1512